Programming ArcObjects with VBA

A Task-Oriented Approach

Programming ArcObjects with VBA

A Task-Oriented Approach

Kang-Tsung Chang

CRC PRESS

Boca Raton   London   New York   Washington, D.C.

Library of Congress Cataloging-in-Publication Data

Chang, Kang-Tsung.
 Programming ArcObjects with VBA : a task-oriented approach / Kang-Tsung Chang.
   p. cm.
 Includes bibliographical references and index.
 ISBN 0-8493-2781-4 (alk. paper)
 1. Microsoft Visual Basic for applications. 2. Graphical user interfaces (Computer
 systems) 3. Geographic information systems. 4. ArcGIS. I. Title.

T58.62.C27 2004
005.1′17—dc22   2004045487

This book contains information obtained from authentic and highly regarded sources. Reprinted material is quoted with permission, and sources are indicated. A wide variety of references are listed. Reasonable efforts have been made to publish reliable data and information, but the author and the publisher cannot assume responsibility for the validity of all materials or for the consequences of their use.

Visit the CRC Press Web site at www.crcpress.com

About the Author

Kang-tsung (Karl) Chang is a professor of geography at the University of Idaho. He received his B.S. in geography from National Taiwan University and his M.A. and Ph.D. from Clark University. He has taught geographic information systems (GIS) and macro programming in GIS since 1988. He has previously published three other books on GIS.

Contents

Introduction

This book is designed for ArcGIS users who want to get a quick start on programming ArcObjects. Both ArcGIS and ArcObjects are products developed and distributed by Environmental Systems Research Institute (ESRI), Inc. ArcObjects is the development platform for ArcGIS, a software package for managing geographic information systems (GIS). Ideally, we should learn ArcObjects before using ArcGIS. But that is not the case in reality. We use ArcGIS first through its toolbars and commands. It is easier to follow the user interface in ArcGIS than to sort out objects, properties, and methods in code. The topic of ArcObjects usually emerges when we realize that programming ArcObjects can actually reduce the amount of repetitive work, streamline the workflow, and even produce functionalities that are not easily available in ArcGIS.

How can we learn programming ArcObjects efficiently and quickly? Perhaps surprising to some, the answer is to apply our knowledge from working with ArcGIS. In fact, this book uses a task-oriented approach in an attempt to directly relate what we already know about ArcGIS to programming ArcObjects.

THE TASK-ORIENTED APPROACH

GIS activities are task-oriented: we use GIS for data integration, data management, data display, data analysis, and so on. Therefore, an efficient way to learn programming ArcObjects is to take a task-oriented approach. The task-oriented approach has at least three main advantages:

First, it connects ArcObjects with what we already know. Take the example of *QueryFilter*. This book first links a *QueryFilter* object to the task of data exploration. After we know that the object can perform the same function as the Select by Attributes command in ArcMap, which we have used many times before, it becomes easy to understand the properties and methods that are associated with the object.

Second, the task-oriented approach groups ArcObjects in a way that is logical to ArcGIS users. With thousands of objects, properties, and methods, it can be difficult, if not impossible, for beginners to navigate the ArcObjects model diagrams. Using the task-oriented approach, we can learn ArcObjects incrementally from one group of tasks to another. Likewise, we can learn programming ArcObjects in an organized fashion.

Third, the task-oriented approach can help us gain a better understanding of ArcGIS with our new knowledge of ArcObjects. For example, as a type of *QueryFilter*, the *SpatialFilter* class has its own properties of geometry and spatial relation in addition to the properties that it inherits from the *QueryFilter* class. This class relationship explains why the Select by Location command in ArcMap can actually accept both attribute and spatial constraints for data query. Perhaps the Select by Location command should be more appropriately named Select by Attributes and Location.

ABOUT THIS BOOK

This book has 14 chapters. The first three chapters introduce ArcObjects, programming basics, and customization. This book adopts Visual Basic for Applications (VBA) for programming ArcObjects. Because VBA is already embedded within ArcMap and ArcCatalog, it is convenient for ArcGIS users to program ArcObjects in VBA. The following major topics are covered in the first three chapters:

- Chapter 1 ArcObjects: Geodatabase, ArcObjects, organization of ArcObjects, and help sources on ArcObjects
- Chapter 2 Programming Basics: Basic elements, writing code, calling subs and functions, Visual Basic Editor, and debugging code
- Chapter 3 Customization of the User Interface: Creating a toolbar with existing commands, adding a new button and tool, adding a form, and making basic templates

Chapters 4 through 14 discuss programming ArcObjects for solving common GIS tasks. Each chapter is organized around a central theme and has three parts. The first part is a quick review of ArcGIS commands on the topic; the second part discusses objects that are related to the theme; and the third part presents sample programs (procedures and modules in VBA) for solving common tasks under the theme. This combination of ArcGIS commands, ArcObjects, and sample programs is aimed at relating our experience of working with ArcGIS to programming ArcObjects.

This book contains 95 sample programs. Each sample program starts with a list of key interfaces and members (i.e., properties and methods) and its usage, followed by the listing and explanation of code. Many programs are divided into two or more parts to better connect the code lines and their explanation. Available on the companion CD to the book, these sample programs are organized by chapter in the *programs* folder, with 3 to 13 in each chapter. Stored as text files, these sample programs can be easily imported to Visual Basic Editor in either ArcMap or ArcCatalog to view and run. The companion CD also includes datasets for test runs.

They are stored in the *data* folder by chapter. ArcGIS 8.3 or a later version is needed to run the programs.

The following major tasks are covered in each chapter:

- Chapter 4 Dataset and Layer Management: Add datasets as layers, manage layers and datasets, and report geographic dataset information
- Chapter 5 Attribute Data Management: List fields, add or delete fields, calculate field values, and join and relate tables
- Chapter 6 Data Conversion: Convert shapefile to geodatabase, convert coverage to geodatabase and shapefile, perform rasterization and vectorization, and add XY data
- Chapter 7 Coordinate Systems: Manipulate on-the-fly projection, define coordinate system, perform geographic transformation, and project datasets
- Chapter 8 Data Display: Display vector data, display raster data, and make a layout page
- Chapter 9 Data Exploration: Perform attribute query, perform spatial query, combine attribute and spatial queries, and derive descriptive statistics
- Chapter 10 Vector Data Analysis: Buffer, perform overlay, join data by location, and manipulate features
- Chapter 11 Raster Data Analysis: Manage raster data, perform local operations, perform neighborhood operations, perform zonal operations, and perform distance measure operations
- Chapter 12 Terrain Mapping and Analysis: Derive contour, slope, aspect, and hill shade, perform viewshed analysis, perform watershed analysis, and create and edit triangulated irregular networks (TINs)
- Chapter 13 Spatial Interpolation: Perform spatial interpolation and compare interpolation methods
- Chapter 14 Binary and Index Models: Build binary and index models, both vector- and raster-based

TYPOGRAPHICAL CONVENTIONS

The following typographical conventions are used in this book:

- Sample programs are set off from the text and appear in a different typeface.
- Names of sample programs are in bold face and italicized.
- ArcObjects, interfaces, properties, and methods appear in italic.
- Names of datasets appear in italic in the text.

CHAPTER 1

ArcObjects

ArcGIS 8.x from Environmental Systems Research Institute (ESRI), Inc. is a different software package from ArcView 3.x and Arc/Info 7.x, two previous packages from the same company. The difference stems mainly from the geodatabase data model and ArcObjects introduced in ArcGIS 8.x. These two new developments are closely related. Together, they provide the foundation for the desktop applications of ArcCatalog and ArcMap in ArcGIS. They also provide the basis for readers of this book to write programs in Visual Basic for Applications (VBA) for new applications in ArcGIS.

The geodatabase data model replaces the georelational data model that has been used for coverages over the past two decades. These two data models differ in how geographic and attribute data are stored. The georelational data model stores geographic and attribute data separately in a split system. Geographic data describing the geometry of features in a coverage are saved as graphic files with such names as arc-node list, arc-coordinate list, left/right polygon list, and polygon/arc list. These graphic files are separate from the polygon attribute table that stores attribute data and keys to link to other tables. By contrast, the geodatabase data model stores geographic and attribute data together in a single system and geographic data in a geometry field.

Another important difference that characterizes the geodatabase data model is the use of object-oriented technology. Object-oriented technology treats a spatial feature as an object and groups spatial features of the same type into a class. A class, and by extension an object in the class, can have properties and methods. A property describes a characteristic or attribute of an object. A method carries out an action by an object. Developers of ArcGIS have already implemented properties and methods on hundreds of classes in ArcGIS. Therefore, when we work in ArcCatalog and ArcMap, we actually interact with these classes and their properties and methods.

This chapter focuses on the geodatabase data model and ArcObjects. To use ArcObjects programmatically, we must understand how spatial data are structured and stored using the geodatabase data model and how classes in ArcObjects are designed and organized. Section 1.1 describes the basics of the geodatabase data model including the types of data that the model covers. Section 1.2 explains the basics of ArcObjects including classes, relationships between classes, interfaces, properties, and methods. Section 1.3 deals with the organization of ArcObjects. Section 1.4 covers the help sources on ArcObjects.

1.1 GEODATABASE

A geographic information system (GIS) manages geographically referenced data. Geographically referenced data are data that describe both the location and characteristics of spatial features such as roads, land parcels, and vegetation stands on the Earth's surface. The locations of spatial features are measured in geographic coordinates (i.e., longitude and latitude values) or projected coordinates (e.g., Universal Transverse Mercator or UTM coordinates). The characteristics of spatial features are expressed as numeric and string attributes. This book uses the term *geographic data* to describe data that include the locations of spatial features, and the term *nongeographic data* to describe data that include only the attributes of spatial features.

A geodatabase uses tables to store geographic data as well as nongeographic data. It is therefore important to distinguish different types of tables. A table consists of rows and columns. Each row corresponds to a feature, and each column or field represents an attribute. A table that contains geographic data has a geometry field, which distinguishes the table from tables that contain only nongeographic data. The following sections describe both geographic and nongeographic data.

1.1.1   Vector Data

The geodatabase data model represents vector-based spatial features as points, polylines, and polygons.[1] A point feature may be a simple point feature or a multipoint feature with a set of points. A polyline feature is a set of line segments, which may or may not be connected. A polygon feature may be made of one or many rings. A ring is a set of connected, closed, nonintersecting line segments.

A geodatabase organizes spatial features into feature classes and feature datasets. A *feature class* is a collection of spatial features with the same type of geometry. A feature class may therefore contain simple point, line, or polygon features. A *feature dataset* is a collection of feature classes based on a common coordinate system. A feature dataset is typically reserved for feature classes that participate in topological relationships with each other such as in a geometric network or a planar (two-dimensional) topology. A feature class is like a shapefile in having simple features. A feature dataset is similar to a coverage in having multiple datasets based on a common coordinate system. However, this kind of analogy does not address other differences between the traditional and geodatabase data models that are driven by advances in computer technology.

In a geodatabase, a feature class can be a standalone feature class or part of a feature dataset. In either case, a feature class is stored as a table. A feature class has a minimum of two fields. One is the object or feature ID and the other is the geometry or shape field. A feature class can have other attribute fields, but the geometry field sets a feature class apart from other tables.

ArcGIS users recognize a feature class as a feature attribute table. When we open the attribute table of a feature layer in ArcMap, we see the two default fields in the table, and we can locate and highlight spatial features on a map only through a feature attribute table.

Features within a feature class can be further segregated by subtype. For example, a road feature class can have subtypes based on average daily traffic volume. The geodatabase data model provides four general validation rules for the grouping of objects: attribute domains, default values, connectivity rules, and relationship rules.[1] An attribute domain limits an attribute's values to a valid range of values or a valid set of values. A default value sets an expected attribute value. Connectivity rules control how features in a geometric network are connected to one another. Relationship rules determine, for example, how many features can be associated with another.

1.1.2 Raster Data

The geodatabase data model represents raster data as a two-dimensional array of equally spaced cells.[1] This concept of using arrays and cells for raster data is the same as the ESRI grid model.

A wide variety of raster data are available in GIS. They include satellite imagery, digital elevation models (DEMs), digital orthophotos, scanned files, graphic files, and software-specific raster data such as ESRI grids.[2] The geodatabase model treats them equally as *raster datasets*. However, a raster dataset may have a single band or multiple bands. An ESRI grid typically contains a single band, whereas a multi-spectral satellite image contains multiple bands.

A multi-band raster dataset may also appear as the output from a raster data operation. For example, a cost distance measure operation can produce results showing the least accumulative cost distance, the back link, and the allocation. These different outputs can be initially saved into a multi-band raster dataset, one band per output, and later extracted to create the proper raster datasets.

1.1.3 Triangulated Irregular Networks (TINs)

The geodatabase data model uses a *TIN dataset* to store a set of nonoverlapping triangles that approximate a surface. Elevation values along with x-, y-coordinates are stored at nodes that make up the triangles. In many instances, a TIN dataset is an alternative to a raster dataset for surface mapping and analysis. The choice between the two depends on data flexibility and computational efficiency.[2]

Inputs to a TIN include DEMs, contour lines, GPS (global positioning system) data, LIDAR (light detection and ranging) data, and survey data. We can also modify and improve a TIN by using linear features such as streams and roads and area features such as lakes and reservoirs. Data flexibility is therefore a major advantage of using a TIN. In addition, the triangular facets of a TIN tend to create a sharper image of the terrain than an elevation raster does.

Computational efficiency is the main advantage of using raster datasets. The simple data structure of arrays and cells makes it relatively easy to perform computations that are necessary for deriving slope, aspect, surface curvature, viewshed, and watershed.

1.1.4   Location Data

The term *location data* refers to data that can be converted to point features. Common examples of location data are tables that contain x-, y-coordinates or street addresses. We can convert a table with x-, y-coordinates directly into a point feature class, with each feature corresponding to a pair of x- and y-coordinates. Using a street network as a reference, we can also convert a list of street addresses into a set of point features.

1.1.5   Nongeographic Data

A table that stores nongeographic data does not have a geometry field. The geodatabase data model defines such table as an *object class*. Examples of object classes include comma delimited text files and dBASE files. These files or tables contain attributes of spatial features and have keys (i.e., relate fields) to link to geographic data in a relational database environment.

1.2 ARCOBJECTS

ArcObjects is the development platform for ArcGIS Desktop. A collection of objects, ArcObjects is behind the menus and icons that we use to perform tasks in ArcGIS. These same objects also allow software developers to access data and to perform tasks programmatically.

1.2.1   Objects and Classes

ArcObjects consists of objects and classes.[3] An *object* represents a spatial feature such as a road or a vegetation stand. In a geodatabase, an object corresponds to a row in a table and the object's attributes appear in columns. A *class* is a set of objects with similar attributes. An ArcObjects class can have built-in interfaces, properties, and methods.

ArcObjects includes three types of classes:

The most common type is the *coclass*. A coclass can be used to create new objects. For example, *FeatureClass* is a coclass that allows new feature class objects to be created as instances of the coclass.

The second type is the *abstract class*. An abstract class cannot be used to create new objects. An abstract class exists so that other classes (i.e., subclasses) can use or share the properties and methods that the class supports. For example, *GeoDataset* is an abstract class. The class exists so that geographic datasets such as feature

classes and raster datasets can all share the properties of extent and spatial reference
that the *GeoDataset* class supports.

The third type is the *class*. A class cannot be used directly to create new objects;
objects of a class can only be created from another class. For example,
EnumInvalidObject is a noncreatable class because an *EnumInvalidObject* can only
be obtained from another object such as a data conversion object. When converting
a shapefile into a geodatabase feature class, for example, a data conversion object
automatically creates an *EnumInvalidObject* to keep track of those objects that
have failed to be converted.

1.2.2   Relationships between Classes

Object-oriented technology has introduced different types of relationships that
can be established between classes. Developers of ArcObjects have generally followed these relationships. A good reference on relationships between classes in
ArcObjects is Zeiler.[3] There are also books such as Larman[4] that deal with this topic
in the general context of object-oriented analysis and design. A basic understanding
of class relationships is important for navigating the object model diagrams and for
programming ArcObjects as well.

Association describes the relationship between two classes. An association uses
multiplicity expressions to define how many instances of one class can be associated
with the other class. Common multiplicity expressions are one (1) and one or more
(1..*). For example, Figure 1.1 shows an association between *Fields* and *Field* and
between *Field* and *GeometryDef*. The multiplicity expressions in Figure 1.1 suggest
that:

One fields object, which represents a collection of fields in a table, can be associated
with one or more field objects.

One field object can be associated with zero or one *GeometryDef* object, which
represents a geometry definition.

A field associated with a geometry definition is the geometry field, and a table can
have one geometry field at most.

Type inheritance defines the relationship between a superclass and a subclass.
A subclass is a member of a superclass and inherits the properties and methods
of the superclass. But a subclass can have additional properties and methods to
separate itself from other members of the superclass. For example, Figure 1.2

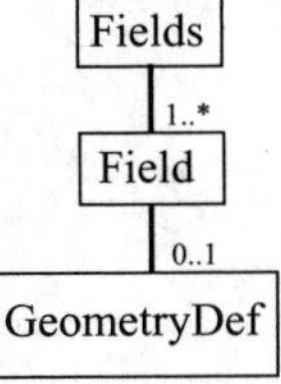

Figure 1.1   The association between *Fields* and *Field* is one or more and between *Field* and
GeometryDef is zero or one.

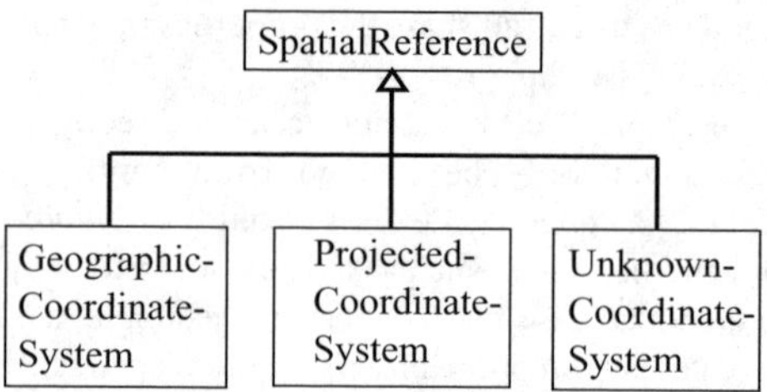

Figure 1.2   *SpatialReference* and its three subclasses.

shows that *GeographicCoordinateSystem* is a type of *SpatialReference*. The *GeographicCoordinateSystem* class shares with *ProjectedCoordinateSystem* and *UnknownCoordinateSystem* the properties and methods that the *SpatialReference* class supports, but the *GeographicCoordinateSystem* class has additional properties and methods that are unique to the geographic coordinate system.

Composition describes the whole–part relationship between classes. Composition is a kind of association except that the multiplicity at the composite end is typically one and the multiplicity at the other end can be zero or any positive integer. For example, a composition describes the relationship between the *Map* class and the *FeatureLayer* class (Figure 1.3). A map object represents a map or a data frame in ArcMap and a feature layer object represents a feature-based layer in a map. A map can be associated with a number of feature layers. Or, to put it the other way, a feature layer is part of a map.

Aggregation, also called shared aggregation, describes the whole–part relationship between classes. Unlike composition, however, the multiplicity at the composite end of an aggregation relationship is typically more than one. For example, Figure 1.4 shows that a *SelectionSet* object can be created from a *QueryFilter* object and a *Table* object. A table and a query filter together at the composite end can create a selection set (i.e., a data subset) at the other end.

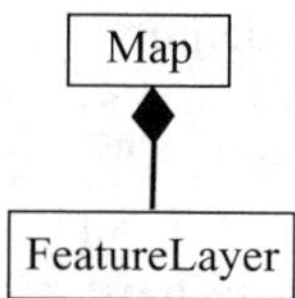

Figure 1.3   A *Map* object comprises zero, one, or more *FeatureLayer* objects.

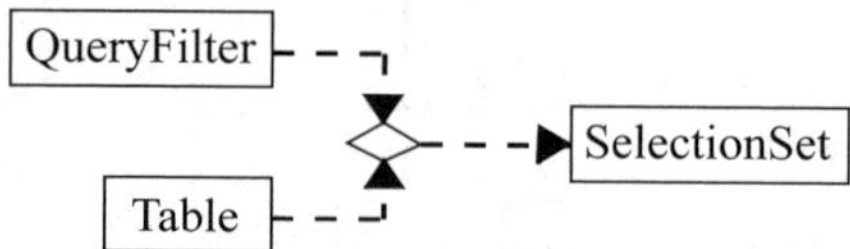

Figure 1.4   A *QueryFilter* object and a *Table* object together can create a *SelectionSet* object.

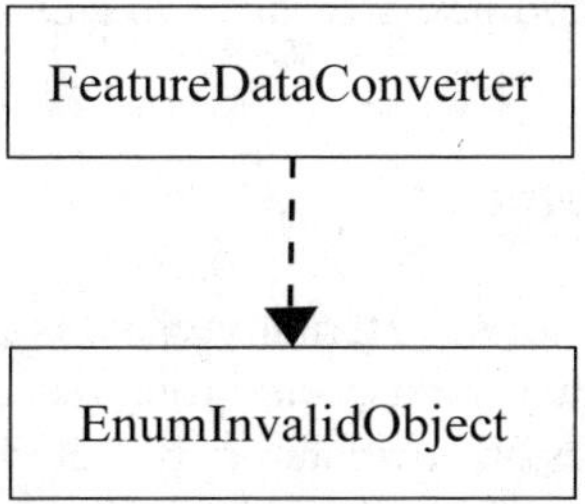

Figure 1.5   An *EnumInvalidObject* can be created by a *FeatureDataConverter* object.

Instantiation means that an object of a class can be created from an object of another class. Figure 1.4 shows that, for example, a selection set can be created from a query filter and a table. Another example is an *EnumInvalidObject*, which, as explained earlier, can be created from a *FeatureDataConverter* object (Figure 1.5).

1.2.3   Interfaces

When programming with objects in ArcObjects, one would never work with the object directly but, instead, would access the object via one of its interfaces. An *interface* represents a set of externally visible operations. For example, a *RasterReclassOp* object implements *IRasterAnalysisEnvironment* and *IReclassOp* (Figure 1.6). We can access a *RasterReclassOp* object via the *IRasterAnalysis-Environment* interface to define the analysis environment such as the output cell size. We can also use the *IReclassOp* interface to perform raster data reclassification.

An object may support two or more interfaces and, additionally, the same object may inherit interfaces from its superclass. Given multiple interfaces, it is possible to access an interface via another interface, or to jump from an interface to another. This technique is called QueryInterface or QI for short. QI simplifies the process of coding. For example, to program a raster data reclassification, we can stay with the same *RasterReclassOp* object. First, we use *IRasterAnalysisEnvironment* to set up the analysis environment. Then, we switch, via QI, to *IReclassOp* to perform data reclassification. Chapter 2 on the basics of programming has a more detailed discussion on the QI technique.

Some objects in ArcObjects have two or more similar interfaces. For example, a *FeatureDataConverter* object implements *IFeatureDataConverter* and *IFeature-DataConverter2*. Both interfaces have methods for data conversion between shapefiles, coverages, and geodatabases. But *IFeatureDataConverter2* has the additional functionality of working with data subsets. Object-oriented technology allows

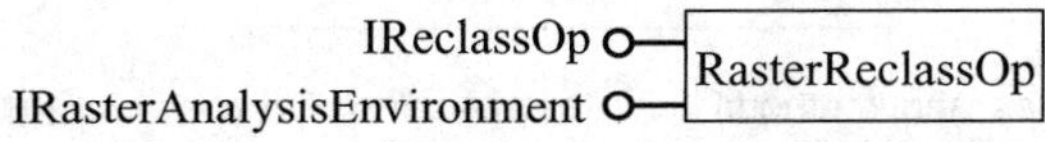

Figure 1.6   A *RasterReclassOp* object supports both *IReclassOp* and *IRasterAnalysis-Environment*.

developers of ArcObjects to add new interfaces to a class without having to remove or update the existing interfaces.

1.2.4   Properties and Methods

An interface represents a set of externally visible operations. More specifically, an interface allows programmers to use the properties and methods that are on the interface. A *property* describes an attribute or characteristic of an object. A *method*, also called behavior, performs a specific action. Figure 1.7, for example, shows the properties and methods on *IRasterAnalysisEnvironment*. These properties and methods are collectively called *members* on the interface.

A property can be for read only, write only, or both read and write. The read property is also called the get property, and the write property the put property. In Figure 1.7, the barbell symbols accompany the properties of *Mask* and *OutWorkspace* on *IRasterAnalysisEnvironment*. The square on the left is for the get property, and the square on the right is for the put property. If the square on the right is open, such as in Figure 1.7, the property is defined as put by reference. If it is solid, the property is defined as put by value. The difference between the two put properties lies in the way a property value is assigned to an object. The put by reference property requires the keyword *Set*, whereas the put by value property does not. For example, to specify an analysis mask through the *Mask* property on *IRaster-AnalysisEnvironment*, we need to use a statement such as: *Set pEnv.Mask = pMask-Dataset*, where *pEnv* represents an analysis environment and *pMaskDataset* represents an analysis mask.

To carry out an action, a method on an interface may require some arguments and may return a value or values. In Figure 1.7, the arrow symbols show the methods of *SetCellSize* and *SetExtent* on *IRasterAnalysisEnvironment*. The syntax of the *SetCellSize* method is object.SetCellSize (envType [,cellSizeProvider]). The method has two arguments, of which the first is required and the second is optional.

An interface may not have both properties and methods. Some interfaces have properties only, while some have methods only. *IReclassOp*, for example, only has methods. Figure 1.8 shows two of the five methods on *IReclassOp*. These methods all perform reclassification of raster data but use different mechanisms. The

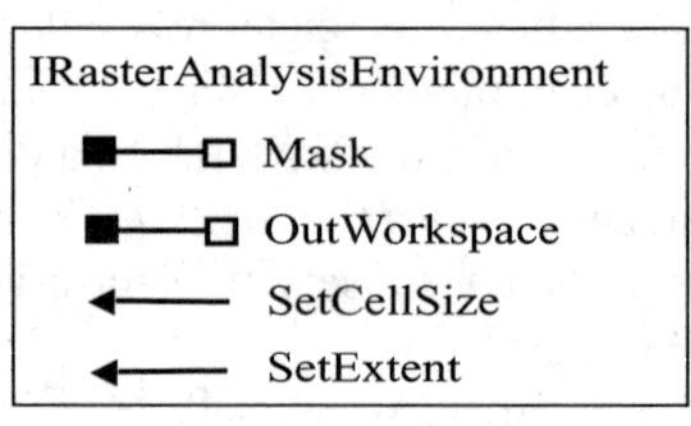

Figure 1.7   Properties and methods on *IRasterAnalysisEnvironment*. Properties are shown with barbell symbols, and methods are shown with arrow symbols. Occasionally in this book, properties and methods are shown with double colon symbols such as *IRasterAnalysisEnvironment::Mask*.

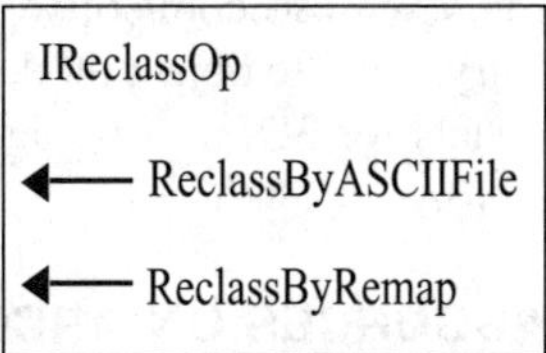

Figure 1.8   Methods on *IReclassOp*.

ReclassByASCIIFile method uses an ASCII file, whereas the *ReclassByRemap* method uses a remap that is built programmatically.

1.3 ORGANIZATION OF ARCOBJECTS

The ArcObjects Developer Help organizes objects in ArcObjects into the following subsystems: 3D Analyst Extension, Application Framework, ArcCatalog, ArcMap, ArcMap Editor, ArcObjects Controls, ArcScan, Display, Geocoding, Geodatabase, Geometry, IMS, Linear Referencing and XY Events, Network, Output, Raster, Spatial Analyst Extension, Spatial Reference, and StreetMap USA Extension. This subsystem organization resembles that of ArcGIS and its applications, thus providing a good starting point for those who are already familiar with operations in ArcGIS.

For example, the Spatial Analyst subsystem organizes objects by type of raster data operation. The subsystem's object model lists *RasterReclassOp*, *Raster-SurfaceOp*, *RasterConversionOp*, *RasterLocalOp*, and other objects that closely resemble different functionalities of the Spatial Analyst extension by name. ArcGIS users who are familiar with the Spatial Analyst extension should have no problems using these objects. Objects in other subsystems such as Geodatabase, ArcMap, and ArcCatalog are more difficult to relate to because many of them represent new object-oriented concepts and methods.

Objects can also be organized by object library. An object library file (*.olb*) is a file that contains a collection of objects needed for a certain application. For example, the ESRI Object Library, also called the Core Object Library, includes objects from ArcCatalog, ArcMap, Geodatabase, and Geometry that are regularly used in programming ArcObjects. The Core Object Library is automatically loaded in VBA. The Spatial Analyst Extension Object Library, on the other hand, has objects that are used by the extension. This object library file is not automatically loaded. We must set a reference to the Spatial Analyst Extension Object Library before using objects in the library for an application.

A recent trend suggests that objects may also be grouped by industry as well. Because real world objects all have different properties and methods, it is impossible to apply, for example, the methods and properties of transportation-related objects to forestry-related objects. ESRI has set up a website that supports the development of object models for address, forestry, transportation, hydro, land parcel, environmental facilities, and other fields (http://www.esri.com/software/arcgisdatamodels/). Parallel with ESRI's promotion of industry-specific object models are increased research activities in developing complex models for 3D, transportation, and other applications.[5–7]

1.4 HELP SOURCES ON ARCOBJECTS

The help sources on ArcObjects include books and documents. ESRI has published two books on ArcObjects: *Exploring ArcObjects*[3], and *Getting to Know ArcObjects: Programming ArcGIS with VBA*.[8] The former is a two-volume reference on ArcObjects, and the latter is a workbook with hands-on exercises. Two other publications deal with topics related to ArcObjects: *ArcGIS Developer's Guide for VBA*[9] covers the basics of developing ArcGIS applications. *Avenue Wraps*[10] is a guide for converting Avenue scripts into VBA code.

This section covers electronic and online help documents. ArcObjects has thousands of classes, interfaces, methods, and properties. One must regularly consult the help documents while programming ArcObjects.

1.4.1   ArcObjects Developer Help

The start up menu of ArcGIS includes the ArcObjects Developer Help. On the Contents tab, the help document lists topics such as Object Model Overviews, ArcObjects Component Help, and Samples. The overviews section describes each subsystem of ArcObjects. The component help section lists the interfaces as well as coclasses and classes. Abstract classes, however, are not included. The samples section provides sample macros by subsystem. On the Index tab, one can type a coclass, a class, an interface, a method, or a property as a search word and find the help information on the search word.

ESRI also has an ArcObjects Online website, http://arconline.esri.com/arcobjectsonline/. This website provides the latest information about ArcObjects including sample code, technical documents, and object model diagrams.

1.4.2   Object Model Diagrams

The Object Model Diagrams of the ArcObjects Developer Help organize classes by subsystem. An object model diagram can be downloaded as a *pdf* file from the help document. Each diagram shows classes, class relationships, interfaces, properties, and methods with different symbols. Figure 1.9, for example, shows a small portion of the Spatial Analyst Object Model.

1.4.3   ESRI Object Browser

The ESRI Object Browser or EOBrowser is a utility for browsing object libraries. The EOBrowser can be installed by running \ArcObjects Developer Kit\Utilities\ EOBrowser.exe of the ArcGIS installation folder. The Object Library References dialog, which can be accessed through the browser's File menu, allows the user to

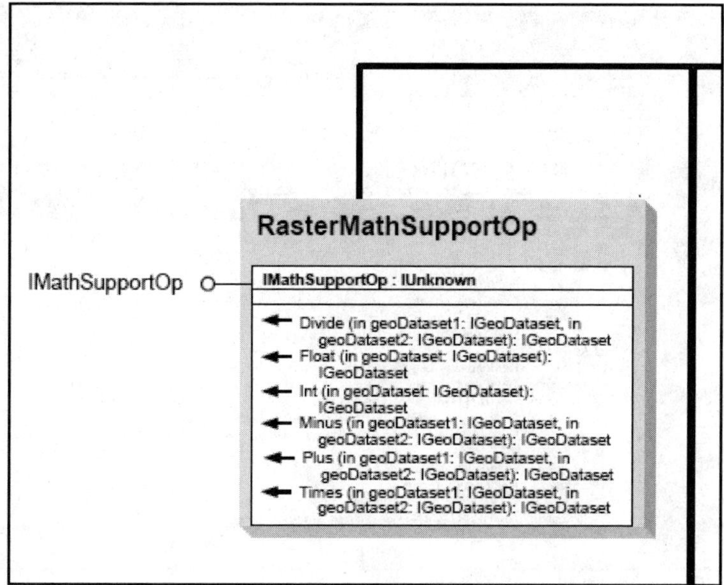

Figure 1.9 A small portion of the Spatial Analyst Object Model diagram.

add and remove object libraries (Figure 1.10). The EOBrowser window has controls so that the user can select all coclasses and all interfaces in an object library for display and browsing (Figure 1.11).

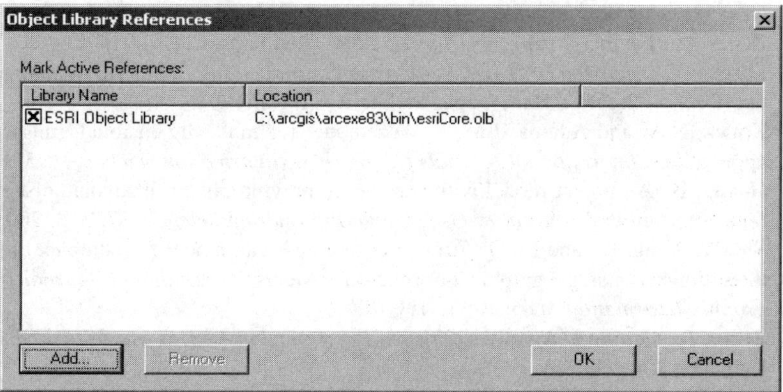

Figure 1.10 The Object Library References dialog lets the user add and remove object libraries.

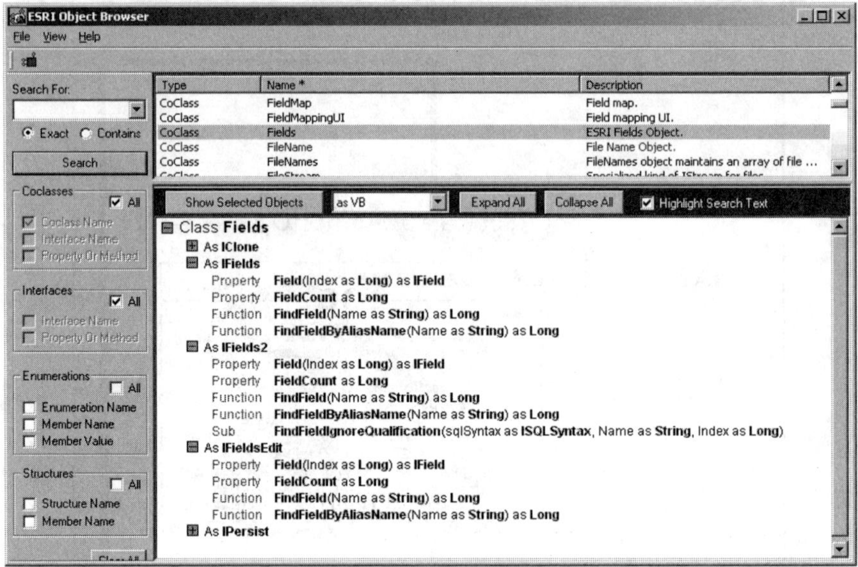

Figure 1.11   The top part of the EOBrowser shows all coclasses in the ESRI Object Library, and the bottom part shows all interfaces that the *Fields* coclass supports.

REFERENCES

1. Zeiler, M., *Modeling Our World: The ESRI Guide to Geodatabase Design*, Environmental Systems Research Institute, Redlands, CA, 1999.
2. Chang, K., *Introduction to Geographic Information Systems*, 2nd ed., McGraw-Hill, New York, 2004.
3. Zeiler, M., Ed., *Exploring ArcObjects*, ESRI, Redlands, CA, 2001.
4. Larman, C., *Applying UML and Patterns: An Introduction to Object-Oriented Analysis and Design*, Prentice Hall, Upper Saddle River, NJ, 1998.
5. Koncz, N.A. and Adams, T.M., A data model for multi-dimensional transportation applications, *International Journal of Geographic Information Science*, 16, 551, 2002.
6. Huang, B., An object model with parametric polymorphism for dynamic segmentation, *International Journal of Geographic Information Science*, 17, 343, 2003.
7. Shi, W., Yang, B., and Li, Q., An object-oriented data model for complex objects in three-dimensional geographic information systems, *International Journal of Geographic Information Science*, 17, 411, 2003.
8. Burke, R., *Getting to Know ArcObjects: Programming ArcGIS with VBA*, ESRI Press, Redlands, CA, 2003.
9. Razavi, A.H., *ArcGIS Developer's Guide For VBA*, OnWord Press/Delmar Learning, Clifton Park, NY, 2002.
10. Tonias, C.N. and Tonias, E.C., *Avenue Wraps*, The CEDRA Press, Rochester, NY, 2002.

CHAPTER 2

Programming Basics

ArcObjects is the development platform for ArcGIS. Because ArcObjects is built using Microsoft's COM (Component Object Model) technology, it is possible to use any COM-compliant development language with ArcObjects to customize applications. This book adopts Visual Basic for Applications (VBA), which is already embedded within ArcMap and ArcCatalog of ArcGIS. Other COM-compliant programming languages include Visual Basic and C++.

Writing application programs for ArcGIS requires knowledge of both VBA and ArcObjects; VBA provides the programming language and ArcObjects provides objects and their built-in properties and methods. It may be of interest to some readers to compare ArcObjects with Avenue and AML (Arc Macro Language), two programming languages previously developed by Environmental Systems Research Institute (ESRI), Inc. Programming ArcObjects is similar to Avenue programming in that both use objects and their built-in properties and methods (called requests in Avenue). The main difference is that we program ArcObjects using VBA, a common programming language available in Microsoft's products. Programming ArcObjects is conceptually different from AML programming because AML is a procedural, rather than an object-oriented, language. However, many objects in ArcObjects resemble ARC/INFO commands in terms of the design of their properties and methods.

This chapter deals with the programming language and code writing, although many examples in the chapter do involve ArcObjects. Section 2.1 discusses the basic elements in VBA programming such as procedures, variables, interfaces, and arrays. Section 2.2 offers common techniques for writing code. Section 2.3 explains how to put together a program as a collection of code blocks. Section 2.4 covers Visual Basic Editor, a medium for preparing, compiling, and running macros. Section 2.5 covers the debugging tools that can help identify mistakes in macros.

2.1 BASIC ELEMENTS

This section covers basic programming elements. Many elements are directly related to VBA. Therefore, additional information on these elements can be found in Microsoft Visual Basic Help, which is accessible through Visual Basic Editor in either ArcMap or ArcCatalog.

2.1.1 Projects, Modules, Procedures, and Macros

Procedures are the basic units in VBA programming. A *procedure* is a block of code that can perform a specific task such as defining the coordinate system of a geographic dataset. Applications developed using VBA are called *macros* in Microsoft's products such as Word, Excel, and Access. A macro is functionally similar to a procedure. A *module* is a collection of procedures, and a *project* is a collection of modules (Figure 2.1). Most sample macros in this book are procedures, but some are modules. For example, modules, each with several procedures, are used to build binary and index models in Chapter 14.

A procedure can be private or public. A private procedure can only be called or used by another procedure in the same module. By contrast, a public procedure is available to different modules that make up a project.

Three types of procedures exist: events, subs, and functions. *Event* procedures are associated with controls on a form, such as command buttons. *Subs* and *functions*, on the other hand, are not directly associated with controls. A function returns a value, whereas a sub does not. This book uses mainly subs and functions. Chapters 3 and 8 have examples that use event procedures to customize the user interface.

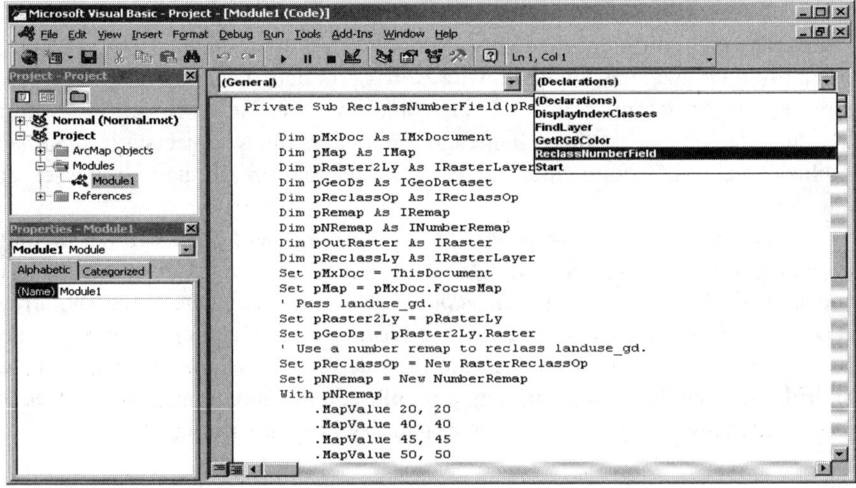

Figure 2.1   On the left of Visual Basic Editor, the Project Explorer shows that Module1 is a module in the Project. On the right, the procedure list shows that ReclassNumberField is the procedure in the Code window.

A procedure starts with the keyword of *Sub* or *Function* and ends with the *End Sub* or *End Function* statement. VBA automatically creates the first and last lines of a new procedure, which are called the wrapper lines.

2.1.2 Variables

A *variable* stores a value that can be accessed or changed by a macro. VBA requires that a variable be declared before it can be used. To make sure that variables are declared explicitly, one can add the *Option Explicit* line to the beginning of a module. Declaration statements can be all placed at the top of a macro or placed wherever they are needed. This book adopts the style of declaring variables at the top of a macro. If a macro is divided into parts, then variables are declared at the top of each part.

How to declare a variable in a macro depends on whether the variable refers to an ArcObjects class or not. The following two lines declare a counter variable *n*, which does not refer to an ArcObjects class, and assign 5 to be its value.

```
Dim n As Integer
n = 5
```

Dim is the most used keyword for declaring a variable. A variable declared with the *Dim* keyword within a procedure is only available in that procedure. But a variable declared with the *Dim* keyword at the head (i.e., the Declarations section) of a module is available to all procedures within the module. Other keywords for declaring variables include *Public* and *Private*. A public variable is available to all modules in a project. A private variable, on the other hand, is available only to the module in which it is declared. A declaration statement usually includes a data type. Integer in the above example represents the data type. Other data types include Boolean, Single, Double, String, and Variant.

If a variable refers to an existing class in ArcObjects, it must be declared by pointing to an interface that the class supports. The properties and methods of an object are hidden according to the encapsulation principle in object-oriented technology. Therefore, the object can only be accessed through the predefined interfaces. Encapsulation also means that interface and object can be interchangeable in discussions and that programming ArcObjects is programming with interfaces.

The following two lines show how to declare a variable by referencing an existing class in ArcObjects.

```
Dim pField As IFieldEdit
Set pField = New Field
```

The first line declares *pField* by pointing the variable to the *IFieldEdit* interface that the *Field* coclass supports (Figure 2.2). The second line creates a new field object by making *pField* an instance of the *Field* class.

The first letter in *pField* stands for pointer. The first letter in *IFieldEdit* stands for interface. *IFieldEdit* has uppercase and lowercase letters for better reading. These

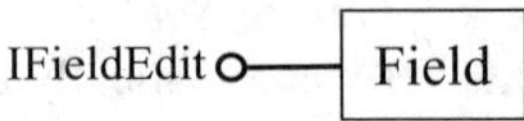

Figure 2.2   A field variable can be declared by pointing to *IFieldEdit* that a *Field* object supports.

are naming conventions in object-oriented programming. The keyword *Set* assigns a value to an object variable.

The next example defines the top layer in an active data frame of the running ArcMap.

```
Dim pMxDoc As IMxDocument
Dim pMap As IMap
Dim pFeatureLayer As IFeatureLayer
Set pMxDoc = ThisDocument
Set pMap = pMxDoc.FocusMap
Set pFeatureLayer = pMap.Layer(0)
```

The *Dim* statements point *pMxDoc* to *IMxDocument*, *pMap* to *IMap*, and *pFeatureLayer* to *IFeatureLayer*. The first *Set* statement assigns ThisDocument to *pMxDoc*. ThisDocument is the predefined name of an *MxDocument* object. When we launch ArcMap, *MxDocument* and *Application* are already in use. The alternative to ThisDocument is Application.Document, which refers to the document of the ArcMap application. The second *Set* statement assigns *FocusMap* or the focus map of the map document to *pMap*. And the third statement assigns *Layer(0)* or the top layer in the focus map to *pFeatureLayer*; the index begins with 0 in VBA. *FocusMap* and *Layer()* are both properties, which are covered in the next section.

2.1.3   Use of Properties and Methods

Properties are attributes of an object. As examples, *FocusMap* is a property on *IMxDocument* and *Layer(0)* is a property on *IMap*. The syntax for using a property is *object.property,* such as *pMxDoc.FocusMap.*

Both *FocusMap* and *Layer()* happen to be get-only, or read-only, properties. The following example shows the put, or write, properties. There are two methods for putting properties: by reference and by value. The second line statement in the example sets *pFeatureClass* to be the feature class of *pFeatureLayer* by reference and the third line statement assigns the string "breakstrm" to be the name of *pFeatureLayer* by value.

```
Dim pFeatureClass As IFeatureClass
Set pFeatureLayer.FeatureClass = pFeatureClass
PFeatureLayer.Name = "breakstrm"
```

The difference between put by reference and put by value is the use of the *Set* keyword. How can we tell which method to use? One approach is to consult the ArcObjects Developer Help. The put by reference property has an open square symbol, whereas the put by value property has a solid square symbol. Another approach is to let the VBA compiler catch the error. The error messages are "Method or data member not found" if the *Set* keyword is missing and "Invalid use of property" if the *Set* keyword is unnecessary.

Methods perform specific actions. A method may or may not return a value. The syntax for calling a method is *object.method.* Many methods require object qualifiers and arguments. The following line, for example, adds a feature layer to a map:

```
PMap.AddLayer pFeatureLayer
```

The *AddLayer* method on *IMap* adds *pFeatureLayer* to *pMap*. The method requires an object qualifier (i.e., *pFeatureLayer*) and does not return a value or an interface.

The next example gets a workspace on disk and then gets a shapefile from the workspace.

```
Dim pFeatureWorkspace As IFeatureWorkspace
Dim pFeatureClass As IFeatureClass
Set pFeatureWorkspace = pWorkspaceFactory. _
OpenFromFile("c:\data\chap2", 0)
Set pFeatureClass = pFeatureWorkspace. _
OpenFeatureClass("emidastrm")
```

The *OpenFromFile* method on *IWorkspaceFactory* returns an interface on the specified workspace (i.e., "c:\data\chap2\"). The code then switches to the *IFeatureWorkspace* interface and uses the *OpenFeatureClass* method to open *emidastrm* in the workspace. Both methods require arguments in their syntax. The first argument for *OpenFromFile* is a workspace, and the second argument of 0 tells VBA to get the ArcMap window handle. The only argument for *OpenFeatureClass* is a string that shows the name of the feature class.

VBA has the automatic code completion feature to work with properties and methods. After an object variable is entered with a dot, VBA displays available properties and methods for the object variable in a dropdown list. We can either scroll through the list to select a property or method or type the first few letters to come to the property or method to use.

2.1.4 QueryInterface

A class object may support two or more interfaces, and each interface may have a number of properties and methods. When we declare a variable, we point the variable to a specific interface. To switch to a different interface, we can use QueryInterface, or QI for short. QI lets the programmer jump from one interface to another.

We can revisit the code fragment from the previous section to get a better understanding of QI.

```
Dim pFeatureWorkspace As IFeatureWorkspace
Dim pFeatureClass As IFeatureClass
Set pFeatureWorkspace = pWorkspaceFactory. _
OpenFromFile("c:\data\chap2", 0)
Set pFeatureClass = pFeatureWorkspace. _
OpenFeatureClass("emidastrm") ' QI
```

The syntax of the *OpenFromFile* method suggests that the method returns the *IWorkspace* interface that a workspace object supports. But to use the *OpenFeatureClass* method, which is on *IFeatureWorkspace*, the code must perform a QI for *IFeatureWorkspace* that a workspace object also supports (Figure 2.3).

The next example shows a code fragment for converting feature data to raster data. A *RasterConversionOp* object supports both *IConversionOp* and *IRasterAnalysisEnvironment* (Figure 2.4). The *IConversionOp* interface has methods for converting feature data to raster data, and the *IRasterAnalysisEnvironment* interface has properties and methods to set the analysis environment. The example uses QI to define the output cell size as 5000 for a vector to raster data conversion.

```
Dim pConversionOp As IConversionOp
Dim pEnv As IRasterAnalysisEnvironment
Set pConversionOp = New RasterConversionOp
Set pEnv = pConversionOp ' QI
PEnv.SetCellSize esriRasterEnvValue, 5000
```

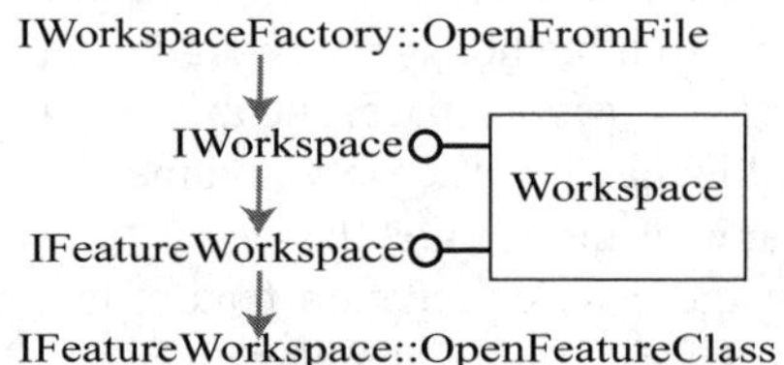

Figure 2.3   The diagram shows how to switch from *IWorkspace* to *IFeatureWorkspace* by using QI.

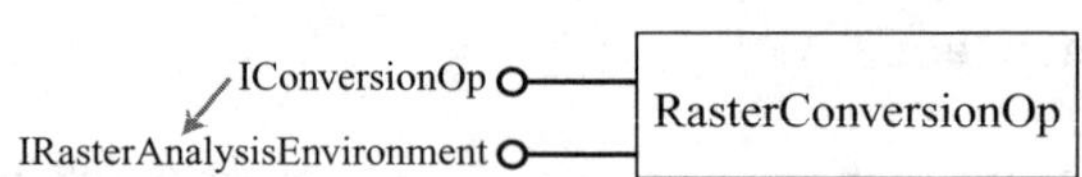

Figure 2.4   Use QI to jump from *IConversionOp* to *IRasterAnalysisEnvironment*.

2.1.5   Comment Lines and Line Continuation

A comment line is a text added to code that explains how the code works. A comment line starts with an apostrophe ('). Except for short comment lines such as QI, which are placed at the end of a statement, this book typically places a comment line before a statement or a group of statements. An underscore (_) at the end of a line statement means line continuation.

2.1.6   Arrays

An *array* is a special type of variable that holds a set of values of the same data type, rather than a single value as in the case of a regular variable. Arrays are declared the same way as other variables, using the *Dim*, *Private*, and *Public* statements. But an array variable must have the additional specification for the size of the array. For example, the following line declares the *AnArray* variable as an array of 11 (0 to 10) integers:

```
Dim AnArray(10) As Integer
```

AnArray in the above example is a static array, meaning that it has a predefined size of 11. The other type of array is a dynamic array. A dynamic array has no fixed size but uses VBA keywords to find out information about the array and to change its size. A dynamic array must be declared at the module level.

2.1.7   Collections

A *collection* consists of a set of ordered objects, which do not have to be of the same data type. Collections are therefore special arrays. A collection can be created as follows:

```
Dim theList As New Collection
```

The following code fragment uses a loop to add the field names to a collection object referenced by *theList*. *Add* is a method of a collection object.

```
Dim theList As New Collection
For ii = 0 To pFields.FieldCount - 1
  Set aField = pFields.Field(ii)
  fieldName = aField.Name
  theList.Add (fieldName)
Next
```

2.2 WRITING CODE

This section covers programming techniques for handling decision-making, branching, repetitive operations, and dialogs.

2.2.1 *If...Then...Else* Statement

A simple approach to decision-making in a macro is to use the *If...Then...Else* Statement. The statement has the following syntax:

```
If condition Then
  [statements]
Else
  [else_statements]
End If
```

If the condition is true, the program executes statements that follow the *Then* keyword. If the condition is false, the program executes statements that follow the *Else* keyword. The *If...Then...Else* statement can therefore handle two possible outcomes. To handle more than two outcomes, one can add the *ElseIf* clause to the statement:

```
If condition Then
  [statements]
ElseIf condition-n Then
  [elseif_statements]
Else
  [else_statements]
End If
```

The example below assigns 5 to the *n* variable if the name of the top layer is idcounty and 3 to *n* if the name is not idcounty.

```
If (pFeatureLayer.Name = "idcounty") Then
  n = 5
Else
  n = 3
End If
```

When the *If...Then...Else* statement is used jointly with the *TypeOf* keyword, the statement can check whether an object supports a specific interface before using the interface. The following code fragment verifies that *pConversionOp* does support *IRasterAnalysisEnvironment* before specifying the output cell size of 5000.

```
Dim pConversionOp As IConversionOp
Dim pEnv As IRasterAnalysisEnvironment
Set pConversionOp = New RasterConversionOp
If TypeOf pConversionOp Is IRasterAnalysisEnvironment Then
  Set pEnv = pConversionOp ' QI
  PEnv.SetCellSize esriRasterEnvValue, 5000
End If
```

Another common use of the *If...Then...Else* statement is to check for a condition that can cause a program error such as division by zero. If such condition is determined to exist, the *Exit Sub* statement placed after *Else* can terminate the execution of a macro immediately.

2.2.2   *Select Case* Statement

The *If...Then...Else* statement can become confusing and untidy if more than three or four possible outcomes exist. An alternative is to use the *Select Case* statement, which has the following syntax:

```
Select Case test_expression
  [Case expression_list-n
  [statements-n]]…
  [Case Else
  [else_statements]]
End Select
```

The following example uses a *Select Case* statement to prepare the data type description of a field. Using ArcObjects, the data type of a field is coded in numeric values from 0 to 8. A *Select Case* statement can translate these numeric values into text strings.

```
Dim fieldType As Integer
Dim typeDes As String
Select Case fieldType
Case 0
  typeDes = "SmallInteger"
Case 1
  typeDes = "Integer"
Case 2
  typeDes = "Single"
Case 3
  typeDes = "Double"
```

```
Case 4
  typeDes = "String"
Case 5
  typeDes = "Date"
Case 6
  typeDes = "OID"
Case 7
  typeDes = "Geometry"
Case 8
  typeDes = "Blob"
End Select
```

2.2.3   *Do...Loop* Statement

A *Do...Loop* statement repeats a block of statements in a macro. There are two types of loops: *Do While* and *Do Until*. A *Do While* loop continues while the condition is true:

```
Do While condition
   [statements]
Loop
```

The following example uses a *Do While* loop to repeat a block of statements as long as the user provides the name of a shapefile. The loop stops when the return from the input box is empty.

```
Dim pInput As String
pInput = InputBox("Enter the name of the input shapefile")
Do While pInput <> ""
   [statements]
Loop
```

A *Do Until* loop continues until the condition becomes true:

```
Do Until condition
   [statements]
Loop
```

The following example uses a *Do Until* loop to count how many cities are in a cursor (i.e., a selection set).

```
Dim pCity As IFeature
Dim intCount As Integer
```

```
Dim pCityCursor As IFeatureCursor
Set pCity = pCityCursor.NextFeature
Do Until pCity is Nothing
  intCount = intCount + 1
  Set pCity = pCityCursor.NextFeature
Loop
```

The *FeatureCursor* object in the above example holds a set of selected features. The *IFeatureCursor* interface has the *NextFeature* method that advances the position of the feature cursor by one and returns the feature at that position. By using the cursor and the *NextFeature* method, the above example increases the *intCount* value by 1 each time *NextFeature* advances a feature. The loop continues until no feature (i.e., Nothing) is advanced.

2.2.4 *For...Next* Statement

Like the *Do...Loop* statement, the *For...Next* statement also repeats a block of statements. The difference is that the *For...Next* statement runs a given number of times as determined by the start, end, and step (with the default of one) values:

```
For counter = start To end [Step step]
  [statements]
Next
```

The following example uses a *For...Next* statement to add the field names of a feature class to an array.

```
Dim pFields As IFields
Dim ii As Long
Dim aField As IField
Dim fieldName As Variant
Dim theList As New Collection
For ii = 0 To pFields.FieldCount - 1
  Set aField = pFields.Field(ii)
  fieldName = aField.Name
  theList.Add (fieldName)
Next
```

The *FieldCount* property on *IFields* returns the number of fields in *pFields*. The code sets the *For...Next* statement to begin with zero and to end with the number of fields minus one so that the *ii* counter corresponds to the index of a field.

The *Exit For* statement provides a way to exit a *For* loop. The following example uses an *Exit For* statement to exit the loop if a layer named idcities is located before reaching a fixed number of loops. The *Exit For* statement transfers control to the statement following the *Next* statement.

```
Dim pMxDoc As IMxDocument
Dim pMap As IMap
Dim pLayer As ILayer
Dim i As Integer
Set pMxDoc = ThisDocument
Set pMap = pMxDoc.FocusMap
For ii = 0 To pMap.LayerCount - 1
  Set pLayer = pMap.Layer(ii)
  If pLayer.Name = "idcities" Then
    i = ii
    Exit For
  End If
Next ii
MsgBox "idcities is at index " & i
```

2.2.5   *For Each...Next* Statement

The *For Each...Next* statement repeats a group of statements for each element in an array or collection:

```
For Each element In group
  [statements]
Next
```

The following code fragment uses a *For Each...Next* statement to print each field name in a collection of field names referenced by *theList*.

```
' Display the list of field names in a message box
For Each fieldName In theList
  MsgBox "The field name is " & fieldName
Next fieldName
```

2.2.6   *With* Statement

The *With* statement lets the programmer perform a series of statements on a single object. The *With* statement has the following syntax:

```
With object
  [statements]
End With
```

The following code fragment uses a *With* block to edit the name, type, and length properties of a new field.

```
Dim pField As IFieldEdit
Set pField = New Field
With pField
  .Name = "pop2000"
  .Type = esriFieldTypeInteger
  .Length = 8
End With
```

The alternative to the *With* block is to use the following line statements:

```
pField.Name = "pop2000"
pField.Type = esriFieldTypeInteger
pField.Length = 8
```

2.2.7 Dialog Boxes

Dialogs in a macro serve the purpose of getting information from and to the user. Dialogs come in a variety of forms. Chapter 3 covers custom dialogs using Visual Basic forms. ArcObjects also offers a number of common dialogs for such purposes as adding data, getting x- and y-coordinates, and reporting progress. A couple of these common dialogs are used in this book. Chapters 4 and 14 have examples of using browser dialogs for selecting datasets, and Chapter 10 has an example of using a progress dialog for reporting the progress of a spatial join operation. This section covers message boxes and input boxes, two simple dialog boxes that are frequently used in VBA macros.

A message box can be used as a statement or a function. As a statement, a message box shows text. For example, the following line displays the quoted text and the value of the *fieldName* variable.

```
MsgBox "The field name is " & fieldName
```

After viewing the field name, the user must acknowledge by clicking the OK button, which is also displayed in the message box. VBA has the following chr$() functions for handling multi-line messages: chr$(13) for a carriage return character and chr$(10) for a linefeed character. Additionally, the constant vbCrLf also functions as chr$(10) in creating a new line.

As a function, a message box returns the ID of the button that the user presses. For example, the following code fragment creates a message box and returns a value to *iAnswer* based on the user's decision. The message box shows a prompt, the Yes and No buttons, a question mark icon, and a title of Continue. The returned value is six for Yes and seven for No.

```
Dim iAnswer As Integer
iAnswer = MsgBox("Do you want to continue?", vbYesNo + _
vbQuestion, "Continue")
MsgBox "The answer is : " & iAnswer
```

An input box displays a prompt in a dialog box and returns a string containing the user's input. For example, the following line displays the prompt of "Enter the name of the input shapefile" in a dialog box and returns the user's input as a string to the *pInput* variable.

```
Dim pInput As String
pInput = InputBox("Enter the name of the input shapefile")
```

2.3 CALLING SUBS AND FUNCTIONS

A procedure, either a sub or a function, can be called by another procedure. VBA actually provides many simple functions that we use regularly in macros. Both message boxes and input boxes are VBA functions. Other examples include CStr and CInt. The CStr function converts a number to a string, and the CInt function returns an integer number.

This section goes beyond simple VBA functions and deals with the topic in a broader context. A procedure, depending on whether it is private or public, can be called by another procedure in the same module or throughout a project. Therefore, we can think of a sub or a function as a tool and build a module as a collection of tools. The major advantage of organizing code into separate subs and functions is that they can be reused in different modules. Other advantages include ease of debugging in smaller blocks of code and a better organization of code.

In the following example, the **Start** sub uses an input box to get a number from the user and then calls the **Inverse** sub to compute and report the inverse of the number.

```
Private Sub Start ()
  Dim n As Integer
  n = InputBox("Type a number")
  ' Call the Inverse sub.
  Inverse n
End Sub
Private Sub Inverse (m)
  Dim d As Double
  d = 1/m
  MsgBox "The inverse of the number is: " & d
End Sub
```

The *Start* sub passes *n* entered by the user as an argument to the *Inverse* sub. *Inverse* uses the passed value of *m* to compute its inverse. Notice that the example does not use the *Call* keyword. If the programmer prefers to use the keyword, the *inverse n* statement can be changed to:

```
Call Inverse (n)
```

The next example lets the *Start* sub call a function instead of a sub to accomplish the same task.

```
Private Sub Start ()
  Dim n As Integer
  Dim dd As Double
  n = InputBox("Type a number")
  ' Call the Inverse function and assign the return
  ' value to dd.
  dd = Inverse (n)
  MsgBox "The inverse of the number is: " & dd
End Sub
Private Function Inverse (m) As Double
  Dim d As Double
  d = 1/m
  Inverse = d ' Return the d value.
End Function
```

A couple of changes are noted when the code calls a function instead of a sub. First, the *Start* sub uses the following line to assign the returned value from the *Inverse* function to *dd*, which has been previously declared as a Double variable:

```
dd = Inverse (n)
```

Second, the code adds the *As Double* clause to the first line of *Inverse*:

```
Private Function Inverse (m) As Double
```

The clause declares *Inverse* to be a Double procedure. Thus the value returned by the function is also of the Double data type.

Third, the following line assigns *d*, which is the inverse of the passed value *m*, to *Inverse*:

```
Inverse = d
```

The *d* value is eventually returned to *Start* and assigned to the *dd* variable.

2.4 VISUAL BASIC EDITOR

Visual Basic Editor is a tool for compiling and running programs. To open Visual Basic Editor in either ArcCatalog or ArcMap, one can click the Tools menu, point to Macros, and select Visual Basic Editor.

Figure 2.5 shows Visual Basic Editor in ArcMap. A menu bar, a tool bar, and windows make up the user interface. Several commands ought to be mentioned at this point. Import File and Export File on the File menu allow the user to import and export macros in text file format. The Debug menu has commands for compiling and debugging macros, and the Run menu has commands for running and resetting macros. The same Run commands of Run Sub/UserForm, Break, and Reset are also available on the toolbar.

Figure 2.5 shows four types of windows: Code, Project, Properties, and Immediate. The Code window is the area for preparing and editing a macro. We can either type a new macro or import a macro. At the top of the Code window are two dropdown lists. On the left is the object list, and on the right is the procedure list. The Project window, also called the Project Explorer, displays a hierarchical list of projects and the contents and references of each project. Normal.mxt is a template for all map documents and is present whenever Visual Basic Editor is launched. Project, on the other hand, is specific to a map document. Macros for specific tasks are typically developed and stored at the current Project level. The Properties window shows the properties of controls such as command buttons and text boxes on a user form. Chapter 3 on customization of the user interface covers the use of the Properties

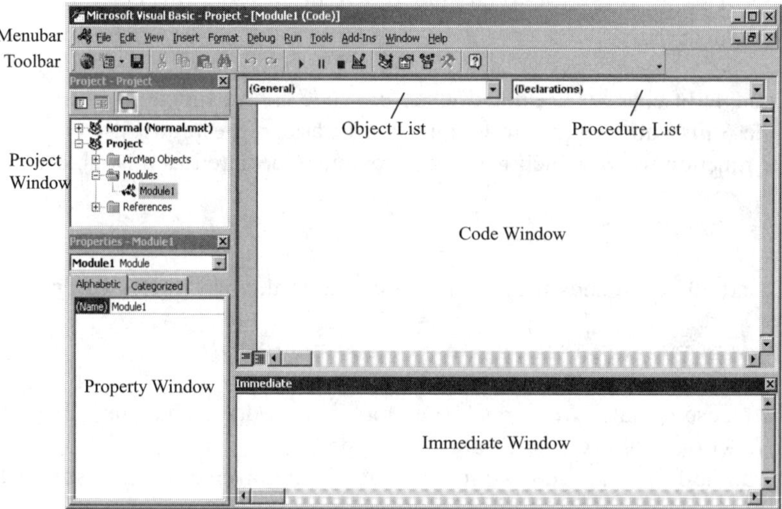

Figure 2.5   Visual Basic Editor consists of a menu bar, a tool bar, the Project window, the Property window, the Code window, the object list, the procedure list, and the Immediate window.

window. The Immediate window is designed for debugging. When used with a *Debug.Print* statement, the window can show the value of a variable.

Visual Basic Editor in ArcCatalog is set up the same way as in ArcMap. The only difference is that the Project Explorer in ArcCatalog contains only Normal.gxt. ArcCatalog does not have documents, and all customizations apply to the application.

The following shows how to use Visual Basic Editor to import and use a sample module on the companion CD of this book.

1. Right-click Project in the Project Explorer in ArcMap and select Import File. (In ArcCatalog, right-click Normal in the Project Explorer and select Import File.)
2. Select All Files from the file type dropdown list in the Import File dialog. Navigate to the sample module in the text file format. Click Open to import the sample module to Visual Basic Editor.
3. Click the plus sign next to the Modules folder in Project to open its contents.
4. Right-click Module1 and select View Code. The code now appears in the Code window, and the procedure list shows the name of the module.
5. Select Compile Project from the Debug menu to make sure that the module compiles successfully. To run the module, simply click on the Run Sub/UserForm button.

Most macros on the companion CD are designed for ArcMap so that the datasets can be displayed and analyzed immediately. Some macros such as those for data conversion can be run in either ArcCatalog or ArcMap.

2.5 DEBUGGING CODE

Every programmer has to deal with programming errors. Some errors are easy to fix, while others may take hours or days to correct. VBA has various debugging tools that can assist programmers in fixing errors. This section covers some of these tools.

2.5.1 Type of Error

There are three possible types of errors in VBA macros: compile, run-time, and logic. VBA stops compiling when it finds a compile error. Compile errors are caused by mistakes with VBA programming syntax. A compile error can occur when a macro misses the *End With* line in a *With* block or the *Loop* keyword in a *Do Until* statement. A compile error can also occur if a macro uses a property or method that is not available on an interface. For example, the *IFeatureWorkspace* interface has the *OpenFeatureClass* method but not *OpenFromFile*. When a macro tries to use *OpenFromFile* to open a feature class, VBA displays a compile error with the message of "Method or data member not found." To make sure that a property or method is available on an interface, one can first highlight the interface in the code window and then press F1. This will open the help page on the interface from the ArcObjects Developer Help.

A run-time error occurs when a macro, which has been compiled successfully, is running. Run-time errors are more difficult than compile errors to fix. For example, the following code is supposed to report the name of each layer in the active map.

```
Private Sub LayerName()
  Dim pMxDoc As IMxDocument
  Dim pMap As IMap
  Dim pFeatureLayer As IFeatureLayer
  Dim ii As Integer
  Set pMxDoc = ThisDocument
  Set pMap = pMxDoc.FocusMap
  ' Loop through each layer, and report its name.
  For ii = 0 To pMap.LayerCount
    Set pFeatureLayer = pMap.Layer(ii)
    MsgBox "The name of layer is: " & pFeatureLayer.name
  Next
End Sub
```

The macro has no compile errors. But it has a run-time error stating "Run-time error '5': Invalid procedure call or argument." VBA expects to have one more layer than those available in the active map. To make the macro run successfully, the *For* statement must be changed to:

```
For ii = 0 To pMap.LayerCount - 1
```

The next example is similar to the module used previously to derive the inverse of a typed number except that it does not pass the typed number *n* as an argument from the calling sub to the function. Therefore, *m* in the **Inverse** function is treated as zero. The error message in this case is "Run-time error '11': Division by Zero."

```
Private Sub Start()
  Dim n As Integer
  Dim dd As Double
  n = InputBox("Type a number")
  ' Call the Inverse function and assign the return
  ' value to dd.
  dd = Inverse()
  MsgBox "The inverse of the number is: " & dd
End Sub
Private Function Inverse() As Double
  Dim d As Double
  d = 1/m
```

```
   Inverse = d ' Return the value d.
End Function
```

Logic errors are even more difficult to correct than run-time errors are. A logic error does not stop a macro from compiling and running but produces an incorrect result. One type of logical error that every programmer dreads is endless loops. Endless loops can be caused by not setting the condition in a *Do...Loop* statement correctly.

2.5.2 *On Error* Statement

VBA has a built-in object called *Err*. The *Err* object has properties that identify the number, description, and source of a run-time error. We can use the *On Error* statement to display the properties of the *Err* object when an error occurs:

```
On Error GoTo line
```

The code below includes the *On Error* statement to trap the run-time error of division by zero.

```
Private Sub Start()
  On Error GoTo ErrorHandler
  Dim n As Integer
  Dim dd As Double
  n = InputBox("Type a number")
  ' Call the Inverse function and assign the return
  ' value to dd.
  dd = Inverse()
  MsgBox "The inverse of the number is: " & dd
  Exit Sub ' Exit to avoid error handler.
ErrorHandler: ' Error-handling routine.
  MsgBox Str(Err.Number) & ": " & Err.Description,, "Error"
End Sub
Private Function Inverse() As Double
  Dim d As Double
  d = 1/m
  Inverse = d ' Return the value d.
End Function
```

When the error occurs, the *ErrorHandler:* routine displays "11: Division by zero" in a message box with the title of Error. Notice that the *On Error* statement is placed at the top of the code. When a run-time error occurs, the code goes to the *ErrorHandler:* routine and displays the error message. Also notice that the *Exit Sub*

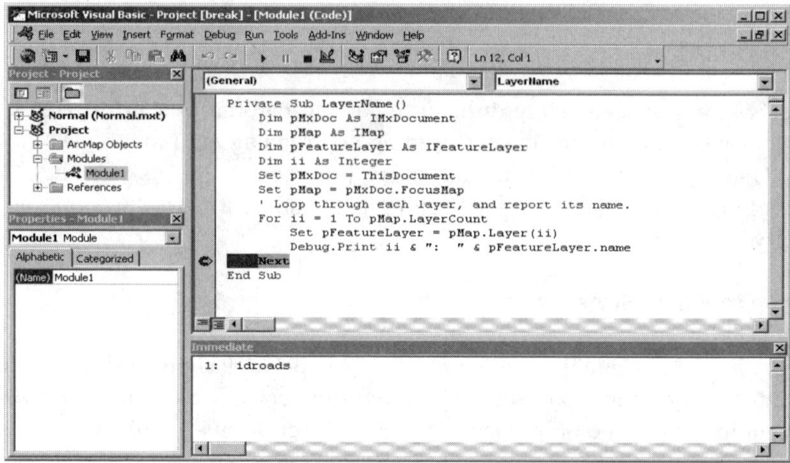

Figure 2.6   The breakpoint at the *Next* line allows the programmer to see that the layer index is one and idroads is the name of the layer.

line is used right before the *ErrorHandler:* routine to avoid the error message if no errors occurred.

2.5.3   Use of Breakpoint and Immediate Window

A breakpoint suspends execution at a specific statement in a procedure. A breakpoint therefore allows the programmer to examine variables and to make sure that the code is working properly. The following code places a breakpoint at the *Next* line of the *For...Next* statement and uses *Debug.Print* (the *Print* method of the *Debug* object) to print the counter value and the layer's name in the Immediate window (Figure 2.6).

```
Private Sub LayerName()
  Dim pMxDoc As IMxDocument
  Dim pMap As IMap
  Dim pFeatureLayer As IFeatureLayer
  Dim ii As Integer
  Set pMxDoc = ThisDocument
  Set pMap = pMxDoc.FocusMap
  ' Loop through each layer, and report its name.
  For ii = 0 To pMap.LayerCount - 1
    Set pFeatureLayer = pMap.Layer(ii)
    Debug.Print ii & ": " & pFeatureLayer.name
  Next
End Sub
```

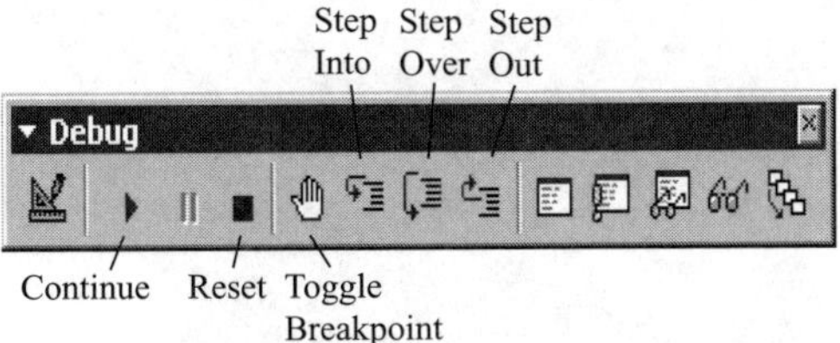

Figure 2.7   The Debug toolbar has the tools of Continue, Reset, Toggle Breakpoint, Step Into, Step Over, and Step Out.

The first time through the loop, the Immediate window shows zero and the name of the top layer in the active map. Click on the Continue button (the same button for Run Macro), and the window will show the next set of values.

Visual Basic Editor has Toggle Breakpoint on the Debug menu, as well as on the Debug toolbar, to add or remove a breakpoint at the current line (Figure 2.7). Other commands on the Debug menu include Step Into for executing code one statement at a time, Step Over for executing a procedure as a unit, Step Out for executing all remaining code in a procedure as if it were a single statement, and Run To Cursor for selecting a statement to stop execution of code.

CHAPTER 3

Customization of the User Interface

As a commercial product, ArcGIS is designed to serve as many users as possible and to meet as many needs as possible. It is no surprise that the software package has a large number of extensions, toolbars, and commands. But most users only use a portion of the available tools at a time. Therefore, a common customization is to simplify the way we interact with ArcGIS. ArcGIS Desktop provides the options to view or hide a toolbar. When working in ArcMap, we typically bring those toolbars that are necessary for a specific task to view and hide the others. Selecting toolbars to view and use is perhaps the easiest form of customization.

Customization can take other forms. One of them is to streamline the workflow. For example, instead of defining a new field and then calculating the field values in separate steps, we may want to combine them into one step. Customization can also reduce the amount of repetitive work. For example, rather than repeating for each dataset the same task of defining a common coordinate system, we may write a macro to complete the entire job with a single button click. Customization can also prevent the user from making unnecessary mistakes. For example, if a project requires distance measures to be in feet, we may choose feet as measurement units in code to prevent use of other units. The above examples show that customization is most useful if a project has a set of well-defined tasks.

This chapter introduces common methods for customizing the user interface. Section 3.1 describes how to create a new toolbar with existing ArcMap commands. Sections 3.2 and 3.3 discuss how to add a new button and a new tool respectively. Section 3.4 demonstrates the procedure for storing a new toolbar in a template. Section 3.5 explains the design and use of a Visual Basic form. Section 3.6 covers the procedure for storing a password-protected form in a template.

3.1 CREATING A TOOLBAR WITH EXISTING ARCMAP COMMANDS

No code writing is required for creating a new toolbar with existing buttons and tools in ArcMap. It is a simple copy-and-paste process. Suppose an application requires the following commands (i.e., buttons and tools) on a new toolbar: Zoom In, Full Extent, Select By Attributes, and Select By Location. The following shows the procedure for completing the task:

1. Select Customize from the Tools menu in ArcMap or double-click on an empty area of a toolbar to open the Customize dialog (Figure 3.1). The Customize dialog has three tabs: Toolbars, Commands, and Options. The Toolbars tab shows all toolbars available in ArcMap. The Commands tab shows all commands available in ArcMap by category. The Options tab has options to lock a customization with a password. The Customize dialog in ArcCatalog is set up the same as in ArcMap.
2. Click New in the Customize dialog. In the New Toolbar dialog, enter Selection for the toolbar name and save the toolbar in Untitled. Click OK to dismiss the New Toolbar dialog. A new toolbar now appears in ArcMap.
3. Click the Commands tab in the Customize dialog. Click the category of Pan/Zoom to view its commands (Figure 3.2). After locating the Full Extent command, drag and drop it onto the new toolbar. Do the same for the Zoom In command.

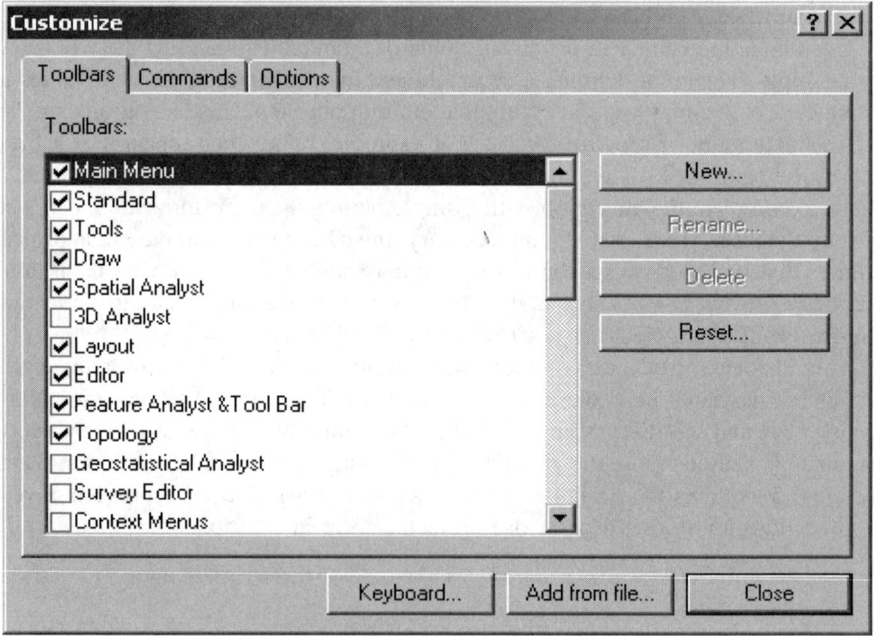

Figure 3.1   The Customize dialog has the three tabs of Toolbars, Commands, and Options.

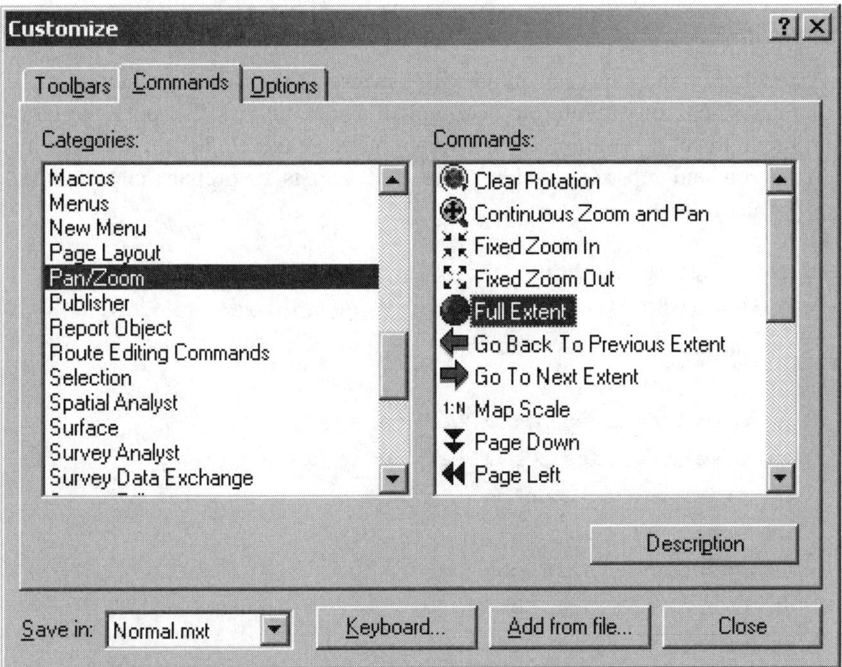

Figure 3.2   On the Commands tab, highlight the category of Pan/Zoom and view existing ArcMap commands in that category.

4. Click the category of Selection. Drag and drop the commands of Select By Attributes and Select By Location onto the new toolbar. As shown in Figure 3.3, the new toolbar now has four commands. These commands can be rearranged in the same way as any graphic elements.
5. Right-click a command on the toolbar to open its context menu. The menu has the options to delete, to change the image icon, to use text only, or to use image and text. However, the option for View Source is not available.

This new toolbar with four commands can now be used by itself or along with other toolbars in ArcMap.

Figure 3.3   The new Selection toolbar has four commands.

3.2 ADDING A NEW BUTTON

A new button must be associated with a macro so that when the button is clicked on, the macro can be executed to accomplish a specific task. Suppose the task is to report the fields of a geographic dataset in a message box. (Chapter 5 covers macros on managing and reporting fields.) The first step is to prepare an event (Click) procedure as follows.

```
Private Sub UIButtonFields_Click ()
  ' Part 1: Get the feature class and its fields.
  Dim pMxDoc As IMxDocument
  Dim pMap As IMap
  Dim pFeatureLayer As IFeatureLayer
  Dim pFeatureClass As IFeatureClass
  Dim pFields As IFields
  Dim count As Long
  Set pMxDoc = ThisDocument
  Set pMap = pMxDoc.FocusMap
  Set pFeatureLayer = pMap.Layer(0)
  Set pFeatureClass = pFeatureLayer.FeatureClass
  Set pFields = pFeatureClass.Fields

  ' Part 2: Prepare a list of fields and display the list.
  Dim ii As Long
  Dim aField As IField
  Dim fieldName As Variant
  Dim theList As New Collection
  Dim NameList As Variant
  ' Loop through each field, and add the field name to
  ' a list.
  For ii = 0 To pFields.FieldCount - 1
    Set aField = pFields.Field(ii)
    fieldName = aField.name
    theList.Add (fieldName)
  Next
  ' Display the list of field names in a message box.
  For Each fieldName In theList
    NameList = NameList & fieldName & Chr(13)
  Next fieldName
  MsgBox NameList,, "Field Names"
End Sub
```

After the macro has been compiled and run successfully, the next step is to link the macro to a button by using the following instructions:

1. Select Customize from the Tools menu in ArcMap.
2. Click New in the Customize dialog. In the New Toolbar dialog, enter Thermal for the toolbar name and save the toolbar in Untitled. Click OK to dismiss the New Toolbar dialog. The Thermal toolbar now appears in ArcMap.
3. On the Commands tab of the Customize dialog, select the category of UIControls and then click the New UIControl button (Figure 3.4). In the next dialog, check the option button for UIButtonControl and click Create (Figure 3.5).

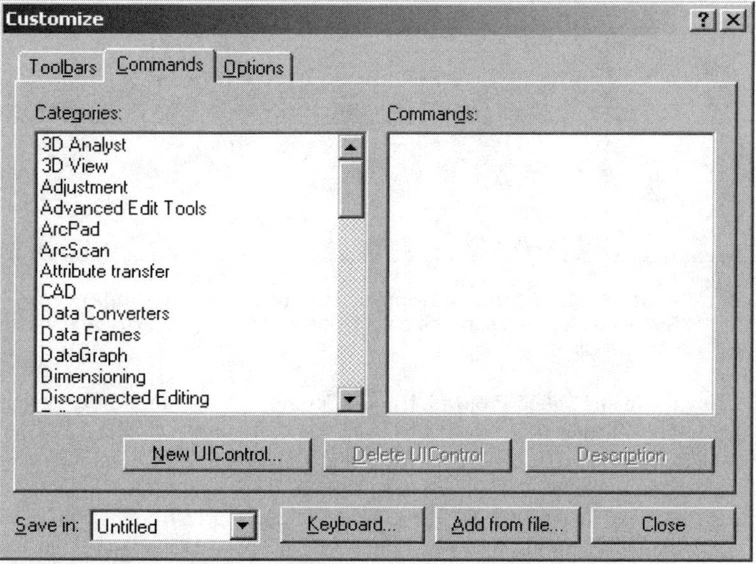

Figure 3.4   The New UIControl button is for creating a new control.

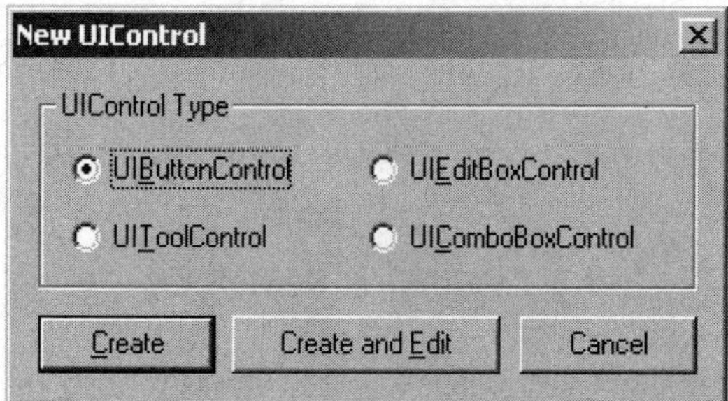

Figure 3.5   The New UIControl dialog shows four types of controls including button and tool.

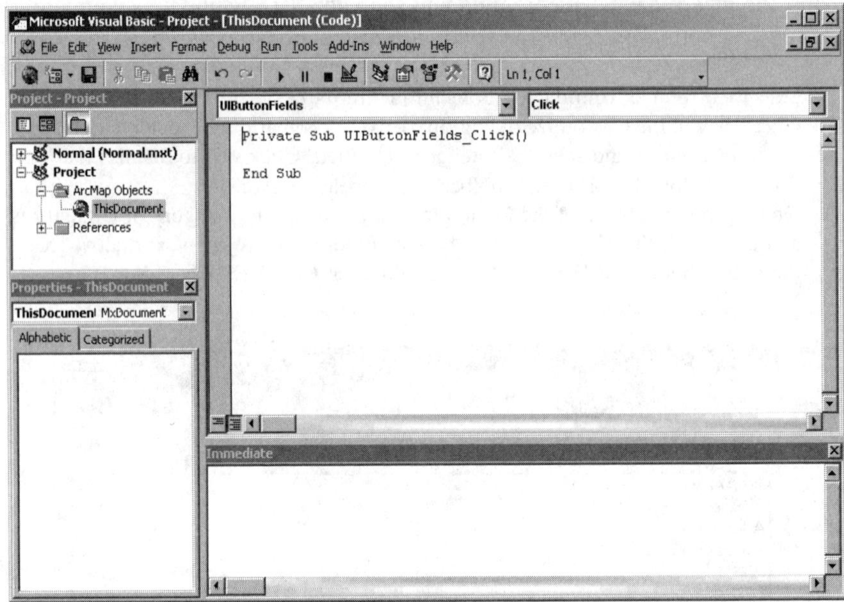

Figure 3.6   Visual Basic Editor automatically adds the wrapper lines of the UIButtonFields_Click sub. The object list shows UIButtonFields, and the procedure list shows Click.

4. A new command called Project.UIButtonControl1 appears in the Customize dialog. UIButtonControl1 is a default name. Click the new command and rename the button Project.UIButtonFields. Drag and drop Project.UIButtonFields onto the new toolbar.

5. Right-click the new button control and select View Source from its context menu. View Source opens Visual Basic Editor. The first and last wrapper lines of the UIButtonFields_Click() Sub are already in the Code window (Figure 3.6). Copy and paste ***UIButtonFields_Click*** on the companion CD to the Code window.

To test how the button works, do the following:

1. Add *thermal.shp* to ArcMap. The shapefile shows thermal springs and wells in Idaho.

2. When the new button is clicked, a message box appears with the fields in *thermal*.

3.3 ADDING A NEW TOOL

A button performs a task as soon as it is clicked on. A tool, on the other hand, requires the user to do something first before the tool can perform a task. The interaction with the user means that a tool has more events to consider and more coding to do than a button does. Events associated with a tool include Select, DblClick, MouseDown, MouseUp, and MouseMove.

This section describes a new tool, which uses a point entered by the user to report the number of features within 16,000 meters of the point. The tool essentially performs a spatial query based on the user's input. The procedure to be associated with the tool is a MouseDown event procedure. A MouseDown event procedure has four variables in its argument list: button, shift, x, and y. The user actually sets the values for these variables. The button value is either 1 or 2: 1 if the user is holding down the left mouse button, and 2 if the user is holding down the right mouse button. The shift value is either 0 or 1: 0 if the Shift key is not depressed, and 1 if the Shift key is depressed. The x and y values represent the location of the mouse pointer on the map display.

The first step is to prepare the event (MouseDown) procedure as follows. (Chapter 9 covers the programming techniques for spatial query.)

```
Private Sub UIToolQuery_MouseDown(ByVal button As Long, _
ByVal shift As Long, ByVal x As Long, ByVal y As Long)
  ' Part 1: Get the point clicked by the user.
  Dim pMxDoc As IMxDocument
  Dim pActiveView As IActiveView
  Dim m_blnMouseDown As Boolean
  Dim pPoint As IPoint
  Set pMxDoc = ThisDocument
  Set pActiveView = pMxDoc.FocusMap
  ' Convert the entered point from display coordinates
  ' to map coordinates.
  Set pPoint = pActiveView.ScreenDisplay. _
DisplayTransformation.ToMapPoint(x, y)
  ' Part 2: Perform a spatial query of features within
  ' 16,000 meters of the entered point.
  Dim pLayer As IFeatureLayer
  Dim pSpatialFilter As ISpatialFilter
  Dim pTopoOperator As ITopologicalOperator
  Dim pSelection As IFeatureSelection
  Dim pElement As IElement
  Dim pSelectionSet As ISelectionSet
  Set pLayer = pMxDoc.FocusMap.Layer(0)
  ' Create a 16000-meter buffer polygon around the clicked
  ' point.
  Set pTopoOperator = pPoint
  Set pElement = New PolygonElement
pElement.Geometry = pTopoOperator.Buffer(16000)
  ' Create a spatial filter for selecting features within
  ' the buffer polygon.
```

```
    Set pSpatialFilter = New SpatialFilter
    pSpatialFilter.SpatialRel = esriSpatialRelContains
    Set pSpatialFilter.Geometry = pElement.Geometry
    ' Refresh the active view.
    pActiveView.PartialRefresh esriViewGeoSelection, _
    Nothing, Nothing
    ' Perform spatial query.
    Set pSelection = pLayer
    pSelection.SelectFeatures pSpatialFilter, _
    esriSelectionResultNew, False
    ' Refresh the active view to highlight the selected
    ' features.
    pActiveView.PartialRefresh esriViewGeoSelection, _
    Nothing, Nothing
    ' Create a selection set and report number of
    ' features in the set.
    Set pSelectionSet = pSelection.SelectionSet
    MsgBox pSelectionSet.Count & " thermals selected"
End Sub
```

After the macro has been compiled successfully, the next step is to link the macro to a tool. In this case, the new tool is added to the Thermal toolbar from the previous section, as follows:

1. Open the Customize dialog.
2. Click Commands in the Customize dialog. Select the category of UIControls and click New UIControl. In the next dialog, select UIToolControl and then click Create. Rename the new control Project.UIToolQuery. Drag and drop Project.UIToolQuery onto the Thermal toolbar.
3. Right-click UIToolQuery and select View Source. View Source opens Visual Basic Editor. The top of the Code window has the object dropdown list on the left and the (event) procedure list on the right. The object list shows UIToolQuery and the procedure list show the default procedure of Select. This application, however, uses the MouseDown event. Click the procedure dropdown arrow and choose MouseDown (Figure 3.7). Visual Basic Editor automatically inserts the wrapper lines for the UIToolQuery_MouseDown Sub in the Code window. Copy and paste ***UIToolQuery_MouseDown*** on the companion CD to the Code window. (The UIToolQuery_Select Sub remains in the Code window. It can be left alone or deleted.) Close Visual Basic Editor.

The Thermal toolbar now has a button and a tool. To test how the new tool works, do the following:

1. Make sure that *thermal.shp* is still in view. Click the new tool. Then click a point on the map.

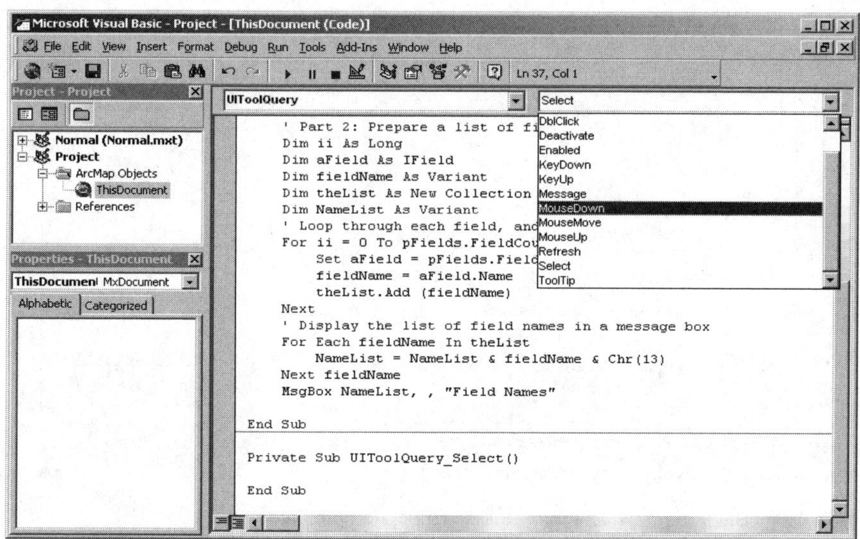

Figure 3.7   The procedure list shows different events including MouseDown.

2. A message box reports how many thermal wells and springs are within 16,000 meters of the entered point. At the same time, the selected thermal wells and springs are highlighted in the map.
3. Click another point on the map. The tool again reports the number of features selected and refreshes the map to show the newly selected features.

3.4 STORING A NEW TOOLBAR IN A TEMPLATE

A customized toolbar such as the Thermal toolbar can be saved for future use or distributed to other users. ArcMap users can save a customized application at three different levels: the Normal template (Normal.mxt), a base template (.mxt), or the current map document (.mxd). Normal.mxt is used every time ArcMap is launched. An mxd file, on the other hand, is available only in a particular map document. A base template represents an intermediate customization between Normal.mxt and the local mxd file. An mxt file is used whenever the user opts to opens the file.

A layout template (e.g., USA.mxt) is one type of template that is familiar to many ArcGIS users; it has a layout design complete with map elements such as a legend and a scale bar. To make a map based on a layout template, we only have to add data, a title, and any other supporting information. A layout template therefore represents a customized application that is available to any ArcMap users who ask for it.

The following shows how to make a template that contains the Thermal toolbar so that the template can be distributed to other users:

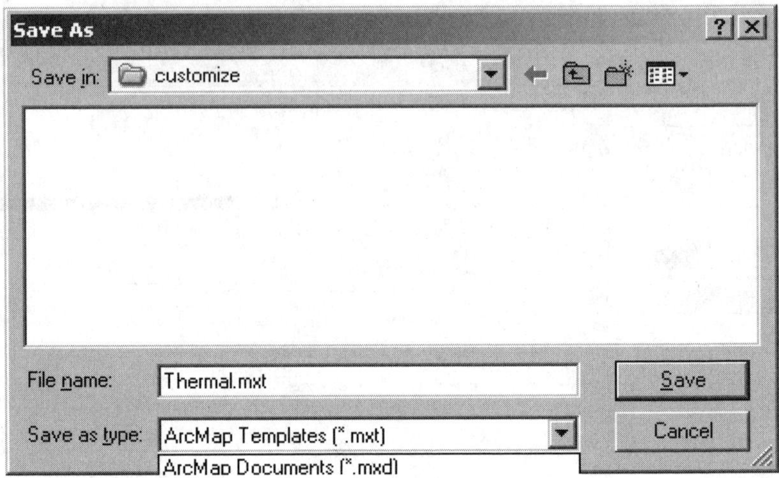

Figure 3.8   Save the new Thermal toolbar as an ArcMap Template.

1. Use *thermal* to test that the commands on the Thermal toolbar work correctly. Remove *thermal*. Select Save As from the File menu in ArcMap. In the Save As dialog, select to save as ArcMap Templates (*.mxt) and enter *Thermal.mxt* for the file name (Figure 3.8). Exit ArcMap.
2. Anyone who has access to *Thermal.mxt* can now use the commands on the Thermal toolbar. Launch ArcMap. Click on *Thermal.mxt* in the ArcMap dialog to open the template. (If *Thermal.mxt* does not show up in the dialog, select Open in the File menu to locate and open the template.) The commands on the Thermal toolbar are ready to work with the top layer in the active map.

3.5 ADDING A FORM

A form is a dialog box that uses controls such as text boxes and command buttons for the user interface. As discussed in Chapter 2, a variety of dialog boxes or forms exist. For example, a message box or an input box is actually a form, albeit with only a couple of controls.

Forms are particularly useful in geographic information systems (GIS) for gathering from the user various inputs needed for an operation. For example, a form can be used to get a numeric field, the number of classes, and the classification method before making a graduated color map. (Chapter 8 has an example of using a form to gather such inputs.)

As an introduction to forms, this section uses a relatively simple form with four controls: a label, a dropdown list with acres and square miles, a command button to run, and a command button to cancel. The user can use the form to convert the area units of a feature class from square meters to either acres or square miles and to save the new area units in a new field.

3.5.1   Designing a Form

Visual Basic Editor provides the environment for designing a user form. The following shows the steps for opening a form and placing controls from the toolbox onto the form in ArcMap:

1. Click the Tools menu in ArcMap, point to Macros, and select Visual Basic Editor.
2. Right-click Project in the Project Explorer, point to Insert, and select UserForm. The Toolbox and UserForm1 now appear in Visual Basic Editor (Figure 3.9). The Properties window shows the default properties of UserForm1. Change the name of the form to *frmAreaUnits*, and change the caption to *New Area Units*. The prefix of frm in *frmAreaUnits* is the recommended naming convention for a form.
3. The Toolbox offers 15 different controls. (If the Toolbox disappears, click the View Object button at the top of the Project Explorer.) The tool tips show that these controls are: Select Object, Label, TextBox, ComboBox, ListBox, CheckBox, OptionButton, ToggleButton, Frame, CommandButton, TabStrip, MultiPage, ScrollBar, SpinButton, and Image. The Microsoft Visual Basic Help covers each control and its usage. The sample form in this section uses a label, a combo box, and two command buttons.
4. This step adds controls to the form. Drag the label control from the toolbox and drop it onto the form. Drag and drop a combo box and two command buttons onto the form. At design time, the controls on the form are graphic elements. Therefore they can be added, removed, resized, and repositioned. Arrange the controls so that the form looks like Figure 3.10.

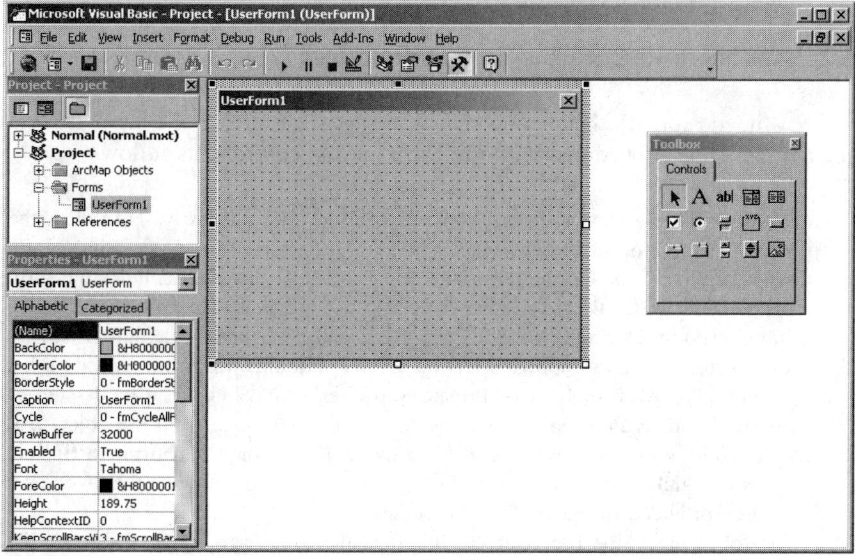

Figure 3.9   Controls in the Toolbox are placed onto UserForm1 to make a form. The Properties window shows the properties of each control including the form.

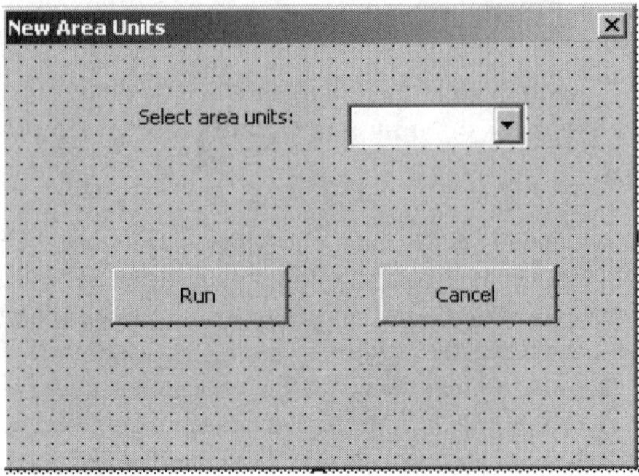

Figure 3.10   The New Area Units form contains four controls.

5. The Properties window allows the user to set the properties of a control at design time. The alternative is to set control properties at run time. This step is to set the properties of each of the controls on *frmAreaUnits* at design time. Click Label1 on the form. Rename the label *lblUnits* and change its caption to Select area units:. Rename the combo box *cboUnits*, and change the Style property to *2 — fmStyleDropDownList* on the dropdown list. Rename the first command button *cmdRun* and change its caption to Run. Rename the second command button *cmdCancel* and change its caption to Cancel. Again, the prefixes lbl, cbo, and cmd are the recommended naming conventions for labels, combo boxes, and command buttons respectively.

3.5.2   Associating Controls with Procedures

After the design of the *frmAreaUnits* form is complete, the next task is to associate the event procedures with the form and its controls, as follows:

1. Double-click the form to open the Code window. (An alternative is to click the View Code button in the Project Explorer.) At the top of the Code window are two dropdown lists. On the left is the object (control) list that includes the form and its controls. On the right is the procedure list. VBA automatically adds Private Sub UserForm_Click() to the Code window because Click is the default procedure for a form. For this sample application, choose Initialize from the procedure list instead. [The wrapper lines of Private Sub UserForm_Click() can be deleted or simply ignored.] Proceed to select *cmdRun* and *cmdCancel* from the object list. The Code window now has the wrapper lines for UserForm_Initialize, cmdRun_Click, and cmdCancel_Click (Figure 3.11). To complete the task, code must be provided for each of the procedures.

2. The following is the procedure to initialize the user form. Copy and paste the **UserForm_Initialize** code on the companion CD into the procedure. The code adds two choices of area units to the combo box at run time.

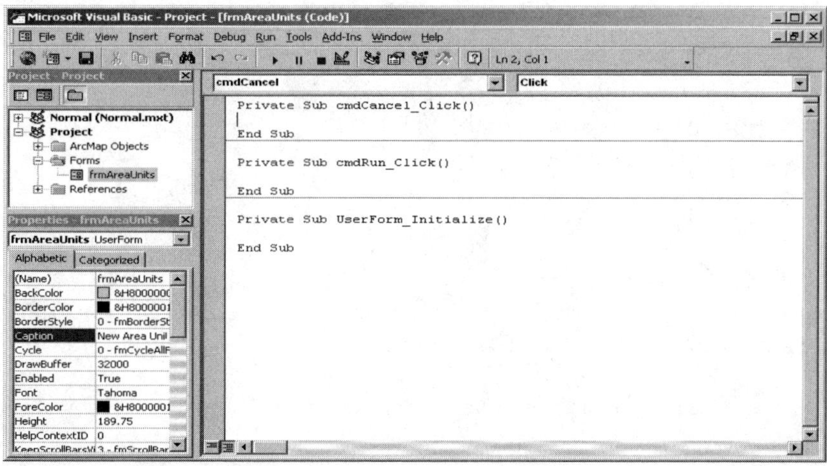

Figure 3.11   The Code window shows the wrapper lines of the three subs to be used for converting area units.

```
Private Sub UserForm_Initialize()
   ' Add items to the dropdown list.
   cboUnits.AddItem "Acres"
   cboUnits.AddItem "SqMiles"
End Sub
```

3. Next is the procedure for cmdRun_Click. Copy and paste the ***cmdRun_Click*** code on the companion CD into the procedure. At a click of the Run command button, the code adds either Acres or SqMiles as a new field to the feature class of the top layer in the active map, prepares a feature cursor, and calculates the new field values. Notice that cboUnits.ListIndex is used in Parts 2 and 3 of the procedure to determine if the user's choice is Acres or SqMiles. If the value of cboUnits.ListIndex is zero, the user's choice is Acres. And if the value is one, the user's choice is SqMiles.

```
Private Sub cmdRun_Click()
   ' Part 1: Define the feature class.
   Dim pMxDoc As IMxDocument
   Dim pFeatureLayer As IFeatureLayer
   Dim pFeatureClass As IFeatureClass
   Set pMxDoc = ThisDocument
   Set pFeatureLayer = pMxDoc.FocusMap.Layer(0)
   Set pFeatureClass = pFeatureLayer.FeatureClass
```

```vba
' Part 2: Add Acres or SqMiles as a new field.
Dim pField As IFieldEdit
Set pField = New Field
pField.Type = esriFieldTypeDouble
If cboUnits.ListIndex = 0 Then
  pField.Name = "Acres"
Else
  pField.Name = "SqMiles"
End If
pFeatureClass.AddField pField

' Part 3: Calculate the new field values.
Dim pCursor As ICursor
Dim pCalculator As ICalculator
' Prepare a cursor with all records.
Set pCursor = pFeatureClass.Update(Nothing, True)
' Define a calculator.
Set pCalculator = New Calculator
Set pCalculator.Cursor = pCursor
' Calculate the field values.
If cboUnits.ListIndex = 0 Then
  pCalculator.Expression = "[Area]/4046.7808"
  pCalculator.Field = "Acres"
  pCalculator.Calculate
Else
  pCalculator.Expression = "([Area]/1000000) * 0.3861"
  pCalculator.Field = "SqMiles"
  pCalculator.Calculate
End If
End Sub
```

4. Finally, the following is the procedure for cmdCancel_Click. Type *End* between the wrapper lines. The *End* statement terminates code execution.

```vba
Private Sub cmdCancel_Click()
  End
End Sub
```

3.5.3   Running a Form

To test how the *frmAreaUnits* form works, do the following:

1. Add *idcounty.shp* to ArcMap. The county shapefile has square meters as area units.
2. Click the Run Sub/UserForm button. The form appears. Select either Acres or SqMiles from the dropdown list. Click Run on the form.
3. Open the attribute table of *idcounty*. A new field has been added to the table and the field values have been calculated.

We can export the *frmAreaUnits* form, after it has been tested successfully, by selecting Export File from the File menu in Visual Basic Editor. The form is saved as a form file with the frm extension. Additionally, an frx file is created to save information about the graphics on the form. (frmAreaUnits_Copy.frm on the companion CD is a copy of the form.)

3.5.4   Linking a Button to a Form

This section shows how to link a customized button to the *frmAreaUnits* form so that when the button is clicked, it will open the form for use:

1. Make sure that the *frmAreaUnits* form is still available in the Project Explorer of Visual Basic Editor. Otherwise, import the form.
2. Select Customize from the Tools menu in ArcMap to open the Customize dialog. Click New in the Customize dialog. Enter Calculate Area Units for the toolbar name and save the toolbar in Untitled.
3. On the Commands tab of the Customize dialog, select the category of UIControls and click the New UIControl button. In the next dialog, check the option button for UIButtonControl and click Create. Change the name of the new command to Project.UIButtonUnits. Drag and drop Project.UIButtonUnits onto the new toolbar.
4. Right-click the new button control and select View Source. Visual Basic Editor opens with the wrapper lines of the UIButtonUnits_Click() Sub in the Code window. Type the following line between the wrapper lines: *frmAreaUnits.Show.* When this line of code runs, the Show method opens the *frmAreaUnits* form.
5. Close Visual Basic Editor. Add *idcounty2.shp* to an active map. Click on the customized button. The New Area Units form appears and is ready for use.

3.6 STORING A FORM IN A TEMPLATE

Similar to a new toolbar with commands, a form and its controls can be stored in the Normal.mxt, a base template, or a map document. The following shows how to store ***frmAreaUnits.frm*** in a base template:

1. Exit ArcMap so that the template to be created will not have datasets from the previous section. Launch ArcMap, and open Visual Basic Editor. Right-click Project in the Project Explorer and select Import File. Import ***frmAreaUnits.frm***.

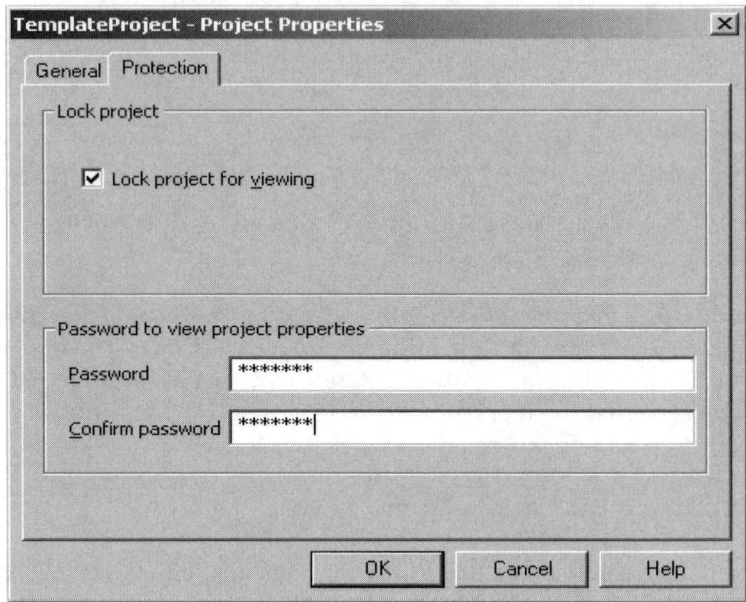

Figure 3.12   The Protection tab of the Template Properties dialog lets the user enter the protection password.

2. Select Save As from the File menu in ArcMap. In the Save As dialog, select to save as ArcMap Templates (*.mxt) and enter *AreaUnits.mxt* for the file name. Exit ArcMap.

3. ArcMap offers the password protection to viewing project properties. This step is to add the password protection to *AreaUnits.mxt*. Launch ArcMap, and open *AreaUnits.mxt*. Open Visual Basic Editor in ArcMap. The Project Explorer lists TemplateProject(AreaUnits.mxt). Select TemplateProject Properties by right-clicking TemplateProject(AreaUnits.mxt). On the Protection tab of the next dialog, choose to lock the project for viewing and enter a password for protection (Figure 3.12). Select Save AreaUnits.mxt from the File menu of Visual Basic Editor.

4. The next time that *AreaUnits.mxt* is opened in ArcMap, a password is required to view the form and its associated procedures.

VBA users can only save a customization in the Normal.mxt, an .mxt file, or an .mxd file. To create a DLL (dynamic-link library) or an EXE (executable) file, we must use standalone Visual Basic, C++, or other programming languages.

CHAPTER 4

Dataset and Layer Management

The geodatabase data model separates geographic data from nongeographic data. Geographic data include the geometry of spatial features, whereas nongeographic data do not. Geographic data include feature-based and raster-based datasets, and nongeographic data include tables in text, dBASE, and other formats.

The first step in many custom applications is to add geographic datasets as layers in ArcMap. A layer is a reference to a geographic dataset. This definition of layer carries two meanings:

First, a layer must be associated with a dataset. A layer can therefore be described as a feature layer if it is associated with a feature-based dataset such as a shapefile, a coverage, or a geodatabase feature class. A raster layer refers to a layer that is associated with a raster dataset.

Second, a layer is a graphic representation of a geographic dataset. We can therefore use different attributes and different symbols to display a layer without affecting the underlying dataset.

ArcMap organizes layers hierarchically. A map document may consist of one or more data frames, and a data frame may have one or more layers. Within a data frame, a layer can be added, deleted, or changed in the drawing order. A layer can also be saved as a layer file.

Nongeographic data are called tables in ArcMap. Tables and layers are managed separately in ArcMap, although tables can be added and deleted in the same way as layers can. The term table can be confusing at times. A dBASE file is a table; so is a feature class. However, a feature class has the geometry field, which sets it apart from a nongeographic table.

This chapter covers management of datasets and layers. Section 4.1 describes use of datasets in ArcGIS. Section 4.2 reviews objects relevant to datasets and layers in ArcObjects. Section 4.3 includes a series of macros for adding different types of datasets in ArcMap. Section 4.4 offers a macro for managing layers in an active map. Section 4.5 discusses macros for copying and deleting datasets. Section 4.6 includes a macro for reporting the spatial reference and area extent of a geographic dataset. All macros start with the listing of key interfaces and key members (i.e., properties and methods) and the usage.

4.1 USING DATASETS IN ARCGIS

ArcCatalog is the ArcGIS Desktop application for managing datasets. The catalog tree groups datasets by the connected folder. Within each folder, different icons represent different types of datasets. The context menu of each dataset, regardless of its type, offers commands to copy, delete, and rename the dataset.

ArcMap is the application for displaying and analyzing datasets. The Add Data button lets the user add geographic datasets as layers and nongeographic datasets as tables to an active data frame. Each data frame has a context menu with commands for removing or activating the data frame. Each data frame also has a table of contents, which lists the datasets it contains. The table of contents has two tabs: Display and Source. The Display tab shows the drawing order of the layers. The Source tab organizes the layers and tables by data source. To list tables that have been added as datasets to a data frame, the table of contents must be on the Source tab.

The context menu of a layer has a command to remove the dataset from an active data frame. It also has a command to save the layer as a layer file. The context menu of a table has commands to remove and open the table.

4.2 ARCOBJECTS FOR DATASETS AND LAYERS

The ArcMap subsystem contains objects for maps and layers. Figure 4.1 shows the hierarchical structure of these objects. At the top of the hierarchy is the *Application*, which in this case represents ArcMap. The *Application* is composed of *MapDocument* objects, a *MapDocument* object is composed of *Map* objects, and a *Map* object is composed of *Layer* objects. A map object represents a data frame in ArcMap. Examples of layers include feature layers, raster layers, and TIN (triangulated irregular network) layers.

The Geodatabase subsystem of ArcObjects contains objects for datasets and data sources. Figure 4.2 shows the hierarchical structure of these objects. At the top of the hierarchy is the *WorkspaceFactory* abstract class. Many coclasses inherit the properties and methods of the *WorkspaceFactory* class. They include *Shapefile-WorkspaceFactory*, *ArcInfoWorkspaceFactory*, *RasterWorkspaceFactory*, and *AccessWorkspaceFactory*. The naming of a workspace factory corresponds to the data source.

A workspace factory object can create a new workspace. The *OpenFromFile* method on *IWorkspaceFactory*, for example, returns an interface (i.e., *IWorkspace*) on a workspace by following the pathname of a file or directory. Using the returned *IWorkspace*, we can QueryInterface (QI) for *IFeatureWorkspace* to open feature-based datasets such as shapefiles or feature classes or QI for *IRasterWorkspace* to

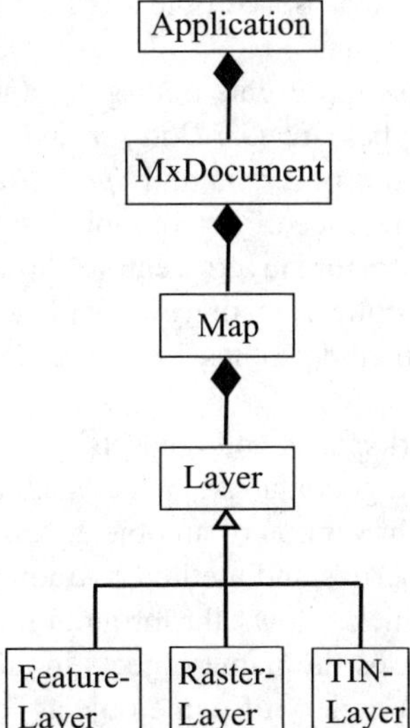

Figure 4.1 The hierarchical structure of the *Application*, *MxDocument*, *Map*, and *Layer* classes in ArcMap.

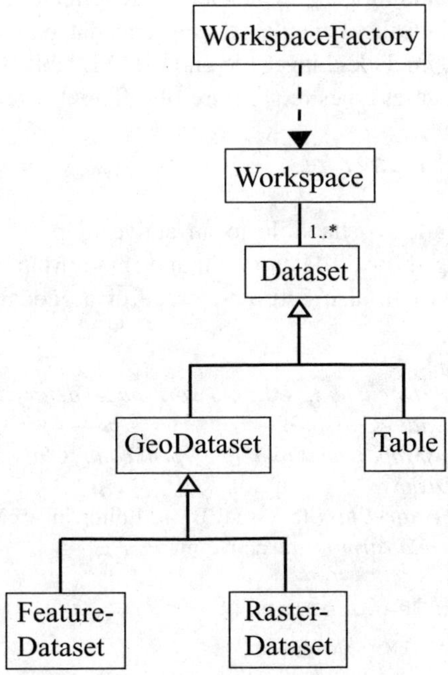

Figure 4.2 The hierarchical structure of the *WorkspaceFactory*, *Workspace*, and *Dataset* classes in Geodatabase.

open raster-based datasets. A workspace object is therefore a container of different types of datasets.

The *Dataset* abstract class represents both geographic and nongeographic data (Figure 4.2). Two *Dataset* types are *GeoDataset* and *Table*. A *GeoDataset* object has the two important properties of *Extent* and *SpatialReference*, which describe the area extent and the spatial reference of a geographic dataset respectively. Examples of geodataset objects include feature layers, feature classes, raster datasets, and raster layers. A *Table* object is a collection of rows with attributes stored in columns. Examples of table objects include tables in text and dBASE formats as well as feature classes.

Macros dealing with workspaces and datasets often use name objects. A *Name* object identifies and locates a geodatabase object such as a workspace or a dataset. A name object is a lightweight version of an object, because a name object typically has a limited number of properties and methods. Among the methods a name object has is the *Open* method, which allows the programmer to get an instance of the actual object. This chapter uses the name objects in adding a nongeographic table in ArcMap. Chapter 6 uses a variety of name objects for data conversion.

4.3 ADDING DATASETS AS LAYERS

This section covers adding geographic and nongeographic datasets in ArcMap. The geographic datasets include the shapefile, coverage, geodatabase feature class, and raster. Nongeographic datasets include a layer file and dBASE table. From the programming perspective, different dataset types require use of different data source objects.

4.3.1 *AddFeatureClass*

AddFeatureClass adds a shapefile to an active map. The macro performs the same function as using the Add Data command in ArcMap. With minor modifications, *AddFeatureClass* can also add a coverage or a geodatabase feature class to an active map.

> **Key Interfaces:** *IMxDocument, IMap, IWorkspaceFactory, IFeatureWorkspace, IFeatureLayer, IFeatureClass.*
> **Key Members:** *FocusMap, OpenFromFile, OpenFeatureClass, FeatureClass, Name, AliasName, AddLayer.*
> **Usage:** Import *AddFeatureClass* to Visual Basic Editor in ArcMap. Run the macro. The macro adds *emidastrm* to the active map.

```
Private Sub AddFeatureClass()
    Dim pMxDoc As IMxDocument
    Dim pMap As IMap
    Dim pWorkspaceFactory As IWorkspaceFactory
    Dim pFeatureWorkspace As IFeatureWorkspace
    Dim pFeatureLayer As IFeatureLayer
```

```
   Dim pFeatureClass As IFeatureClass
   ' Specify the workspace and the feature class.
   Set pWorkspaceFactory = New ShapefileWorkspaceFactory
   Set pFeatureWorkspace = pWorkspaceFactory.OpenFromFile _
   ("c:\data\chap4", 0)
   Set pFeatureClass =  pFeatureWorkspace. _
   OpenFeatureClass("emidastrm")
   ' Prepare a feature layer.
   Set pFeatureLayer = New FeatureLayer
   Set pFeatureLayer.FeatureClass = pFeatureClass
   pFeatureLayer.Name = pFeatureLayer.FeatureClass. _
   AliasName
   ' Add the feature layer to the active map.
   Set pMxDoc = ThisDocument
   Set pMap = pMxDoc.FocusMap
   pMap.AddLayer pFeatureLayer
   ' Refresh the active view.
   PMxDoc.ActiveView.Refresh
End Sub
```

The macro first creates *pWorkspaceFactory* as an instance of the *Shapefile-WorkspaceFactory* class. Next the code uses the *OpenFromFile* method on *IWorkspaceFactory* to return an *IWorkspace*, performs a QI for the *IFeatureWorkspace* interface, and uses the *OpenFeatureClass* method to open a feature class. The feature class is *emidastrm*, which is referenced by *pFeatureClass*. Using *pFeatureClass* and its name, the code creates *pFeatureLayer* as an instance of the *FeatureLayer* class. The last part of the macro adds *pFeatureLayer* to an active map. The code sets *pMxDoc* to be ThisDocument and *pMap* to be the focus map of *pMxDoc*. (This- Document is the predefined name of the *MxDoc* object, which, along with the *Application* object, is already in use when ArcMap is launched.) The *AddLayer* method on *IMap* adds *pFeatureLayer* to the active map, before refreshing the view. Figure 4.3 illustrates the process of adding a shapefile to an active map, which involves objects from the Geodatabase and ArcMap subsystems.

With two minor changes, we can use **AddFeatureClass** to add a coverage to an active map. Suppose we want to add the arcs of the *breakstrm* coverage. The first change relates to the workspace factory and the path to the feature class:

```
   ' Specify the workspace and the feature class.
   Set pWorkspaceFactory = New ArcInfoWorkspaceFactory
   Set pFeatureWorkspace = pWorkspaceFactory. _
   OpenFromFile("c:\data\chap4\", 0)
   Set pFeatureClass = pFeatureWorkspace. _
   OpenFeatureClass("breakstrm:Arc")
```

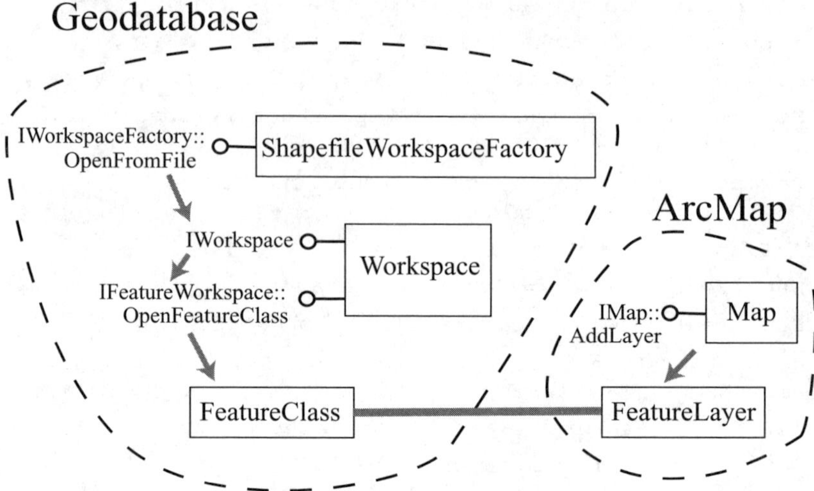

Figure 4.3   The process of adding a shapefile from its data source to an active map involves objects in Geodatabase and ArcMap.

ArcInfoWorkspaceFactory is the class that creates workspaces for coverages. Also, the argument for the *OpenFeatureClass* method must specify arc for the feature class.

The second change relates to the name property of *pFeatureLayer*, which includes a prefix of breakstrm:. Without the prefix, the feature layer would be named simply as Arc.

```
' Add the prefix to the layer name.
pFeatureLayer.Name = "breakstrm: " & pFeatureLayer. _
FeatureClass.AliasName
```

With one minor change, we can also use ***AddFeatureClass*** to add a geodatabase feature class to an active map. The feature class can be either standalone or part of a feature dataset. For example, to add the *emidastrum* feature class in *emida.mdb*, we need to make the following change in ***AddFeatureClass***:

```
' Specify the workspace and the feature class.
Set pWorkspaceFactory = New AccessWorkspaceFactory
Set pFeatureWorkspace = pWorkspaceFactory. _
OpenFromFile("c:\data\chap4\emida.mdb", 0)
Set pFeatureClass = pFeatureWorkspace. _
OpenFeatureClass("emidastrm")
```

AccessWorkspaceFactory is the class that creates workspaces for geodatabases. Also, the path to the feature workspace must include the geodatabase (i.e., *emida.mdb*). Each feature class, whether it is standalone or part of a feature dataset, has a unique name. Therefore, the name can be used in the *OpenFeatureClass* method to open the feature class regardless of its type.

4.3.2 *AddFeatureClasses*

AddFeatureClasses lets the user select shapefiles from a dialog box and adds them to an active map. The dialog box is similar to the Add Data tool in ArcMap, except that it only shows shapefiles.

> **Key Interfaces:** *IGxDialog, IGxObjectFilter, IEnumGxObject, IGxDataset.*
> **Key Members:** *AllowMultiSelect, ButtonCaption, ObjectFilter, StartingLocation, Title, DoModalOpen, Next, Refresh, UpdateContents.*
> **Usage:** Import *AddFeatureClasses* to Visual Basic Editor in ArcMap. Run the macro. A dialog box with the Add Shapefiles caption appears. Choose *idcities.shp* and *idcounty.shp* and click Add. The macro adds the two shapefiles, which are based on the same coordinate system, to the active map.

```
Private Sub AddFeatureClasses()
  ' Part 1: Prepare an Add Shapefiles dialog.
  Dim pGxDialog As IGxDialog
  Dim pGxFilter As IGxObjectFilter
  Set pGxDialog = New GxDialog
  Set pGxFilter = New GxFilterShapefiles
  ' Define the dialog's properties.
  With pGxDialog
    .AllowMultiSelect = True
    .ButtonCaption = "Add"
    Set.ObjectFilter = pGxFilter
    .StartingLocation = "c:\data\chap4"
    .Title = "Add Shapefiles"
  End With
```

Part 1 prepares an Add Shapefiles dialog. The code first creates *pGxDialog* as an instance of the *GxDialog* class and *pGxFilter* as an instance of the *GxFilter-Shapefiles* class. Both *GxDialog* and *GxFilter* are ArcCatalog classes. COM (Component Object Model) technology allows ArcCatalog objects to be used in ArcMap. A *GxDialog* object is basically a form that has been coded by ArcGIS developers to accept the datasets selected by the user and to add them to ArcMap. A *GxFilter* object filters the type of data to be displayed in a *GxDialog* object. *GxFilter* is an abstract class with more than 30 different types. Part 1 uses the *GxFilterShapefiles* class, thus limiting the data sources to only shapefiles. The rest of Part 1 uses a *With* block to define the properties of *pGxDialog*: the title is Add Shapefiles, the button caption is Add, the object filter is *pGxFilter*, and the starting location is the path to the data sources. The *AllowMultiSelect* property is set to be true, meaning that the user can select multiple datasets. If false, then the user can only select a single dataset at a time.

```
    ' Part 2: Get the datasets from the dialog and add them
    ' to the active map.
    Dim pGxObjects As IEnumGxObject
    Dim pMxDoc As IMxDocument
    Dim pMap As IMap
    Dim pGxDataset As IGxDataset
    Dim pLayer As IFeatureLayer
    Set pMxDoc = ThisDocument
    Set pMap = pMxDoc.FocusMap
    ' Open the dialog.
    pGxDialog.DoModalOpen 0, pGxObjects
    Set pGxDataset = pGxObjects.Next
    ' Exit sub if no dataset has been added.
    If pGxDataset Is Nothing Then
      Exit Sub
    End If
    ' Step through the datasets and add them as layers to
    ' the active map.
    Do Until pGxDataset Is Nothing
      Set pLayer = New FeatureLayer
      Set pLayer.FeatureClass = pGxDataset.Dataset
      pLayer.Name = pLayer.FeatureClass.AliasName
      pMap.AddLayer pLayer
      Set pGxDataset = pGxObjects.Next
    Loop
    ' Refresh the map and update the table of contents.
    pMxDoc.ActivatedView.Refresh
    pMxDoc.UpdateContents
  End Sub
```

Part 2 gets the datasets selected by the user and adds them as layers to the active map. The *DoModalOpen* method on *IGxDialog* opens the dialog box and saves the selected shapefiles into a collection. In the code, the first argument for *DoModalOpen* is set to be zero (i.e., to use the ArcMap window) and the second is *pGxObjects*, a reference to an *EnumGxObject*. An *EnumGxObject* represents a collection of ordered objects. The code uses the *Next* method on *IEnumGxObject* to advance one object at a time and assigns the object to the *pGxDataset* variable, a reference to a *GxDataset* object. A type of *GxObject*, a *GxDataset* object represents a dataset (Figure 4.4). If the *Next* method advances nothing the first time, exit the sub. If the

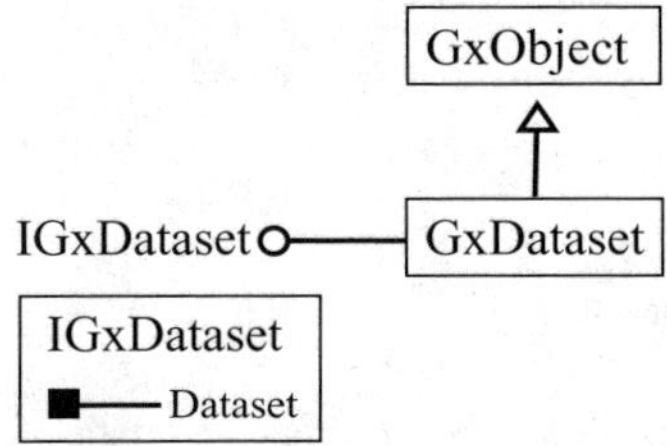

Figure 4.4 A type of GxObject, a GxDataset object represents a dataset, which can be read via the *Dataset* property on *IGxDataset*.

collection contains selected shapefiles, then the code uses a *Do...Loop* to step through each of them. Within each loop, the code creates *pLayer* as an instance of the *FeatureLayer* class, assigns the dataset of *pGxDataset* to be the feature class of *pLayer*, and adds the layer to the active map. Finally, the code refreshes the view and updates the table of contents of the map document.

4.3.3 *AddRaster*

AddRaster adds a raster dataset to an active map. The macro performs the same function as using the Add Data command in ArcMap.

Key Interfaces: *IWorkspaceFactory, IRasterWorkspace, IRasterLayer, IRaster-Dataset.*
Key Members: *OpenFromFile, OpenRasterDataset, CreateFromDataset, AddLayer.*
Usage: Import *AddRaster* to Visual Basic Editor in ArcMap. Run the macro. The macro adds *emidalat* to the active map.

```vb
Private Sub AddRaster()
  Dim pMxDoc As IMxDocument
  Dim pMap As IMap
  Dim pWorkspaceFactory As IWorkspaceFactory
  Dim pRasterWorkspace As IRasterWorkspace
  Dim pRasterDS As IRasterDataset
  Dim pRasterLayer As IRasterLayer
  ' Specify the workspace and the raster dataset.
  Set pWorkspaceFactory = New RasterWorkspaceFactory
  Set pRasterWorkspace = pWorkspaceFactory. _
  OpenFromFile("c:\data\chap4\", 0)
  Set pRasterDS = pRasterWorkspace.OpenRasterDataset _
  ("emidalat")
  ' Prepare a raster layer.
  Set pRasterLayer = New RasterLayer
  pRasterLayer.CreateFromDataset pRasterDS
  ' Add the raster layer to the active map.
```

```
  Set pMxDoc = ThisDocument
  Set pMap = pMxDoc.FocusMap
  pMap.AddLayer pRasterLayer
  pMxDoc.ActiveView.Refresh
End Sub
```

The macro creates *pWorkspaceFactory* as an instance of the *RasterWorkspace-Factory* class and uses the *OpenFromFile* method on *IWorkspaceFactory* to open a raster-based workspace referenced by *pRasterWorkspace*. Next the code uses the *OpenRasterDataset* method on *IRasterWorkspace* to open a raster dataset named *emidalat* and referenced by *pRasterDS*. Then the code creates *pRasterLayer* as an instance of the *RasterLayer* class and defines its dataset. Finally, the code uses the *AddLayer* method on *IMap* to add *pRasterLayer* to the active map.

The *CreateFromDataset* method is one of the three methods on *IRasterLayer* for creating a raster layer. The other two methods are *CreateFromFilePath* and *CreateFromRaster* (Figure 4.5). The following macro uses the *CreateFromFilePath* method to complete the same task as **AddRaster**.

```
Private Sub AddRaster_2()
  Dim pMxDocument As IMxDocument
  Dim pMap As IMap
  Set pMxDocument = ThisDocument
  Set pMap = pMxDocument.FocusMap
  Dim pRasterLayer As IRasterLayer
  Set pRasterLayer = New RasterLayer
  pRasterLayer.CreateFromFilePath "c:\data\chap4\emidalat"
  pMap.AddLayer pRasterLayer
End Sub
```

4.3.4   *AddLayerFile*

AddLayerFile adds a layer file to an active map. The macro performs the same function as using the Add Data command in ArcMap.

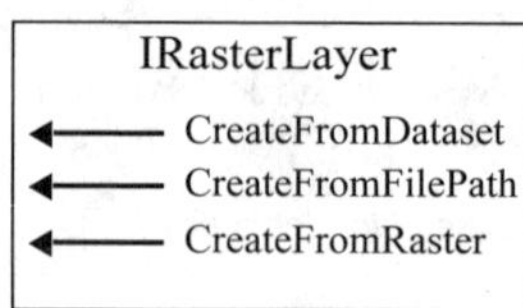

Figure 4.5   Methods on *IRasterLayer.*

Key Interfaces: *IGxFile, IGxLayer.*
Key Members: *Path, Layer, AddLayer.*
Usage: Import *AddLayerFile* to Visual Basic Editor in ArcMap. Run the macro. The macro adds *emidalat.lyr* to the active map.

```
Private Sub AddLayerFile()
  Dim pMxDoc As IMxDocument
  Dim pMap As IMap
  Dim pGxLayer As IGxLayer
  Dim pGxFile As IGxFile
  ' Get the layer file.
  Set pGxLayer = New GxLayer
  Set pGxFile = pGxLayer
  pGxFile.Path = "c:\data\chap4\emidalat.lyr"
  ' Add the layer file to the active map.
  Set pMxDoc = ThisDocument
  Set pMap = pMxDoc.FocusMap
  pMap.AddLayer pGxLayer.Layer
  pMxDoc.ActiveView.Refresh
End Sub
```

The macro creates *pGxLayer* as an instance of the *GxLayer* class. Next the code performs a QI for the *IGxFile* interface and uses the *Path* property to define the path to *emidalat.lyr* (Figure 4.6). Then the code uses the *AddLayer* method on *IMap* to add the layer associated with *pGxLayer* to the active map. Both *GxLayer* and *GxFile* are ArcCatalog objects.

4.3.5 *AddTable*

AddTable adds a nongeographic table to an active map and opens the table. The macro performs the same function as using the Add Data command in ArcMap to add a table and using the Open command to open the table. *AddTable* has three parts: Part 1 uses the name objects to define the input table, Part 2 adds the table to the active map, and Part 3 uses a table window to open the table.

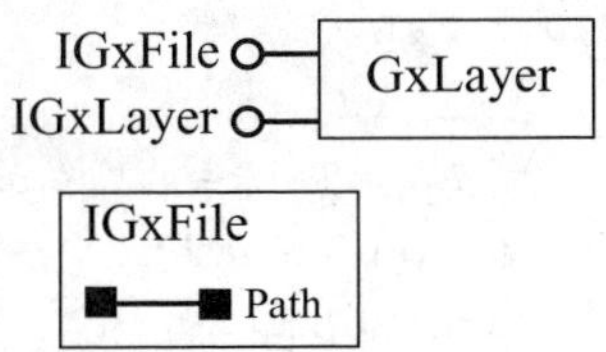

Figure 4.6 *GxLayer* and the interfaces that a GxLayer object supports.

Key Interfaces: *IWorkspaceName, IDatasetName, IName, ITable, ITableCollection, ITableWindow.*
Key Members: *WorkspaceFactoryProgID, PathName, Name, WorkspaceName, AddTable, UpdateContents, Table, Application, Show.*
Usage: Import ***AddTable*** to Visual Basic Editor in ArcMap. Run the macro. The macro opens the *comp.dbf* table and adds the table to the active map. Click on the Source tab in the table of contents to see *comp.dbf*.

```
Private Sub AddTable()
  ' Part 1: Define the input table.
  Dim pWSName As IWorkspaceName
  Dim pDatasetName As IDatasetName
  Dim pName As IName
  Dim pTable As ITable
  ' Get the dbf file by specifying its workspace and name.
  Set pDatasetName = New TableName
  Set pWSName = New WorkspaceName
  pWSName.WorkspaceFactoryProgID = _
  "esriCore.ShapefileWorkspaceFactory"
  pWSName.PathName = "c:\data\chap4"
  pDatasetName.Name = "comp.dbf"
  Set pDatasetName.WorkspaceName = pWSName
  Set pName = pDatasetName
  ' Open the dbf table.
  Set pTable = pName.Open
```

Part 1 first creates *pDatasetName* as an instance of the *TableName* class. Next the code defines the workspace name and name properties of *pDatasetName* by using members on *IWorkspaceName* and *IDatasetName* (Figure 4.7). Notice that the program ID of the workspace factory is esriCore.ShapefileWorkspaceFactory because the dataset to be added is a dBASE file. (If the dataset to add is a text file, one would opt for esriCore.TextfileWorkspaceFactory.) The code then switches to the *IName* interface and uses the *Open* method to open *pTable*.

```
  ' Part 2: Add the table to the active map.
  Dim pMxDoc As IMxDocument
  Dim pMap As IMap
  Dim pTableCollection As ITableCollection
  Set pMxDoc = Application.Document
  Set pMap = pMxDoc.FocusMap
  Set pTableCollection = pMap ' QI
  pTableCollection.AddTable pTable
  pMxDoc.UpdateContents
```

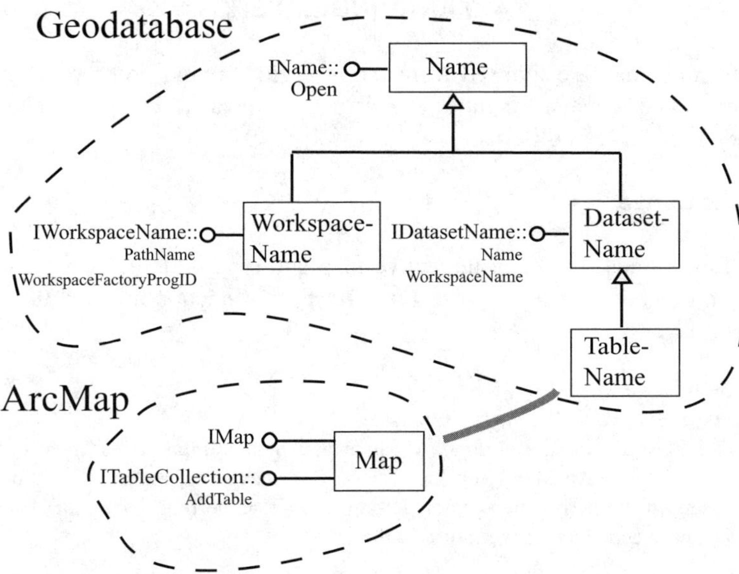

Figure 4.7   Adding a table to an active map involves objects in the Geodatabase and ArcMap subsystems. See text for explanation.

Part 2 first sets *pMap* to be the active map. The code then switches to the *ITable-Collection* interface and uses the *AddTable* method to add *pTable* to *pMap*. The *UpdateContents* method on *IMxDocument* updates ArcMap's table of contents.

```
' Part 3: Open the table in a table window.
Dim pTableWindow As ITableWindow
' Create a table window and specify its properties and
' methods.
Set pTableWindow = New TableWindow
With pTableWindow
  Set.Table = pTable
  Set.Application = Application
  .Show True
End With
End Sub
```

Part 3 creates *pTableWindow* as an instance of the *TableWindow* class. Next the code uses a *With* block to define the table window for view. The *Application* property is set to be Application, which represents ArcMap in this case. If the *Application* property is not set, the macro will crash.

4.4 MANAGING LAYERS

Layers in a map are indexed from the top with the base of zero. A common application of layer management is to locate a particular layer by its index so that the layer can be accessed in code.

4.4.1 *FindLayer*

FindLayer finds a layer in the active map and reports its index value. Sample modules in this book frequently use ***FindLayer*** as a function and call the function to access a layer.

> **Key Interfaces:** *IMap.*
> **Key Members:** *LayerCount, Layer(), name.*
> **Usage:** Add *emidalat* and *emidastrm.shp* to an active map. Import ***FindLayer*** to Visual Basic Editor in ArcMap. Run the macro. The macro first reports the number of datasets in the active map. After getting a layer name (e.g., *emidalat*) from the user, the macro reports the index of the layer.

```vba
Private Sub FindLayer()
  Dim pMxDoc As IMxDocument
  Dim pMap As IMap
  Dim FindDoc As Variant
  Dim aLName As String
  Dim name As String
  Dim i As Long
  Set pMxDoc = ThisDocument
  Set pMap = pMxDoc.FocusMap
  MsgBox "The active map has " & pMap.LayerCount & _
  " datasets."
  ' Use an input box to get a layer name.
  name = InputBox("Enter a layer name:", "")
  ' Loop through layers in the active map and match the
  ' entered name with the layer name in uppercase.
  For i = 0 To pMap.LayerCount - 1
    aLName = UCase(pMap.Layer(i).name)
    ' Given a match, assign the counter to FindDoc.
    If (aLName = (UCase(name))) Then
      FindDoc = i
    End If
  Next
  MsgBox name & " is at index " & FindDoc
End Sub
```

The macro first uses the *LayerCount* property on *IMap* to report the number of layers in the active map. Next the code gets a layer name from the user and assigns it to the *name* variable. Using a *For...Next* loop, the code steps through each layer, converts the name of the layer to its uppercase, and matches the name with the uppercase of the input name. When a match is found, its index is assigned to the *FindDoc* variable. A message box then reports the name of the layer and its index. UCase is a Visual Basic function that converts a string to the uppercase.

4.5 MANAGING DATASETS

This section covers dataset management such as copying and deleting datasets programmatically. Before deleting a dataset, the layer that uses the dataset should be first removed from ArcMap.

4.5.1 *CopyDataset*

CopyDataset copies the dataset of a layer in the active map and saves the copied dataset in a specified workspace. The macro performs the same function as using the Copy command in ArcCatalog.

Key Interfaces: *IWorkspaceFactory, IFeatureWorkspace, IDataset, IFeatureClass.*
Key Members: *OpenFromFile, Copy.*
Usage: Add *emidastrm.shp* to an active map. Import *CopyDataset* to Visual Basic Editor in ArcMap. Run the macro. The macro copies *emidastrm.shp* and saves the copy as *emidastrmCopy.shp*.

```
Private Sub CopyDataset()
  ' Part 1: Define the top layer as the dataset to be copied.
  Dim pMxDocument As IMxDocument
  Dim pMap As IMap
  Dim pFeatureLayer As IFeatureLayer
  Dim pFeatureClass As IFeatureClass
  Set pMxDocument = ThisDocument
  Set pMap = pMxDocument.FocusMap
  Set pFeatureLayer = pMap.Layer(0)
  Set pFeatureClass = pFeatureLayer.FeatureClass
```

Part 1 defines *pFeatureClass* as the feature class of the top layer in the active map. This feature class is the geographic dataset to be copied.

```
  ' Part 2: Copy the dataset and add it to the active map.
  Dim pWorkspaceFactory As IWorkspaceFactory
  Dim pFeatureWorkspace As IFeatureWorkspace
```

```
    Dim pDataset As IDataset
    Dim pCopyFC As IFeatureClass
    Dim CopyDSName As String
    ' Define the workspace for the copied dataset.
    Set pWorkspaceFactory = New ShapefileWorkspaceFactory
    Set pFeatureWorkspace = pWorkspaceFactory. _
    OpenFromFile("c:\data\chap4\", 0)
    ' Copy the dataset.
    Set pDataset = pFeatureClass
    CopyDSName = pFeatureLayer.name & "Copy"
    Set pCopyFC = pDataset.Copy _
    (CopyDSName, pFeatureWorkspace)
End Sub
```

Part 2 creates *pWorkspaceFactory* as an instance of the *ShapefileWorkspace-Factory* class and uses the *OpenFromFile* method to open a feature-based workspace. Then the code performs a QI for the *IDataset* interface, and uses the *Copy* method to make a copy of *pFeatureClass*.

4.5.2   *DeleteDataset*

DeleteDataset removes a layer from an active map and deletes the layer's dataset. The macro performs the same function as using the Remove command in ArcMap to remove a layer first and then using the Delete command in ArcCatalog to delete the layer's dataset.

> **Key Interfaces:** *IMap, IDataset, IActiveView.*
> **Key Members:** *DeleteLayer, Delete, Refresh.*
> **Usage:** Add *emidastrmCopy.shp* to an active map. Import **DeleteDataset** to Visual Basic Editor in ArcMap. Run the macro. The macro removes the *emidastrmCopy* layer from the active map and deletes *emidastrmCopy.shp*.

```
Private Sub DeleteDataset()
  Dim pMxDocument As IMxDocument
  Dim pMap As IMap
  Dim pFeatureLayer As IFeatureLayer
  Dim pFeatureClass As IFeatureClass
  Dim pDataset As IDataset
  Dim pActiveView As IActiveView
  Set pMxDocument = ThisDocument
  Set pMap = pMxDocument.FocusMap
  ' Define the dataset to be deleted.
  Set pFeatureLayer = pMap.Layer(0)
```

```
    Set pFeatureClass = pFeatureLayer.FeatureClass
    ' Remove the layer from the active map.
    pMap.DeleteLayer pFeatureLayer
    ' Delete the dataset.
    Set pDataset = pFeatureClass
    pDataset.Delete
    ' Refresh the map.
    Set pActiveView = pMap
    pActiveView.Refresh
End Sub
```

The macro sets *pFeatureLayer* to be the top layer in the active map and *pFeatureClass* to be its feature class. Next the code uses the *DeleteLayer* method on *IMap* to remove *pFeatureLayer* from the active map. Then the code switches to *IDataset* and uses the *Delete* method to delete *pFeatureClass*. Finally, the code refreshes the map.

4.6 REPORTING GEOGRAPHIC DATASET INFORMATION

This section shows how we can report the spatial reference and area extent properties of a geographic dataset object by using a macro.

4.6.1 *SpatialRef*

SpatialRef reports the spatial reference and the extent of a geographic dataset. The macro performs the same function as looking up the metadata of a geographic dataset in ArcCatalog or getting the information on the Source tab of the Layer Properties dialog in ArcMap.

> **Key Interfaces:** *IGeoDataset, ISpatialReference, IEnvelope.*
> **Key Members:** *SpatialReference, Extent, Xmin, YMin, XMax, YMax.*
> **Usage:** Add *emidastrm.shp* to an active map. Import *SpatialRef* to Visual Basic Editor in ArcMap. Run the macro. The macro reports the dataset's coordinate system and area extent.

```
Private Sub SpatialRef()
    Dim pMxDocument As IMxDocument
    Dim pMap As IMap
    Dim pFeatureLayer As IFeatureLayer
    Dim pGeoDataset As IGeoDataset
    Dim pSpatialRef As ISpatialReference
    Dim pEnvelope As IEnvelope
    Dim MinX, MaxX, MinY, MaxY As Double
```

```
Set pMxDocument = ThisDocument
Set pMap = pMxDocument.FocusMap
' Define the input geodataset.
Set pFeatureLayer = pMap.Layer(0)
Set pGeoDataset = pFeatureLayer 'QI
' Derive the spatial reference and extent of the geodataset.
Set pSpatialRef = pGeoDataset.SpatialReference
Set pEnvelope = pGeoDataset.Extent
' Get minx, miny, maxx, and maxy.
MinX = pEnvelope.XMin
MinY = pEnvelope.YMin
MaxX = pEnvelope.XMax
MaxY = pEnvelope.YMax
' Report the geodataset information.
MsgBox "The layer's spatial reference is: " & _
pSpatialRef.Name
MsgBox "Minimum X is: " & MinX & " Minimum Y is: " & _
MinY & Chr$(10) & "Maximum X is: " & MaxX & _
" Maximum Y is: " & MaxY
End Sub
```

The macro first defines *pFeatureLayer* as the top layer in the active map. Next the code switches to the *IGeoDataset* interface and derives the *SpatialReference* and *Extent* properties of *pFeatureLayer*. The extent of a geographic dataset is an envelope or a rectangular object. The code assigns the *XMin, YMin, XMax,* and *YMax* values of the envelope to the Double variables of *MinX, MinY, MaxX,* and *MaxY*, respectively. Finally, the code uses the dialog boxes to report the spatial information of *pFeatureLayer*.

CHAPTER 5

Attribute Data Management

A geographic information system (GIS) involves both geographic and attribute data: geographic data relate to the geometry of spatial features, whereas attribute data describe the characteristics of the features. The geodatabase data model uses tables to store both types of data in a relational database environment. A table with a geometry field is a feature class, a feature attribute table, or simply a geographic dataset. A table with attribute data only is a nongeographic dataset.

A table, either a feature class or a nongeographic table, consists of rows and columns. Each row represents a feature, and each column represents a characteristic. A row is also called a record, and a column a field. The intersection of a column and a row shows the value of a particular characteristic for a particular feature.

Attribute data management takes place at either the field level or the table level. At the field level, common tasks include deriving the field information, adding fields, deleting fields, and calculating the field values. Many of these tasks require working with the properties of a field such as name, type, and length. At the table level, common tasks typically involve joining and relating tables in a relational database environment. A join brings together two tables. A relate connects two tables but keeps the tables separate. Both operations use keys and relationships to link tables.

This chapter covers attribute data management. Section 5.1 reviews management of attribute data in ArcGIS. Section 5.2 discusses objects relevant to tables, fields, and relationship classes. Section 5.3 includes macros for listing fields and the field properties. Section 5.4 offers a macro for adding and deleting fields. Section 5.5 offers a macro for calculating the field values. Section 5.6 discusses macros for joining and relating tables. All macros start with the listing of key interfaces and key members (i.e., properties and methods) and the usage.

5.1 MANAGING ATTRIBUTE DATA IN ARCGIS

An ArcGIS user can add and delete fields in either ArcCatalog or ArcMap. To add a field, we must first define the field properties. Depending on the field type, the definition may include length, precision, and scale. Length is the maximum length, in bytes, reserved for the field. Precision is the number of digits reserved for a numeric field. Scale is the number of decimal digits reserved for a field of the Double data type.

The Field Calculator available through a field's context menu in ArcMap is a tool for calculating the field values. To use the Field Calculator, we must prepare a calculation expression with fields and mathematical functions.

Joins and relates are available through the context menu or the properties of a feature layer or a table in ArcMap. To add a join or relate, we must specify a table to join or relate and the fields on which the join or relate is based. The fields used in a join or relate are called keys. A primary key represents a field whose values can uniquely identify a record in a table. Its counterpart in another table for the purpose of linkage is a foreign key. Joins and relates can only be built one at a time, but existing joins and relates can be removed individually or as a group.

There are four possible relationships, also called cardinalities, in connecting two tables. The one-to-one relationship means that one and only one record in a table is related to one and only one record in another table. The one-to-many relationship means that one record in a table may be related to many records in another table. The many-to-one relationship means that many records in a table may be related to one record in another table. And the many-to-many relationship means that many records in a table may be related to many records in another table.

Joins are usually recommended for the one-to-one or many-to-one relationship. Given a one-to-one relationship, two tables are joined by record. Given a many-to-one relationship, many records in the base table have the same value from a record in the other table. Joins are inappropriate with the one-to-many relationship because only the first matching record value from the other table is assigned to a record in the base table. Relates, on the other hand, are appropriate for all four relationships.

5.2 ARCOBJECTS FOR ATTRIBUTE DATA MANAGEMENT

This section covers objects that are related to tables, fields, field, and relationship.

5.2.1 Tables

Figure 5.1 shows the hierarchical structure of *Table*, *ObjectClass*, and *FeatureClass*. The *ObjectClass* is a type of the *Table* class whose rows represent entities, and the *FeatureClass* is a type of the *ObjectClass* whose rows represent features. A feature class object has two default fields: the shape field that stores the geometry of features, and the FID field that stores the feature IDs.

Chapter 4 has shown how to access a feature class through its data source. Another way to access a feature class is through a feature layer that is already present

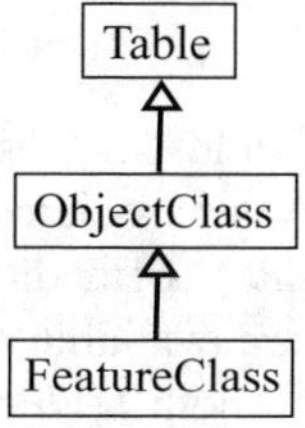

Figure 5.1 *FeatureClass* is a type of *ObjectClass*, and *ObjectClass* is a type of *Table*.

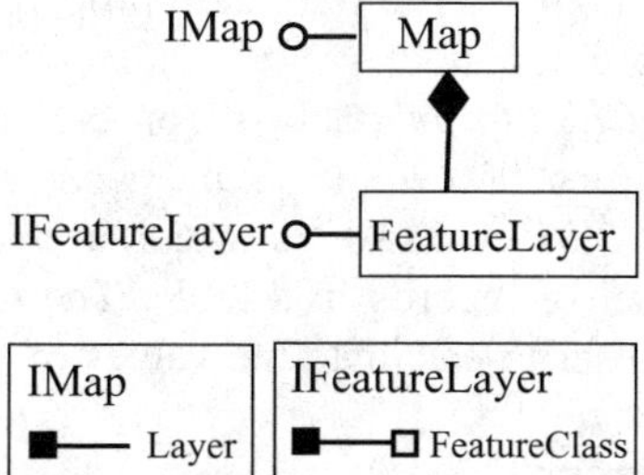

Figure 5.2 A feature class can be accessed through a feature layer in an active map.

in an active map. For example, we can access a feature layer by using the *Layer()* property of *IMap* and then access the layer's feature class by using the *FeatureClass* property on *IFeatureLayer* (Figure 5.2).

Chapter 4 has also shown how to access a dBASE file or a text file through its data source. An alternative for accessing a file is through a table that is already present in an active map, called a standalone table. A *StandaloneTable* object is not associated with a feature class but is based on a nongeographic table. Figure 5.3 shows how to access the table underlying a standalone table. First, use the *IStandaloneTableCollection* interface that a map object supports to access the standalone table. Second, use the *Table* property on *IStandaloneTable* to access the table. Conceptually, a standalone table is like a feature layer, and the *Table* property of a standalone table is like the *FeatureClass* property of a feature layer.

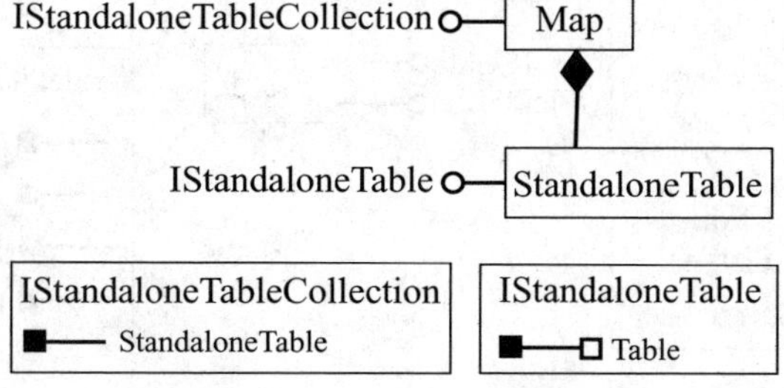

Figure 5.3 A table can be accessed through a standalone table in an active map.

5.2.2   Fields and Field

A *Fields* object is a collection of fields such as a feature class or a nongeographic table. In either case, a fields object is associated with a table object (Figure 5.4). We can therefore add, delete, or find a field through a fields or table object. The *AddField* and *DeleteField* methods are available on both *ITable* and *IFieldsEdit*, and the *FindField* method is available on both *ITable* and *IFields*.

A fields object consists of one or more *Field* objects (Figure 5.4). Each field has an index or a numbered position. A field object implements *IField* and *IFieldEdit*. *IField* has read-only field properties such as name, length, precision, and type. *IFieldEdit*, on the other hand, has write-only field properties and is therefore useful for the purpose of defining a new field.

ArcObjects has the *Calculator* coclass for calculating the field values. *ICalculator* has the properties of *Cursor*, *Expression*, and *Field* as well as the *Calculate* method (Figure 5.5). A cursor is a data-access object, which allows a macro to step through a set of records in a table. The *Calculate* method uses an expression defined by the user to calculate the values of a specified field.

5.2.3   Relationship Classes

In ArcObjects, a join or relate is defined as a relationship class linking two tables. Relationship classes can be stored in a geodatabase or, as in this chapter, built in code between tables that are in use. To build a join or relate in code requires that the relationship class be prepared as a memory (i.e., virtual) relationship class first (Figure 5.6). The *MemoryRelationshipClassFactory* coclass implements *IMemoryRelationshipClassFactory*, which has the *Open* method to create memory relationship class objects.

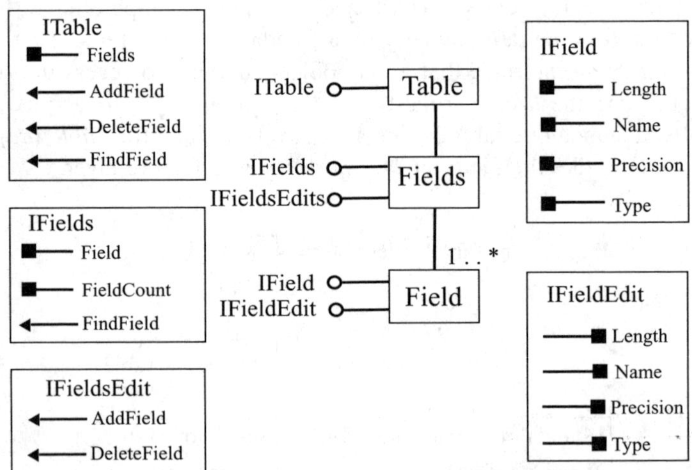

Figure 5.4   The diagram shows the relationship between the *Table*, *Fields*, and *Field* objects as well as the properties and methods of these objects.

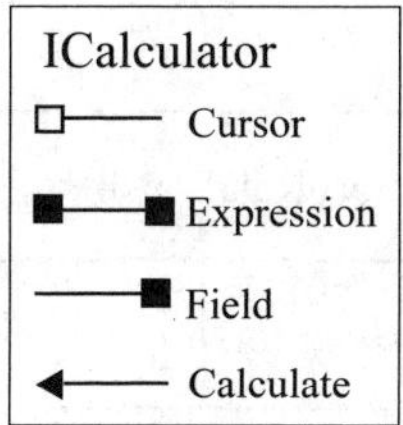

Figure 5.5   Properties and methods on *ICalculator.*

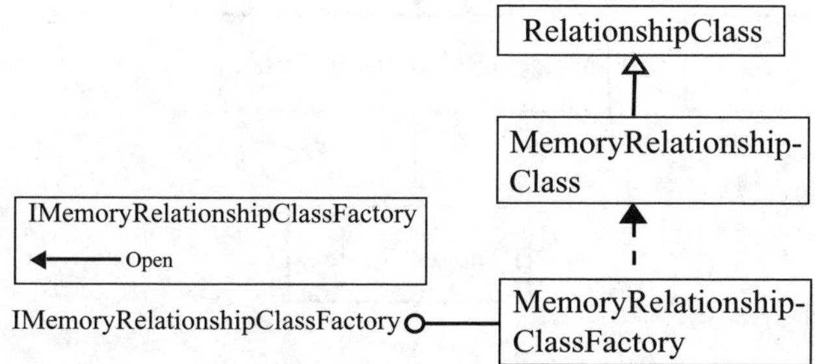

Figure 5.6   A *MemoryRelationshipClassFactory* can create a *MemoryRelationshipClass* object, which is a type of *RelationshipClass.*

After a memory relationship class is opened, a join or relate can be set up through a feature layer object. The method differs, however, between a join and a relate. A feature layer object implements *IDisplayRelationshipClass*, which has the *DisplayRelationshipClass* method to set up a join between two tables and to get the joined table ready for use (Figure 5.7).

To join more than two tables, ArcObjects stipulates that a *RelQueryTable* object be created from the first join and the object be used as the source to create another *RelQueryTable* object for the second join (Figure 5.8). Representing a joined pair of tables, a *RelQueryTable* object is obtained through the *RelQueryTableFactory* coclass. *IRelQueryTableFactory* provides the *Open* method, which can create a new *RelQueryTable* object.

A feature layer object also implements *IRelationshipClassCollection*, and *IRelationshipClassCollectionEdit* for managing relates (Figure 5.9). *IRelationshipClassCollection* has members for deriving and finding relates, and *IRelationshipClassCollectionEdit* has methods for adding or removing a relate.

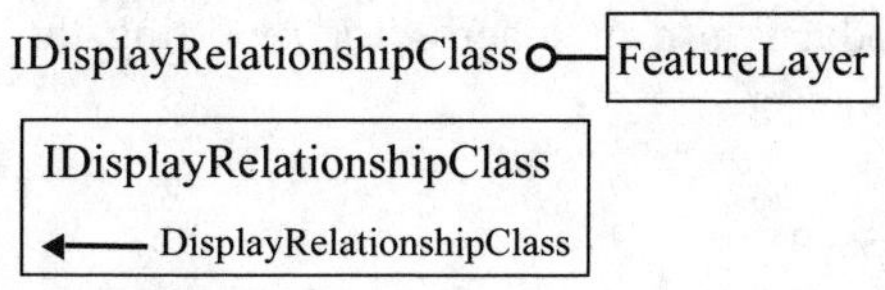

Figure 5.7   *IDisplayRelationshipClass* can set up a join for a feature layer.

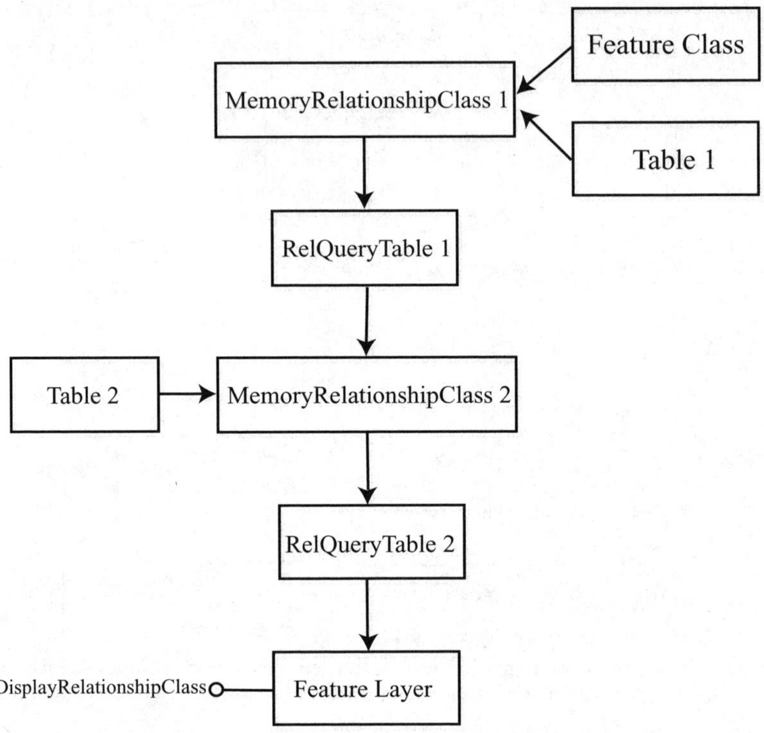

Figure 5.8   The diagram shows a flow chart for joining two tables to a feature class. See text for explanation.

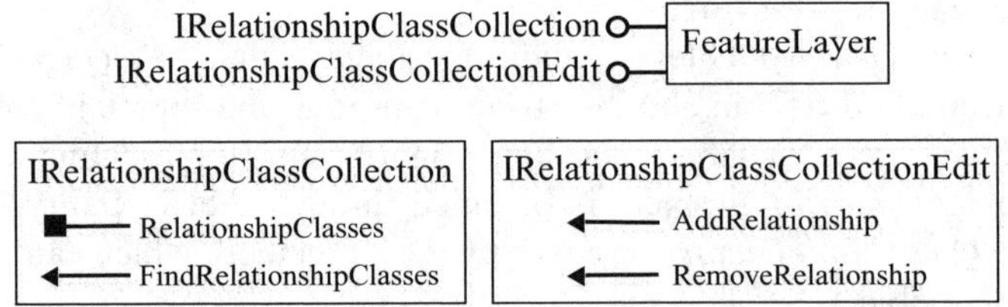

Figure 5.9   A *FeatureLayer* object supports interfaces that work with relates.

5.3 LISTING FIELDS AND FIELD PROPERTIES

This section introduces use of macros for reporting the number of fields and field properties in a dataset.

5.3.1 *ListOfFields*

ListOfFields reports the number of fields in a feature class and the field names. The macro performs the same function as using the Fields tab in a feature layer's

Properties dialog. ***ListOfFields*** has two parts. Part 1 gets the feature class and reports the number of fields in the feature class. Part 2 steps through each field, gets the field name, adds the field name to a list, and reports the list.

Key Interfaces: *IFields, IField.*
Key Members: *Fields, FieldCount, Field( ), name, Add.*
Usage: Add *idcounty.shp* to an active map. *idcounty* shows 44 counties in Idaho and has some demographic attributes. Import ***ListOfFields*** to Visual Basic Editor. Run the macro. The first message box reports the number of fields in *idcounty*, and the second message lists the field names.

```
Private Sub ListOfFields()
  ' Part 1: Get the feature class and its fields.
  Dim pMxDoc As IMxDocument
  Dim pMap As IMap
  Dim pFeatureLayer As IFeatureLayer
  Dim pFeatureClass As IFeatureClass
  Dim pFields As IFields
  Dim count As Long
  Set pMxDoc = ThisDocument
  Set pMap = pMxDoc.FocusMap
  Set pFeatureLayer = pMap.Layer(0)
  Set pFeatureClass = pFeatureLayer.FeatureClass
  Set pFields = pFeatureClass.Fields
  ' Get the number of fields.
  count = pFields.FieldCount
  MsgBox "There are " & count & " fields"
```

Part 1 sets *pFeatureClass* to be the feature class of the top layer in the active map. Next the code sets *pFields* to be the fields of *pFeatureClass*, and the *count* variable to be the *FieldCount* property on *IFields*. A message box then reports the number of fields.

```
  ' Part 2: Prepare a list of fields and display the list.
  Dim ii As Long
  Dim aField As IField
  Dim fieldName As Variant
  Dim theList As New Collection
  Dim NameList As Variant
  ' Loop through each field, and add the field name to a list.
  For ii = 0 To pFields.FieldCount - 1
    Set aField = pFields.Field(ii)
    fieldName = aField.name
```

```
    theList.Add (fieldName)
  Next
  ' Display the list of field names in a message box.
  For Each fieldName In theList
    NameList = NameList & fieldName & Chr(13)
  Next fieldName
  MsgBox NameList,, "Field Names"
End Sub
```

Part 2 uses a *For...Next* statement to step through each field in *pFields*. Because the loop starts with the base of zero, the predefined number of iterations is based on the *FieldCount* value minus one. Each time through the loop, the code assigns the name of the field to the *fieldName* variable and adds the name to a collection referenced by *theList*. Finally, the code uses a *For Each...Next* statement to display each field name in a message box. The constant Chr(13) adds a carriage return.

5.3.2 *ListFieldProps*

ListFieldProps reports the field name, type, length, precision, and scale of each field in a dataset. The macro performs the same function as using the Fields tab in a feature layer's Properties dialog. *ListFieldProps* has two parts. Part 1 gets the feature class and its fields. Part 2 gets the field properties of each field and reports them.

> **Key Interfaces:** *IFields, IField.*
> **Key Members:** *Fields, FieldCount, Field( ), name, Type, Length, Precision, Scale, Add.*
> **Usage:** Add *idcounty.shp* to an active map. Import *ListFieldProps* to Visual Basic Editor. Run the macro. The message box lists the properties of each field in *idcounty.*

```
Private Sub ListFieldProps()
  ' Part 1: Get the feature class and its fields.
  Dim pMxDoc As IMxDocument
  Dim pMap As IMap
  Dim pFeatureLayer As IFeatureLayer
  Dim pFeatureClass As IFeatureClass
  Dim pFields As IFields
  Dim count As Long
  Set pMxDoc = ThisDocument
  Set pMap = pMxDoc.FocusMap
  Set pFeatureLayer = pMap.Layer(0)
  Set pFeatureClass = pFeatureLayer.FeatureClass
  Set pFields = pFeatureClass.Fields
```

Part 1 is the same as ***ListOfFields***. The code derives the number of fields from the feature class of the top layer and saves the fields in *pFields*.

```
' Part 2: Get the field properties for each field, and
' report them.
Dim ii As Long
Dim aField As IField
Dim fieldName As String
Dim fieldType As Integer
Dim fieldLength As Integer
Dim fieldPrecision As Integer
Dim fieldScale As Integer
Dim typeDes As String
Dim out As Variant
Dim theList As New Collection
Dim NameList As Variant
' Set up a do loop.
For ii = 0 To pFields.FieldCount - 1
  Set aField = pFields.Field(ii)
  ' Derive properties of the field.
  fieldName = aField.name
  fieldType = aField.Type
  fieldLength = aField.Length
  fieldPrecision = aField.Precision
  fieldScale = aField.Scale
  ' Determine the field type.
  Select Case fieldType
    Case 0
     typeDes = "SmallInteger"
    Case 1
     typeDes = "Integer"
    Case 2
     typeDes = "Single"
    Case 3
     typeDes = "Double"
    Case 4
     typeDes = "String"
    Case 5
       typeDes = "Date"
```

```
      Case 6
        typeDes = "OID"
      Case 7
        typeDes = "Geometry"
      Case 8
        typeDes = "Blob"
    End Select
    ' Save the field properties to the variable out.
    out = fieldName & " " & typeDes & " " & fieldLength _
      & " " & fieldPrecision & " " & fieldScale
    ' Add the variable out to a list.
    theList.Add out
  Next
  ' Display the list in a message box.
  For Each out In theList
    NameList = NameList & out & Chr(13)
  Next out
  MsgBox NameList,, _
    "Field Name, Type, Length, Precision, & Scale"
End Sub
```

Part 2 uses a *For...Next* statement to step through each field in *pFields* and to derive its field property values. Besides the field name, the code works with the additional field properties of type, length, precision, and scale. Because the field type value can range from zero to eight, the code uses a *Select Case* statement to translate the value into a description such as integer or double. The field length is the maximum length in bytes. The field precision is the number of digits reserved for a numeric field. The field scale is the number of decimal digits reserved for a double field. After the field properties are derived, they are strung together and assigned to the variant variable *out*. A variant variable can take any type of data. The code then adds *out* to *theList*. Finally, a message box reports *theList*, one field at a time.

5.3.3   *UseFindLayer*

UseFindLayer consists of a sub and a function. The **FindLayer** function uses an input box to get the name of a layer from the user and returns the index of the layer to the **Start** sub. **Start** then reports the number of fields in the layer.

Key Interfaces: *IFields, IField.*
Key Members: *Fields, FieldCount, Field, name, Add.*
Usage: Add *idcounty.shp* and *idcities.shp* to an active map. Import *UseFindLayer* to Visual Basic Editor. Run the module. Enter the name of a layer in the input box. The module reports the number of fields in the layer.

```
Private Sub Start()
  Dim pMxDoc As IMxDocument
  Dim pMap As IMap
  Dim i As Long
  Dim pFeatureLayer As IFeatureLayer
  Dim pFeatureClass As IFeatureClass
  Dim pFields As IFields
  Dim count As Long
  Set pMxDoc = ThisDocument
  Set pMap = pMxDoc.FocusMap
  ' Run the FindLayer function.
  i = FindLayer()
  ' Use the returned Id to locate the layer.
  Set pFeatureLayer = pMap.Layer(i)
  Set pFeatureClass = pFeatureLayer.FeatureClass
  Set pFields = pFeatureClass.Fields
  ' Get the number of fields.
  count = pFields.FieldCount
  MsgBox "There are " & count & " fields"
End Sub
```

Start assigns the returned value from *FindLayer* to *i*, and uses *i* as the index to locate *pFeatureLayer*. Next the code sets *pFeatureClass* to be the feature class of *PFeatureLayer* and *pFields* to be the fields of *pFeatureClass*. A message box reports the number of fields in *pFields*.

```
Private Function FindLayer() As Long
  Dim pMxDoc As IMxDocument
  Dim pMap As IMap
  Dim FindDoc As Variant
  Dim aLName As String
  Dim name As String
  Dim i As Long
  Set pMxDoc = ThisDocument
  Set pMap = pMxDoc.FocusMap
  ' Use an input box to get a layer name.
  name = InputBox("Enter a layer name:", "")
  ' Locate the layer in the active map.
  For i = 0 To pMap.LayerCount - 1
    aLName = UCase(pMap.Layer(i).name)
```

```
    If (aLName '= (UCase(name))) Then
      FindDoc = i
    End If
  Next
  FindLayer = FindDoc
End Function
```

FindLayer receives a layer name from an input box and assigns it to the *name* variable. Next the code steps through each layer in the active map and matches the upper case of *name* with the upper case of the layer's name. When a match is found, the code assigns the layer's index value *i* to the *FindDoc* variable. The function then returns *FindDoc* as the value of **FindLayer** to the **Start** sub.

5.4 ADDING OR DELETING FIELDS

This section covers use of a macro for adding fields to and deleting fields from a dataset.

5.4.1 *AddDeleteField*

AddDeleteField adds two new fields to, and deletes a field from, a feature attribute table. The macro performs the same function as using a dataset's Properties dialog in ArcCatalog to add and delete fields. **AddDeleteField** is organized into three parts. Part 1 defines the feature class. Part 2 defines two new fields and adds the fields to the feature class. Part 3 deletes a field from the feature class.

> **Key Interfaces:** *IFeatureClass, IFieldEdit, IFields, IField.*
> **Key Members:** *FeatureClass, name, Type, Length, AddField, Fields, FindField, Field(), DeleteField.*
> **Usage:** Add *idcounty2.shp* to an active map. Open the attribute table of *idcounty2*. The table shows a field named Pop94, which will be deleted by the macro. Two new fields, Pop1990 and Pop2000, will be added to the table. Import **AddDelete-Field** to Visual Basic Editor. Run the macro. Check the attribute table of *idcounty2* again to make sure that the task is done correctly.

```
Private Sub AddDeleteField()
  ' Part 1: Get a handle on the feature class.
  Dim pMxDoc As IMxDocument
  Dim pFeatureLayer As IFeatureLayer
  Dim pFeatureClass As IFeatureClass
  Set pMxDoc = ThisDocument
  Set pFeatureLayer = pMxDoc.FocusMap.Layer(0)
  Set pFeatureClass = pFeatureLayer.FeatureClass
```

Part 1 sets *pFeatureClass* to be the feature class of the top layer in the active map.

```
' Part 2: Add two new fields.
Dim pField1 As IFieldEdit
Dim pField2 As IFieldEdit
' Define the first new field.
Set pField1 = New Field
pField1.name = "Pop1990"
pField1.Type = esriFieldTypeInteger
pField1.Length = 8
' Add the first new field.
pFeatureClass.AddField pField1
' Define the second new field.
Set pField2 = New Field
pField2.name = "Pop2000"
pField2.Type = esriFieldTypeInteger
pField2.Length = 8
' Add the second new field.
pFeatureClass.AddField pField2
```

Part 2 creates *pField1* as an instance of the *Field* class and uses the *IFieldEdit* interface to define its field properties of name, type, and length. Next the code uses the *AddField* method on *IFeatureClass* to add *pField1* to *pFeatureClass*. The code uses the same procedure to add *pField2* to *pFeatureClass*.

```
' Part 3: Delate a field.
Dim pFields As IFields
Dim ii As Integer
Dim pField3 As IField
Set pFields = pFeatureClass.Fields
ii = pFields.FindField("Pop94")
Set pField3 = pFields.Field(ii)
pFeatureClass.DeleteField pField3
End Sub
```

Part 3 first sets *pFields* to be the fields of *pFeatureClass*. Next the code uses the *FindField* method on *IFields* to find the index of Pop94 among the fields, assigns the index value to the *ii* variable, and sets *pFields3* to be the field at the index *ii*. Then the code uses the *DeleteField* method on *IFeatureClass* to delete *pField3*.

5.5 CALCULATING FIELD VALUES

This section focuses on the field value. The first macro shows how to use an expression to calculate the field values programmatically. The second macro shows how to update the field values of a data subset.

5.5.1 *CalculateField*

CalculateField calculates the values of a field by using the values of two existing fields in a feature class. The macro performs the same function as using the Field Calculator. *CalculateField* has two parts. Part 1 finds the field to be calculated. Part 2 calculates the field value for each record (feature) in a cursor.

> **Key Interfaces:** *IFeatureClass, IFields, ICursor, ICalculator.*
> **Key Members:** *FeatureClass, Fields, FindField, Update, Cursor, Expression, Field, Calculate.*
> **Usage:** Add *idcounty3.shp* to an active map. *idcounty3*'s attribute table contains three fields: Pop1990, county population in 1990; Pop2000, county population in 2000; and Change. Import *CalculateField* to Visual Basic Editor. Run the macro. The macro calculates the field values of Change from Pop1990 and Pop2000.

```
Private Sub CalculateField()
  ' Par 1: Find the field to be calculated.
  Dim pMxDoc As IMxDocument
  Dim pFeatLayer As IFeatureLayer
  Dim pFeatureClass As IFeatureClass
  Dim pFields As IFields
  Dim ii As Integer
  Set pMxDoc = ThisDocument
  Set pFeatLayer = pMxDoc.FocusMap.Layer(0)
  Set pFeatureClass = pFeatLayer.FeatureClass
  Set pFields = pFeatureClass.Fields
  ii = pFields.FindField("Change")
```

Part 1 sets *pFeatureClass* to be the feature class of the top layer in the active map. The code then sets *pFields* to be the fields in *pFeatureClass* and uses the *FindField* method on *IFields* to find the field Change.

```
  ' Part 2: Calculate the field values by using a cursor.
  Dim pCursor As ICursor
  Dim pCalculator As ICalculator
  ' Prepare a cursor with all records.
  Set pCursor = pFeatureClass.Update(Nothing, True)
```

```
  ' Define a calculator.
  Set pCalculator = New Calculator
  With pCalculator
    Set.Cursor = pCursor
    .Expression = _
    "(([Pop2000] - [Pop1990])/[Pop1990]) * 100"
    .Field = "Change"
  End With
  ' Calculate the field values.
  pCalculator.Calculate
End Sub
```

Part 2 first creates a cursor from *pFeatureClass* by using the *Update* method on *IFeatureClass*. Because the *Update* method uses Nothing in place of a query filter object, all records are included in the cursor. (Chapter 9 covers use of query filter objects.) Next the code creates *pCalculator* as an instance of the *Calculator* class and defines its properties in a *With* statement. The cursor is set to be *pCursor*, the expression is an equation to calculate the percent change of county population between 1990 and 2000, and the field is Change. Finally, the code uses the *Calculate* method on *ICalculator* to complete the task.

5.5.2 *UpdateValue*

UpdateValue differs from *CalculateField* in two aspects. First, *UpdateValue* calculates the field values for a data subset instead of every record. Second, the macro does not use *ICalculator* to calculate the field values.

UpdateValue performs the same function as using the Select by Attributes command in the Selection menu to first select a data subset, and then using the Field Calculator to calculate the field values. The macro has three parts: Part 1 defines the feature class, Part 2 selects a data subset, and Part 3 calculates the field value for each record in the data subset.

Key Interfaces: *IFeatureClass, IFields, IQueryFilter, IFeatureCursor, IFeature.*

Key Members: *FeatureClass, WhereClause, Update, FindField, NextFeature, Value, UpdateFeature.*

Usage: Add *idcounty4.shp* to an active map. *idcounty4*'s attribute table contains the field Change, which shows the rate of population change between 1990 and 2000 for Idaho counties. Import *UpdateValue* to Visual Basic Editor. Run the macro. The macro populates the field Class with the value of one for high-growth counties (i.e., Change > 30%) and zero for other counties.

```
Private Sub UpdateValue()
  ' Par 1: Define the feature class.
  Dim pMxDoc As IMxDocument
  Dim pFeatLayer As IFeatureLayer
```

```
Dim pFeatureClass As IFeatureClass
Dim pFields As IFields
Dim ii As Integer
Set pMxDoc = ThisDocument
Set pFeatLayer = pMxDoc.FocusMap.Layer(0)
Set pFeatureClass = pFeatLayer.FeatureClass
```

Part 1 sets *pFeatureClass* to be the feature class of the top layer in the active map.

```
' Part 2: Prepare a feature cursor.
Dim pQFilter As IQueryFilter
Dim pUpdateFeatures As IFeatureCursor
' Prepare a query filter.
Set pQFilter = New QueryFilter
pQFilter.WhereClause = "Change > 30"
' Create a feature cursor for updating.
Set pUpdateFeatures = pFeatureClass.Update _
(pQFilter, False)
```

Part 2 creates *pQFilter* as an instance of the *QueryFilter* class and defines its *WhereClause* condition as "Change > 30." The purpose of a query filter object is to filter a data subset that meets its *WhereClause* expression. The code then uses *pQFilter* as an object qualifier and the *Update* method on *IFeatureClass* to create a cursor of selected features. A feature cursor is a data-access object designed for the updating, deleting, and inserting of features. A feature cursor object has the *FindField*, *NextFeature*, and *UpdateFeature* methods (Figure 5.10).

```
' Part 3: Calculate the Class value.
Dim indexClass As Integer
Dim pFeature As IFeature
indexClass = pUpdateFeatures.FindField("Class")
```

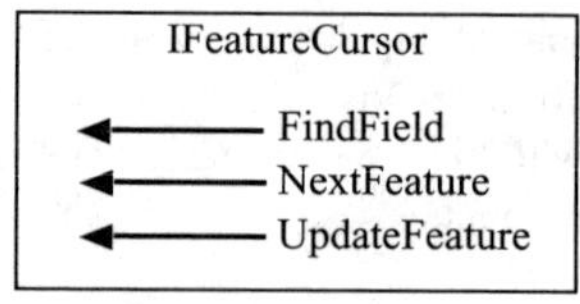

Figure 5.10 *IFeatureCursor* has methods for managing fields and features.

```
 Set pFeature = pUpdateFeatures.NextFeature
 ' Loop through each feature and update its Class value.
 Do Until pFeature Is Nothing
   pFeature.Value(indexClass) = 1
   pUpdateFeatures.UpdateFeature pFeature
   Set pFeature = pUpdateFeatures.NextFeature
 Loop
End Sub
```

Part 3 first uses the *FindField* method on *IFeatureCursor* to find the index of the field Class and assigns the index value to the *indexClass* variable. The rest of the code uses a *Do...Loop* to step through each selected feature, assigns one as the value to the field at *indexClass*, and uses the *UpdateFeature* method on *IFeatureCursor* to update the feature.

5.6 JOINING AND RELATING TABLES

This section covers joins and relates with four sample macros, two on join and two on relate. By having two macros on each type of relationship class, this section shows how to link a nongeographic table to a feature class as well as how to link two or more nongeographic tables to a feature class.

5.6.1 *JoinTableToLayer*

JoinTableToLayer joins a dBASE file to a feature class. The macro performs the same function as using Join in a feature layer's context menu in ArcMap. *JoinTableToLayer* has three parts. Part 1 defines the feature class. Part 2 defines the attribute table to be joined. Part 3 joins the attribute table to the feature class.

Key Interfaces: *IFeatureClass, ITable, IStandaloneTableCollection, IStandaloneTable, IMemoryRelationshipClassFactory, IRelationshipClass, IDisplayRelationshipClass.*

Key Members: *FeatureClass, StandaloneTable(), Table, Open, DisplayRelationship-Class.*

Usage: Add *wp.shp* and *wpdata.dbf* to an active map. *wp* is a shapefile of vegetation stands, and *wpdata* is a dBASE file containing attributes of the vegetation stands. The common field in *wp*'s attribute table and *wpdata* is ID. Import *JoinTableToLayer* to Visual Basic Editor. Run the macro. Enter ID for the name of the join field. The macro joins the attributes in *wpdata* to *wp*'s attribute table. To verify that the macro works, open the attribute table of *wp*. The table should have two sets of attributes: one set has the prefix of wp and the other set has the prefix of wpdata.

```
Private Sub JoinTableToLayer()
  ' Part 1: Get a handle on the layer's attribute table.
  Dim pMxDoc As IMxDocument
  Dim pMap As IMap
  Dim pFeatureLayer As IFeatureLayer
  Dim pFeatureClass As IFeatureClass
  Set pMxDoc = ThisDocument
  Set pMap = pMxDoc.FocusMap
  Set pFeatureLayer = pMap.Layer(0)
  Set pFeatureClass = pFeatureLayer.FeatureClass
```

Part 1 sets *pFeatureClass* to be the feature class of the top layer in the active map.

```
  ' Part 2: Define the table to be joined.
  Dim pTabCollection As IStandaloneTableCollection
  Dim pStTable As IStandaloneTable
  Dim pFromTable As ITable
  Set pTabCollection = pMap
  Set pStTable = pTabCollection.StandaloneTable(0)
  Set pFromTable = pStTable.Table
```

Part 2 first performs a QueryInterface (QI) for the *IStandaloneTableCollection* interface and sets *pStTable* to be the first standalone table in *pMap*. The code then sets *pFromTable* to be the table underlying *pStTable*.

```
  ' Part 3: Join the table to the layer's attribute table.
  Dim strJnField As String
  Dim pMemRelFact As IMemoryRelationshipClassFactory
  Dim pRelClass As IRelationshipClass
  Dim pDispRC As IDisplayRelationshipClass
  ' Prompt for the join field.
  strJnField = InputBox("Provide the name of the join field: ", _
  "Joining a table to a layer", "")
  ' Create a memory relationship class.
  Set pMemRelFact = New MemoryRelationshipClassFactory
  ' The underscore _ means continuation of a line statement.
  Set pRelClass = pMemRelFact.Open("TableToLayer", _
  pFromTable, strJnField, pFeatureClass, strJnField, _
  "wp", "wpdata", esriRelCardinalityOneToOne)
  ' Perform a join.
  Set pDispRC = pFeatureLayer
```

```
pDispRC.DisplayRelationshipClass pRelClass, _
esriLeftOuterJoin
End Sub
```

Part 3 creates a memory relationship class (i.e., virtual join) before joining the two tables. The code first gets the join field from an input box. Next the code creates *pMemRelFact* as an instance of the *MemoryRelationshipClassFactory* class and uses the *Open* method on *IMemoryRelationshipClassFactory* to create a relationship class referenced by *pRelClass*. The *Open* method uses eight object qualifiers and arguments. The two object qualifiers of *pFromTable* and *pFeatureClass* have been defined. The name of the relationship class is "TableToLayer." The origin primary key and the origin foreign key are both *strJnField*. The forward path label and the backward path label are wp and wpdata respectively. And the cardinality or the type of relationship is specified as one-to-one. Finally, the code switches to the *IDisplay-RelationshipClass* interface and uses the *DisplayRelationshipClass* method to perform the join. Besides *pRelClass*, *DisplayRelationshipClass* uses the argument of *esriLeftOuterJoin*, which stipulates that the join operation includes all rows, rather than the matched rows only.

5.6.2 *JoinMultipleTables*

JoinMultipleTables joins two dBASE tables to a feature class. Conceptually, *JoinMultipleTables* is similar to *JoinTableToLayer*. But programmatically, *Join-MultipleTables* requires use of *IRelQueryTable* objects to carry out two joins. The macro performs the same function as using the Join command in a feature layer's context menu in ArcMap. *JoinMultipleTables* has five parts. Part 1 defines the feature class. Part 2 defines the two tables to be joined. Part 3 creates the first virtual join, followed by the second virtual join in Part 4. Part 5 performs the join operation.

> **Key Interfaces:** *IStandaloneTableCollection, IStandaloneTable, ITable, IMemory-RelationshipClassFactory, IRelationshipClass, IRelQueryTableFactory, IRelQueryTable, IDisplayRelationshipClass.*
>
> **Key Members:** *FeatureClass, StandaloneTable(), Table, Open, DisplayRelationship-Class.*
>
> **Usage:** Add *idcounty4.shp, change.dbf,* and *population.dbf* to an active map. *idcounty4* is a shapefile showing 44 Idaho counties. *change* and *population* are two dBASE files that contain demographic data of Idaho counties. (This macro will not work with text files.) The keys for joining *change* to *idcounty* are both co_name. The keys for joining *population* to the joined table are change.co_name and co_name. The keys have been hard coded in the macro. Import *JoinMultipleTables* to Visual Basic Editor. Run the macro. The macro joins the attributes in *change* and *population* to *idcounty4*'s attribute table. To verify that the macro works, open the attribute table of *idcounty4*. The table should include attributes from *change* and *population*. Each field in the table should have a prefix to identify its source.

```
Private Sub JoinMultipleTables()
   ' Part 1: Define the feature class.
   Dim pMxDoc As IMxDocument
   Dim pMap As IMap
   Dim pFeatureLayer As IFeatureLayer
   Dim pFeatureClass As IFeatureClass
   Set pMxDoc = ThisDocument
   Set pMap = pMxDoc.FocusMap
   Set pFeatureLayer = pMap.Layer(0)
   Set pFeatureClass = pFeatureLayer.FeatureClass
```

Part 1 sets *pFeatureClass* to be the feature class of the top layer in the active map.

```
   ' Part 2: Define the two tables to be joined.
   Dim pTabCollection As IStandaloneTableCollection
   Dim pStTable1 As IStandaloneTable
   Dim pFromTable1 As ITable
   Dim pStTable2 As IStandaloneTable
   Dim pFromTable2 As ITable
   Set pTabCollection = pMap
   ' Define the first table.
   Set pStTable1 = pTabCollection.StandaloneTable(0)
   Set pFromTable1 = pStTable1.Table
   ' Define the second table.
   Set pStTable2 = pTabCollection.StandaloneTable(1)
   Set pFromTable2 = pStTable2.Table
```

Part 2 performs a QI for the *IStandaloneTableCollection* interface and sets *pStTable1* and *pStTble2* to be the first and second standalone tables respectively in *pMap*. The code then defines *pFromTable1* and *pFromTable2* to be the underlying tables of *pStTable1* and *pStTble2* respectively.

```
   ' Part 3: Join the first table to the feature class.
   Dim pMemRelFact As IMemoryRelationshipClassFactory
   Dim pRelClass1 As IRelationshipClass
   Dim pRelQueryTableFact As IRelQueryTableFactory
   Dim pRelQueryTab1 As IRelQueryTable
   ' Create the first virtual join.
   Set pMemRelFact = New MemoryRelationshipClassFactory
```

```
Set pRelClass1 = pMemRelFact.Open("Table1ToLayer", _
pFeatureClass, "co_name", pFromTable1, "co_name", _
"forward", "backward", esriRelCardinalityOneToOne)
' Create the first relquerytable.
Set pRelQueryTableFact = New RelQueryTableFactory
Set pRelQueryTab1 = pRelQueryTableFact.Open(pRelClass1, _
True, Nothing, Nothing, "", True, True)
```

Part 3 creates *pMemRelFact* as an instance of the *MemoryRelationshipClass-Factory* class and uses the *Open* method to create a memory relationship class object referenced by *pRelClass1*. The two tables specified for the *Open* method are *pFeatureClass* and *pFromTable1*. Next the code creates *pRelQueryTableFact* as an instance of the *RelQueryTableFactory* class and uses the *Open* method on *IRelQueryTableFactory* and *pRelClass1* as an argument to create *pRelQueryTab1*, a reference to an *IRelQueryTable* object.

```
' Part 4: Join the second table to the joined table.
Dim pRelClass2 As IRelationshipClass
Dim pRelQueryTab2 As IRelQueryTable
' Create the second virtual join.
Set pMemRelFact = New MemoryRelationshipClassFactory
Set pRelClass2 = pMemRelFact.Open("Table2ToLayer", _
pRelQueryTab1, "change.co_name", pFromTable2, _
"co_name", "forward", "backward", _
esriRelCardinalityOneToOne)
' Create the second relquerytable.
Set pRelQueryTableFact = New RelQueryTableFactory
Set pRelQueryTab2 = pRelQueryTableFact.Open(pRelClass2, _
True, Nothing, Nothing, ,"" True, True)
```

Part 4 follows the same procedure as in Part 3: use *pRelQueryTab1* and *pFromTable2* as inputs to create *pRelClass2*, and use *pRelClass2* as an input to create *pRelQueryTab2*.

```
' Part 5: Perform the join operation.
Dim pDispRC2 As IDisplayRelationshipClass
Set pDispRC2 = pFeatureLayer
pDispRC2.DisplayRelationshipClass _
pRelQueryTab2.RelationshipClass, esriLeftOuterJoin
End Sub
```

Part 5 switches to the *IDisplayRelationshipClass* interface and uses the *Display-RelationshipClass* method to perform the join operation. Notice that the method uses *RelQueryTab2* as an object qualifier.

5.6.3 *RelateTableToLayer*

RelateTableToLayer creates a relate between a dBASE file and a feature class. The macro performs the same function as using Relate in a feature layer's context menu in ArcMap. *RelateTableToLayer* has three parts. Part 1 defines the feature class. Part 2 defines the nongeographic table for a relate. Part 3 asks for the keys to establish a relate, creates a virtual relate, and performs a relate.

> **Key Interfaces:** *IFeatureClass, ITable, IStandaloneTableCollection, IStandalone-Table, IMemoryRelationshipClassFactory, IRelationshipClass, IRelationship-ClassCollectionEdit.*
>
> **Key Members:** *FeatureClass, StandaloneTable(), Table, Open, AddRelationshipClass.*
>
> **Usage:** Add *wp.shp* and *wpdata.dbf* to an active map. ID in both *wp*'s attribute table and *wpdata* can be used as the key. Import *RelateTableToLayer* to Visual Basic Editor. Run the macro. Enter ID in both input boxes. The macro creates a relate between *wpdata* and *wp*'s attribute table. To verify that the macro works, open *wp*'s attribute table and select some records from the table. Click Options in *wp*'s attribute table and point to Related Tables. TableToLayer: wpdata should appear on the side bar. Click on the side bar, and *wpdata* will appear with highlighted records that correspond to the selected records in *wp*.

```
Private Sub RelateTableToLayer()
    ' Part 1: Define the feature class.
    Dim pMxDoc As IMxDocument
    Dim pMap As IMap
    Dim pFeatureLayer As IFeatureLayer
    Dim pFeatureClass As IFeatureClass
    Set pMxDoc = ThisDocument
    Set pMap = pMxDoc.FocusMap
    Set pFeatureLayer = pMap.Layer(0)
    Set pFeatureClass = pFeatureLayer.FeatureClass
```

Part 1 sets *pFeatureClass* to be the feature class of the top layer in the active map.

```
    ' Part 2: Define the table for a relate.
    Dim pTabCollection As IStandaloneTableCollection
    Dim pStTable As IStandaloneTable
    Dim pFromTable As ITable
    Set pTabCollection = pMap
    Set pStTable = pTabCollection.StandaloneTable(0)
    Set pFromTable = pStTable.Table
```

Part 2 defines *pFromTable* to be the underlying table of the first standalone table in the active map.

```
' Part 3: Perform the relate.
Dim strLayerField As String
Dim strTableField As String
Dim pMemRelFact As IMemoryRelationshipClassFactory
Dim pRelClass As IRelationshipClass
Dim pRelClassCollEdit As IRelationshipClassCollectionEdit
' Prompt for the keys.
strLayerField = InputBox _
("Provide the key from the layer: ", _
"Relating a table to a layer", "")
strTableField = InputBox _
("Provide the key from the table: ", _
"Relating a table to a layer", "")
' Create a virtual relate.
Set pMemRelFact = New MemoryRelationshipClassFactory
Set pRelClass = pMemRelFact.Open("TableToLayer", _
pFromTable, strTableField, pFeatureClass, strLayerField, _
"wp", "wpdata", esriRelCardinalityOneToOne)
' Add the relate to the collection.
Set pRelClassCollEdit = pFeatureLayer
pRelClassCollEdit.AddRelationshipClass pRelClass
End Sub
```

Part 3 starts by prompting for the keys for establishing the relate. Next the code creates *pMemRelFact* as an instance of the *MemoryRelationshipClassFactory* class. The code then uses the *Open* method on *IMemoryRelationshipClassFactory* to create a virtual relate referenced by *pRelClass*. Finally, the code switches to the *IRelationship- ClassCollectionEdit* interface and uses the *AddRelationshipClass* method to add *pRelClass* to the relation class collection of the feature layer.

5.6.4   *RelationalDatabase*

RelationalDatabase creates relates between a feature class and three dBASE files from a relational database. The macro performs the same function as using Relate in ArcMap three times. *RelationalDatabase* has four parts. Part 1 defines the feature class. Part 2 defines the three dBASE files for relates. Part 3 creates three virtual relates. Part 4 adds the relates to the relationship class collection of the feature layer.

> **Key Interfaces:** *IFeatureClass, ITable, IStandaloneTableCollection, IStandalone-Table, IMemoryRelationshipClassFactory, IRelationshipClass, IRelationship-ClassCollectionEdit.*
>
> **Key Members:** *FeatureClass, StandaloneTable( ), Table, Open, AddRelatoinshipClass.*
>
> **Usage:** Add *mosoils.shp, comp.dbf, forest.dbf,* and *plantnm.dbf* to an active map. The macro assumes that the order of the three tables is *comp, forest,* and *plantnm* from top to bottom in the table of contents. *mosoils* is a soils shapefile, and the three dBASE files contain soil attributes from a relational database. The key relating *mosoils* and *comp* is MUSYM, the key relating *comp* and *forest* is MUID, and the key relating *forest* and *plantnm* is PLANTSYM. Import **RelationalDatabase** to Visual Basic Editor. Run the macro. Enter the proper key in each of the three input boxes. To verify that the macro works, select some records from the attribute table of *mosoils* and then open *comp* to see the corresponding records.

```
Private Sub RelationalDatabase()
  ' Part 1: Get a handle on the mosoils attribute table.
  Dim pMxDoc As IMxDocument
  Dim pMap As IMap
  Dim pFeatureLayer As IFeatureLayer
  Dim pFeatureClass As IFeatureClass
  Set pMxDoc = ThisDocument
  Set pMap = pMxDoc.FocusMap
  Set pFeatureLayer = pMap.Layer(0)
  Set pFeatureClass = pFeatureLayer.FeatureClass
```

Part 1 sets *pFeatureClass* to be the feature class of the top layer in the active map.

```
  ' Part 2: Define the tables for relates.
  Dim pTabCollection As IStandaloneTableCollection
  Dim pStTable1 As IStandaloneTable
  Dim pCompTable As ITable
  Dim pStTable2 As IStandaloneTable
  Dim pForestTable As ITable
  Dim pStTable3 As IStandaloneTable
  Dim pPlantnmTable As ITable
  Set pTabCollection = pMap
  ' Define the first table.
  Set pStTable1 = pTabCollection.StandaloneTable(0)
  Set pCompTable = pStTable1.Table
  ' Define the second table.
  Set pStTable2 = pTabCollection.StandaloneTable(1)
  Set pForestTable = pStTable2.Table
  ' Define the third table.
```

```
Set pStTable3 = pTabCollection.StandaloneTable(2)
Set pPlantnmTable = pStTable3.Table
```

Part 2 sets the tables underlying the three standalone tables as *pCompTable*, *pForestTable*, and *pPlantnmTable* respectively.

```
' Part 3: Create three virtual relates.
Dim strField1 As String
Dim pMemRelFact As IMemoryRelationshipClassFactory
Dim pRelClass1 As IRelationshipClass
Dim strField2 As String
Dim pRelClass2 As IRelationshipClass
Dim strField3 As String
Dim pRelClass3 As IRelationshipClass
' Create the first virtual relate.
strField1 = InputBox _
("Provide the common field in Mosoils and Comp: ", _
"Relate Mosoils to Comp", "")
Set pMemRelFact = New MemoryRelationshipClassFactory
Set pRelClass1 = pMemRelFact.Open _
("Mosoils-Comp", pCompTable, strFeild1, _
pFeatureClass, strField1, "Mosoils", "Comp", _
esriRelCardinalityManyToMany)
' Create the second virtual relate.
strField2 = InputBox _
"Provide the common field in Comp and Forest: ", _
"Relate Comp to Forest", "")
Set pRelClass2 = pMemRelFact.Open _
("Comp-Forest", pForestTable, strField2, _
pCompTable, strField2, "Comp", "Forest", _
esriRelCardinalityManyToMany)
' Create the third virtual relate.
strField3 = InputBox _
("Provide the common field in Forest and Plantnm: ", _
"Relate Forest to Plantnm", "")
Set pRelClass3 = pMemRelFact.Open _
("Forest-Plantnm", pPlantnmTable, strField3, _
pForestTable, strField3, "Forest", "Plantnm", _
esriRelCardinalityManyToMany)
```

Part 3 uses the proper keys and tables to create three memory relationship classes, referenced by *pRelClass1*, *pRelClass2*, and *pRelClass3*, respectively. Notice that the *Open* method specifies many-to-many for the type of relationship (i.e., cardinality) argument.

```
' Part 4: Add the relates to the collection.
Dim pRelClassCollEdit As IRelationshipClassCollectionEdit
Set pRelClassCollEdit = pFeatureLayer
pRelClassCollEdit.AddRelationshipClass pRelClass1
pRelClassCollEdit.AddRelationshipClass pRelClass2
pRelClassCollEdit.AddRelationshipClass pRelClass3
End Sub
```

Part 4 performs a QI for the *IRelationshipClassCollectionEdit* interface and uses its *AddRelationshipClass* method to add the three virtual relates to the relationship class collection of *pFeatureLayer*.

CHAPTER 6

Data Conversion

A major application of geographic information systems (GIS) is integration of data from different sources and in different formats. To allow for data integration, a GIS package must be able to read different data formats and to convert data from one format into another. Data conversion can take place in a variety of ways, of which the three most common are:

Between vector data of different types
Between vector and raster data
From x-, y-coordinates to point data

ESRI has introduced a new vector data model with the release of each major product for the past 20 years: coverages with ARC/INFO, shapefiles with ArcView, and geodatabases with ArcGIS. Although ArcGIS users can use all three types of vector data, many are or have been converting traditional coverages and shapefiles into the new geodatabases. This conversion allows users to take advantage of object-oriented technology and new developments from Environmental Systems Research Institute (ESRI), Inc.

Conversion between vector and raster data is important for digitizing (e.g., tracing from scanned files) and data analysis. Vector-to-raster data conversion, or rasterization, converts points, lines, and polygons to cells and fills the cells with values from an attribute. Raster-to-vector data conversion, or vectorization, extracts points and lines from cells and usually requires generalization and weeding of the extracted features.

Tables that record the location of weather stations or a hurricane track typically contain x-, y-coordinates. Representing the geographic locations, these x and y coordinates can be converted to point features. It is an alternative to digitizing the point locations on a digitizing tablet.

This chapter deals with data conversion of the above three types. Section 6.1 reviews data conversion operations in ArcGIS. Section 6.2 discusses objects related to data conversion. Section 6.3 includes macros for converting shapefiles into standalone feature classes or feature classes in a feature dataset. Section 6.4 provides macros for converting coverages into shapefiles or geodatabases. Section 6.5 discusses macros for rasterization and vectorization. Section 6.6 presents a macro for converting x-, y-coordinates into point data. All macros start with the listing of key interfaces and key members (i.e., properties and methods) and the usage.

6.1 CONVERTING DATA IN ARCGIS

ArcGIS has data conversion commands in all three applications of ArcCatalog, ArcMap, and ArcToolbox. ArcToolbox is probably the first choice among many users because it provides a large set of data conversion tools in one place. ArcToolbox organizes these tools into the following 13 categories:

Export from CAD (Computer-aided design)
Export from coverage
Export from geodatabase
Export from raster
Export from shapefile
Export from table
Export from TIN
Import to coverage
Import to geodatabase
Import to raster
Import to shapefile
Import to table
Import to TIN (triangulated irregular network)

ArcCatalog incorporates data conversion commands into context menus. The context menu of a personal geodatabase has the Import command for importing shapefiles, coverages, and tables to geodatabase. The context menu of a shapefile or a coverage has the Export command for exporting the dataset to geodatabase and other formats. The context menu of a table has the Create Features command, which can create point features from x- and y-coordinates.

In ArcMap, conversion of vector data from one format into another is available through the Data/Export Data command in the dataset's context menu. Conversion between raster and vector data, on the other hand, is available through the Convert command in Spatial Analyst and 3D Analyst. The Tools menu in ArcMap has the Add XY Data command for converting a table with x-, y-coordinates into point features.

6.2 ARCOBJECTS FOR DATA CONVERSION

This section covers objects for feature data conversion, rasterization and vectorization, and *XY* event.

6.2.1 Objects for Feature Data Conversion

The principal component in ArcObjects for converting data between geodatabases, shapefiles, and coverages is the *FeatureDataConverter* class (Figure 6.1). A feature data converter object implements *IFeatureDataConverter* and *IFeatureDataConverter2*. Both interfaces have methods for converting feature classes, feature datasets, and tables. *IFeatureDataConverter2* has the additional functionality of working with data subsets. An alternative to *IFeatureDataConverter* is *IExportOperation*, which, although with fewer options, also has methods for converting vector data of different formats (Figure 6.2).

Three name objects appear frequently in this chapter. A *WorkspaceName* object can identify and locate a workspace. To create a workspace name object, one must define the type of workspace factory and the path name (Figure 6.3). The *WorkspaceFactoryProgID* property on *IWorkspaceName* lets the programmer specify the type of workspace factory. A path name specifies the path of a workspace, which can be coded as a string or based on a generic *PropertySet* object. *IPropertySet* has methods that can set and hold properties for objects such as workspaces.

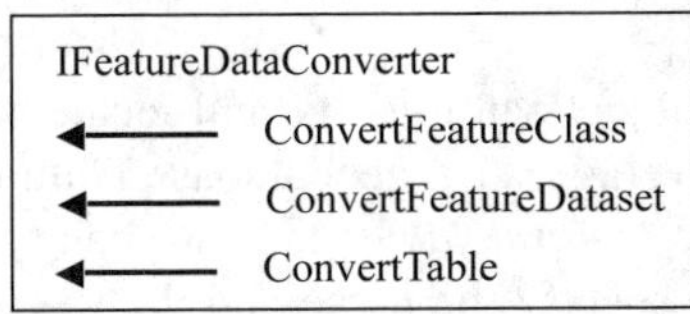

Figure 6.1 *IFeatureDataConverter* has methods for converting feature data.

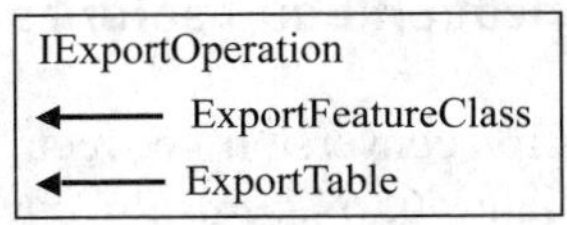

Figure 6.2 *IExportOperation* has methods for exporting feature data.

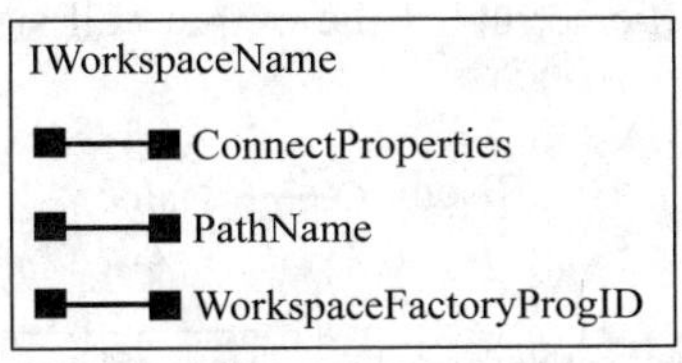

WorkspaceFactoryProgID:
esricore.AccessWorkspaceFactory
esricore.ArcInfoWorkspaceFactory
esricore.RasterWorkspaceFactory
esricore.ShapefileWorkspaceFactory

Figure 6.3 *IWorkspaceName* has properties for identifying workspace factory and path.

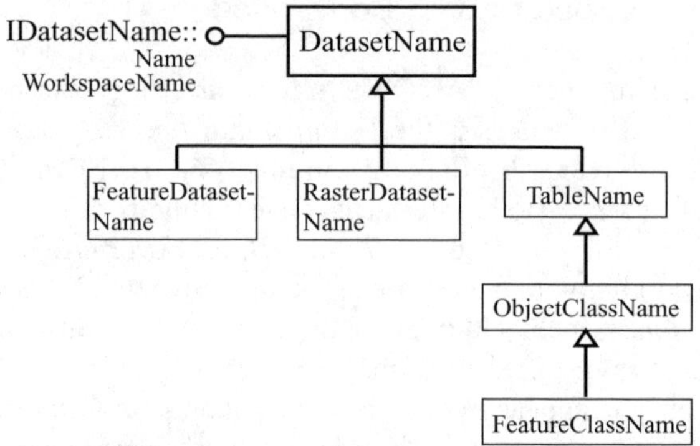

Figure 6.4   Types of *DatasetName* can share the properties of *Name* and *WorkspaceName* on *IDatasetName*.

A *DatasetName* object can represent various types of datasets (Figure 6.4). *IDatasetName* has the properties of *Name* and *WorkspaceName* that let the programmer set the name (i.e., name of the dataset) and the workspace name of a dataset name object.

A *FeatureClassName* object can identify and locate a feature class, which may represent a shapefile, a coverage, or a geodatabase feature class. Because *FeatureClassName* is a type of the *DatasetName* class, we can QueryInterface (QI) for the *IDatasetName* interface to specify the name and the workspace of the feature class (Figure 6.5). Likewise, because *FeatureClassName* is a type of the *Name* class, we can QI for the *IName* interface and use the *Open* method to open the feature class.

6.2.2   Objects for Rasterization and Vectorization

The principal component for conversion between vector and raster data is the *RasterConversionOp* class (Figure 6.6). A *RasterConversionOp* object implements *IConversionOp* and *IRasterAnalysisEnvironment*. *IConversionOp* offers methods for converting raster data to point, line, and polygon feature data as well as for converting vector data to raster data. *IRasterAnalysisEnvironment* has members for the conversion environment such as the set up of the output cell size.

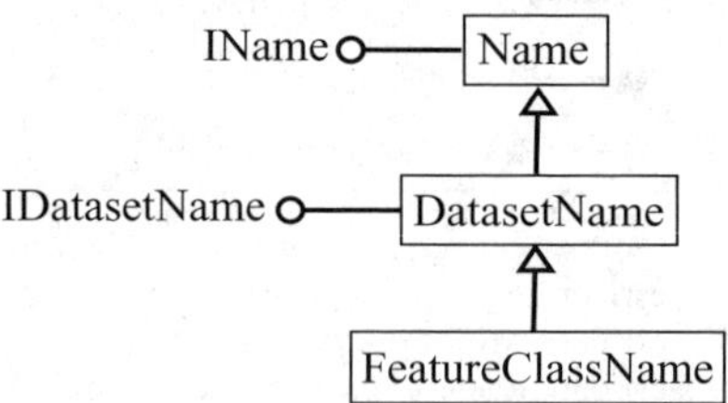

Figure 6.5   A *FeatureClassName* object inherits *IName* and *IDatasetName*.

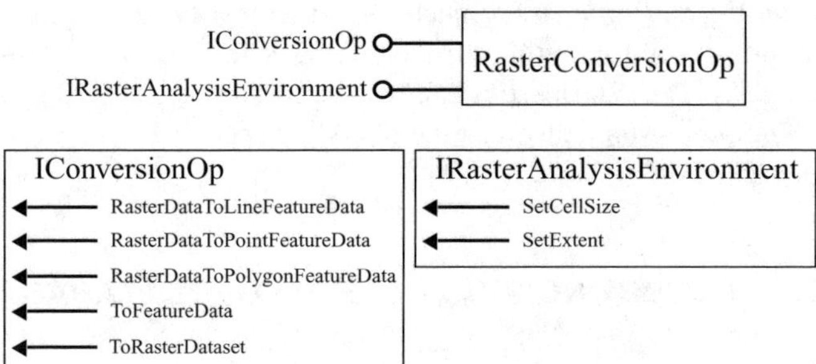

Figure 6.6   A *RasterConversionOp* object supports *IConversionOp* and *IRasterAnalysis-Environment*.

A *RasterDatasetName* object is the raster equivalent of a *FeatureClassName* object. But macros for rasterization or vectorization do not use raster dataset name objects because methods on *IConversionOp* require use of the actual datasets rather than the name objects.

6.2.3   Objects for XY Event

XYEventSource, *XYEventSourceName*, and *XYEvent2FieldsProperties* are the primary components for converting x-, y-coordinates into point features (Figure 6.7). An XY event source object is unique in several ways. First, an XY event source object is created through an XY event source name object. Second, an XY event source object is a type of point feature class. Third, the point feature class represented by an XY event source object is dynamic, meaning that the feature class is generated from a table rather than a physical dataset. Once a dynamic feature class is created, however, it can be exported to a shapefile or a geodatabase feature class.

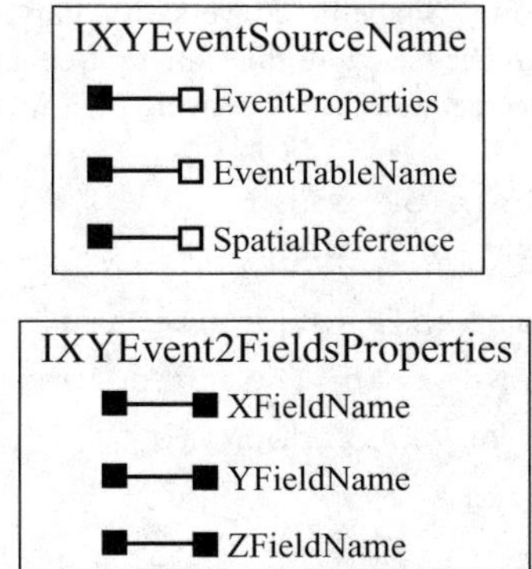

Figure 6.7   *IXYEventSourceName* and *IXYEvent2FieldsProperties* have properties that can define an *XYEventSource* object.

An *XYEvent2FieldsProperties* object implements *IXYEvent2FieldsProperties*, which has the properties for setting up the *x* field, *y* field, and *z* field (optional) of a point feature class. These properties can be passed on to an XY event source name object. *IXYEventSourceName* also has members for specifying the source table and the spatial reference.

6.3 CONVERTING SHAPEFILE TO GEODATABASE

A shapefile contains one set of spatial data, which may represent point, line, or area features. When converted into a geodatabase, a shapefile becomes a feature class, which can be a standalone feature class or part of a feature dataset. This section covers both types of conversion and also shows two ways of getting multiple shapefiles for conversion: one uses a dialog box and the other uses the input box.

6.3.1 *ShapefileToAccess*

ShapefileToAccess converts a shapefile to a feature class and saves the output as a standalone feature class in a new geodatabase. The macro performs the same function as creating a new personal geodatabase in ArcCatalog and using the Import/Shapefile to Geodatabase command to import a shapefile to the geodatabase. *ShapefileToAccess* is organized into three parts. Part 1 defines the output including its workspace and name. Part 2 defines the input. Part 3 performs the data conversion. *ShapefileToAccess* uses the lightweight name objects throughout the code.

> **Key Interfaces:** *IWorkspaceName, IFeatureClassName, IDatasetName, IFeature-DataConverter.*
> **Key Members:** *WorkspaceFactoryProgID, PathName, Name, WorkspaceName, ConvertFeatureClass.*
> **Usage:** Import *ShapefileToAccess* to Visual Basic Editor in ArcCatalog. Run the macro. The macro reports "Shapefile conversion complete," when the conversion is done. The macro adds *Trial.mdb* to the Catalog tree and *Soils* as a feature class in the geodatabase. The feature class is converted from *soils.shp*.

```
Private Sub ShapefileToAccess()
    ' Part 1: Define the output.
   Dim pWorkspaceName As IWorkspaceName
   Dim pFeatureClassName As IFeatureClassName
   Dim pDatasetName As IDatasetName
   ' Define the workspace.
   Set pWorkspaceName = New WorkspaceName
   pWorkspaceName.WorkspaceFactoryProgID = _
   "esricore.AccessWorkspaceFactory"
   pWorkspaceName.PathName = "c:\data\chap6\Trial.mdb"
    ' Define the dataset.
```

```
    Set pFeatureClassName = New FeatureClassName
    Set pDatasetName = pFeatureClassName
    Set pDatasetName.WorkspaceName = pWorkspaceName
    pDatasetName.name = "Soils"
```

Part 1 first creates *pWorkspaceName* as an instance of the *WorkspaceName* class and defines its *WorkspaceFactoryProgID* and *PathName* properties. Next the code creates *pFeatureClassName* as an instance of the *FeatureClassName* class and switches to *IDatasetName* to set its *WorkspaceName* and *Name* properties.

```
    ' Part 2: Define the input.
    Dim pInShpWorkspaceName As IWorkspaceName
    Dim pFCName As IFeatureClassName
    Dim pShpDatasetName As IDatasetName
    ' Define the workspace.
    Set pInShpWorkspaceName = New WorkspaceName
    pInShpWorkspaceName.PathName = "c:\data\chap6"
    pInShpWorkspaceName.WorkspaceFactoryProgID = _
    "esriCore.ShapefileWorkspaceFactory"
    ' Define the dataset.
    Set pFCName = New FeatureClassName
    Set pShpDatasetName = pFCName
    pShpDatasetName.name = "soils.shp"
    Set pShpDatasetName.WorkspaceName = pInShpWorkspaceName
```

Part 2 creates *pInShpWorkspaceName* as an instance of the *WorkspaceName* class and specifies its *PathName* and *WorkspaceFactoryProgID* properties. Next the code creates *pFCName* as an instance of the *FeatureClassName* class and switches to *IDatasetName* to specify its *Name* and *WorkspaceName* properties.

```
    ' Part 3: Perform data conversion.
    Dim pShpToFC As IFeatureDataConverter
    Set pShpToFC = New FeatureDataConverter
    pShpToFC.ConvertFeatureClass pFCName, Nothing, Nothing, _
    pFeatureClassName, Nothing, Nothing, "", 1000, 0
    MsgBox "Shapefile conversion complete!"
End Sub
```

Part 3 creates *pShpToFC* as an instance of the *FeatureDataConverter* class and uses the *ConvertFeatureClass* method on *IFeatureDataConverter* to convert *pFCName* into *pFeatureClassName*. Besides the two object qualifiers, the *Convert-FeatureClass* method specifies 1000 for the flush interval. The flush interval dictates the interval for committing data, which may be important for loading large amounts of data into a geodatabase. An alternative to *IFeatureDataConverter* for data conversion is *IExportOperation*, which also has methods for exporting feature classes and tables. For converting a shapefile to a geodatabase feature class, *IExportOperation* functions in nearly the same way as *IFeatureDataConverter*. The following bit of code using *IExportOperation* can replace Part 3 of **ShapefileToAccess**:

```
' Part 3: Use a new ExportOperation for comparison.
Dim pShpToFC As IExportOperation
Set pShpToFC = New ExportOperation
pShpToFC.ExportFeatureClass pFCName, Nothing, Nothing, _
Nothing, pFeatureClassName, 0
```

6.3.2 *MultipleShapefilesToAccess*

MultipleShapefilesToAccess converts two or more shapefiles into standalone feature classes and saves the feature classes in a new geodatabase. The macro performs the same function as creating a new personal geodatabase in ArcCatalog and using the Import/Shapefile to Geodatabase command to import two or more shapefiles to the geodatabase. *MultipleShapefilesToAccess* has three parts. Part 1 defines the output and input workspace name objects. Part 2 uses a dialog box to get the shapefiles to be converted. Part 3 uses a *Do...Loop* to convert shapefiles into standalone feature classes.

> **Key Interfaces:** *IWorkspaceName, IGxDialog, IGxObjectFilter, IEnumGxObject, IGxDataset, IFeatureClassName, IDatasetName, IFeatureDataConverter.*
> **Key Members:** *WorkspaceFactoryProgID, PathName, AllowMultiSelect, ButtonCaption, ObjectFilter, StartingLocation, Title, DoModalOpen, Next, Name, WorkspaceName, ConvertFeatureClass.*
> **Usage:** Import *MultipleShapefilesToAccess* to Visual Basic Editor in ArcCatalog. Run the macro. Select *landuse.shp, sewers.shp,* and *soils.shp* from the dialog box for conversion. The macro adds *SiteAnalysis.mdb* to the Catalog tree and *landuse, sewers,* and *soils* as feature classes in the database. The feature classes in the geodatabase are converted from the shapefiles. The macro uses the prefix of the shapefile (e.g., landuse) to name the output feature class.

```
Private Sub MultipleShapefilesToAccess()
  ' Part 1: Define the output and input workspaces.
  Dim pWorkspaceName As IWorkspaceName
  Dim pInShpWorkspaceName As IWorkspaceName
  ' Define the output workspace.
  Set pWorkspaceName = New WorkspaceName
  pWorkspaceName.WorkspaceFactoryProgID = _
  "esricore.AccessWorkspaceFactory"
  pWorkspaceName.PathName = _
  "c:\data\chap6\SiteAnalysis.mdb"
```

```
' Define the input workspace.
Set pInShpWorkspaceName = New WorkspaceName
pInShpWorkspaceName.pathname = "c:\data\chap6"
pInShpWorkspaceName.WorkspaceFactoryProgID = _
"esriCore.ShapefileWorkspaceFactory"
```

Part 1 creates *pWorkspaceName* as an instance of the *WorkspaceName* class and defines its *WorkspaceFactoryProgID* and *PathName* properties. The code creates and defines *pInShpWorkspaceName* in the same way.

```
' Part 2: Prepare a dialog box for selecting shapefiles
' to convert.
Dim pGxDialog As IGxDialog
Dim pGxFilter As IGxObjectFilter
Dim pGxObjects As IEnumGxObject
Dim pGxDataset As IGxDataset
Set pGxDialog = New GxDialog
Set pGxFilter = New GxFilterShapefiles
' Define the dialog's properties.
With pGxDialog
  .AllowMultiSelect = True
  .ButtonCaption = "Add"
  Set.ObjectFilter = pGxFilter
  .StartingLocation = pInShpWorkspaceName.PathName
  .Title = "Select Shapefiles to Convert"
End With
' Open the dialog.
pGxDialog.DoModalOpen 0, pGxObjects
' Exit sub if no shapefile has been selected.
Set pGxDataset = pGxObjects.Next
If pGxDataset Is Nothing Then
  Exit Sub
End If
```

Part 2 creates *pGxDialog* as an instance of the *GxDialog* class and *pGxFilter* as an instance of the *GxFilterShapefiles* class. Next the code defines the properties of *pGxDialog* to allow for multiple selections, to start at the input workspace location, and to show only shapefiles. The code then uses the *DoModalOpen* method on *IGxDialog* to open the dialog box and to save the selected shapefiles into a collection. If no shapefile has been selected, exit the sub.

```vba
' Part 3: Loop through each selected shapefile and
' convert it to a feature class.
Dim pFeatureClassName As IFeatureClassName
Dim pDatasetName As IDatasetName
Dim pFCName As IFeatureClassName
Dim pShpDatasetName As IDatasetName
Dim pShpToFC As IFeatureDataConverter
Do Until pGxDataset Is Nothing
  ' Define the output dataset.
  Set pFeatureClassName = New FeatureClassName
  Set pDatasetName = pFeatureClassName
  Set pDatasetName.WorkspaceName = pWorkspaceName
  pDatasetName.name = pGxDataset.Dataset.name
  ' Define the input dataset.
  Set pFCName = New FeatureClassName
  Set pShpDatasetName = pFCName
  pShpDatasetName.name = pGxDataset.Dataset.name & ".shp"
  Set pShpDatasetName.WorkspaceName = pInShpWorkspaceName
  ' Perform data conversion.
  Set pShpToFC = New FeatureDataConverter
  pShpToFC.ConvertFeatureClass pFCName, Nothing, _
  Nothing, pFeatureClassName, Nothing, Nothing, _
  "", 1000, 0
  Set pGxDataset = pGxObjects.Next
Loop
MsgBox "Shapefile conversions complete!"
End Sub
```

Part 3 uses a *Do...Loop* to step through each shapefile selected for conversion. The code performs a QI for the *IDatasetName* interface to specify the *Name* and *Workspace- Name* properties of the output dataset and the input dataset. Then the code uses the *ConvertFeatureClass* method to convert the input dataset into the output dataset. The loop stops when nothing is advanced from *pGxObjects*.

6.3.3   *ShapefilesToFeatureDataset*

ShapefilesToFeatureDataset converts one or more shapefiles into feature classes and saves the feature classes in a specified feature dataset of a geodatabase. The output from the macro therefore follows a hierarchical structure of geodatabase, feature dataset, and feature class. Because all feature classes in a feature dataset must share the same spatial reference (i.e., the same coordinate system and extent), a feature dataset is typically reserved for feature classes that participate in topological relationships with each other.

ShapefilesToFeatureDataset performs the same function as creating a new personal geodatabase, creating a new feature dataset, and using the Import/Shapefile to Geodatabase command to import one or more shapefiles to the geodatabase. By default, a feature dataset uses the extent of the first shapefile as its area extent. If the first shapefile has a smaller area extent, then portions of the other shapefiles will disappear after conversion.

ShapefilesToFeatureDataSet is organized into two parts. Part 1 defines the output workspace, the output feature dataset, and the input workspace. Part 2 uses the input box and a *Do...Loop* statement to get shapefiles for conversion.

> **Key Interfaces:** *IWorkspaceName, IFeatureDatasetName, IDatasetName, IFeatureClassName, IFeatureDataConverter.*
>
> **Key Members:** *WorkspaceFactoryProgID, PathName, Name, WorkspaceName, ConvertFeatureClass.*
>
> **Usage:** Import *ShapefilesToFeatureDataset* to Visual Basic Editor in ArcCatalog. Run the macro. Enter *landuse, sewers,* and *soils* sequentially in the input box. Click Cancel in the input box to dismiss the dialog. The macro adds to the Catalog tree a hierarchy of *SiteAnalysis2.mdb, StudyArea1*, and *landuse, sewers,* and *soils.* The feature classes in the geodatabase are converted from the shapefiles.

```
Private Sub ShapefilesToFeatureDataset()
  ' Part 1: Define the output and input workspaces.
  Dim pWorkspaceName As IWorkspaceName
  Dim pFeatDSName As IFeatureDatasetName
  Dim pDSName As IDatasetName
  Dim pInShpWorkspaceName As IWorkspaceName
  ' Define the output workspace.
  Set pWorkspaceName = New WorkspaceName
  pWorkspaceName.WorkspaceFactoryProgID = _
  "esricore.AccessWorkspaceFactory"
  pWorkspaceName.PathName = _
  "c:\data\chap6\SiteAnalysis2.mdb"
  ' Specify the output feature dataset.
  Set pFeatDSName = New FeatureDatasetName
  Set pDSName = pFeatDSName
  Set pDSName.WorkspaceName = pWorkspaceName
  pDSName.Name = "StudyArea1"
  ' Specify the input workspace.
  Set pInShpWorkspaceName = New WorkspaceName
  pInShpWorkspaceName.pathname = "c:\data\chap6"
  pInShpWorkspaceName.WorkspaceFactoryProgID = _
  "esriCore.ShapefileWorkspaceFactory"
```

Part 1 defines the output and input workspaces in the same way as *Multiple-ShapefilesToAccess.* The change is with the feature dataset for the output. The code creates *pFeatDSName* as an instance of the *FeatureDatasetName* class and switches to *IDatasetName* to define its *WorkspaceName* and *Name* properties.

```
' Part 2: Loop through every shapefile to be converted.
Dim pFeatureClassName As IFeatureClassName
Dim pDatasetName As IDatasetName
Dim pFCName As IFeatureClassName
Dim pShpDatasetName As IDatasetName
Dim pShpToFC As IFeatureDataConverter
Dim pInput As String
' Use the input box to get an input shapefile.
pInput = InputBox("Enter the name of the input shapefile")
' Loop through each input shapefile while the input box
' is not empty.
Do While pInput <> ""
    ' Define the output dataset.
    Set pFeatureClassName = New FeatureClassName
    Set pDatasetName = pFeatureClassName
    Set pDatasetName.WorkspaceName = pWorkspaceName
    pDatasetName.Name = pInput
    ' Define the input dataset.
    Set pFCName = New FeatureClassName
    Set pShpDatasetName = pFCName
    pShpDatasetName.Name = pInput & ".shp"
    Set pShpDatasetName.WorkspaceName = pInShpWorkspaceName
    ' Perform data conversion.
    Set pShpToFC = New FeatureDataConverter
    pShpToFC.ConvertFeatureClass pFCName, Nothing, _
    pFeatDSName, pFeatureClassName, Nothing, _
    Nothing, "", 1000, 0
    pInput = InputBox("Enter the name of the input shapefile")
Loop
MsgBox "Shapefile conversions complete!"
End Sub
```

Part 2 uses the *InputBox* function and a *Do...Loop* statement to loop through every shapefile to be converted. The *ConvertFeatureClass* method uses an additional object qualifier of *pFeatDSName* for the feature dataset.

6.4 CONVERTING COVERAGE TO GEODATABASE AND SHAPEFILE

This section covers conversion of traditional coverages into geodatabase feature class and shapefiles. A coverage consists of different datasets such as arc, tic, and labels. A coverage based on the regions or dynamic segmentation data model has even more datasets. Therefore, conversion of a coverage must start by identifying the dataset to be converted.

6.4.1 *CoverageToAccess*

CoverageToAccess converts a coverage to a feature class and saves the feature class in a specified geodatabase. The macro performs the same function as creating a new personal geodatabase in ArcCatalog and using the Import/Coverage to Geodatabase command to import a coverage into the geodatabase. *CoverageTo-Access* has three parts. Part 1 defines the output including its workspace and name. Part 2 defines the input including the workspace and the specific dataset of the coverage to be converted. Part 3 performs the data conversion.

Key Interfaces: *IPropertySet, IWorkspaceFactory, IWorkspaceName, IFeatureClass-Name, IDatasetName, IFeatureDataConverter.*

Key Members: *SetProperty, Create, WorkspaceName, Name, PathName, Workspace-FactoryProgID, ConvertFeatureClass.*

Usage: Import *CoverageToAccess* to Visual Basic Editor in ArcCatalog. Run the macro. The macro adds *emida.mdb* to the Catalog tree and *breakstrm* as a feature class in the geodatabase. The feature class in the geodatabase is converted from the arcs of the *breakstrm* coverage.

```
Private Sub CoverageToAccess()
    ' Part 1: Define the output.
    Dim pPropset As IPropertySet
    Dim pOutAcFact As IWorkspaceFactory
    Dim pOutAcWorkspaceName As IWorkspaceName
    Dim pOutAcFCName As IFeatureClassName
    Dim pOutAcDSName As IDatasetName
    ' Set the property named database.
    Set pPropset = New PropertySet
    pPropset.SetProperty "Database", "c:\data\chap6"
    ' Define the workspace.
    Set pOutAcFact = New AccessWorkspaceFactory
    Set pOutAcWorkspaceName = pOutAcFact.Create _
    ("c:\data\chap6", "emida", pPropset, 0)
    ' Define the dataset.
    Set pOutAcFCName = New FeatureClassName
```

```
  Set pOutAcDSName = pOutAcFCName
  Set pOutAcDSName.WorkspaceName = pOutAcWorkspaceName
  pOutAcDSName.name = "breakstrm"
```

Part 1 defines the input. The code first creates *pPropset* as an instance of the *PropertySet* class. The *Create* method on *IWorkspaceFactory* uses *pPropset* and other arguments to create *pOutAcWorkspaceName*. Next the code creates *pOutAc-FCName* as an instance of the *FeatureClassName* class and performs a QI for the *IDatasetName* interface to set its *WorkspaceName* and *Name* properties.

```
  ' Part 2: Specify the Input.
  Dim pInCovWorkspaceName As IWorkspaceName
  Dim pFCName As IFeatureClassName
  Dim pCovDatasetName As IDatasetName
  ' Define the workspace.
  Set pInCovWorkspaceName = New WorkspaceName
  pInCovWorkspaceName.PathName = "c:\data\chap6"
  pInCovWorkspaceName.WorkspaceFactoryProgID = _
  "esriCore.ArcInfoWorkspaceFactory.1"
  ' Define the dataset of the coverage to be converted.
  Set pFCName = New FeatureClassName
  Set pCovDatasetName = pFCName
  pCovDatasetName.name = "breakstrm:arc"
  Set pCovDatasetName.WorkspaceName = pInCovWorkspaceName
```

Part 2 specifies the input. The code creates *pInCovWorkspaceName* as an instance of the *WorkspaceName* class and defines its *PathName* and *WorkspaceFactoryProgID* properties. Then the code creates *pFCName* as an instance of the *FeatureClassName* class and switches to the *IDatasetName* interface to set the *Name* and *Workspace-Name* properties. Notice that the name is "*breakstrm:arc*," which refers to the arcs of *breakstrm*.

```
  ' Part 3: Perform data conversion.
  Dim pCovtoFC As IFeatureDataConverter
  Set pCovtoFC = New FeatureDataConverter
  pCovtoFC.ConvertFeatureClass pFCName, Nothing, Nothing, _
  pOutAcFCName, Nothing, Nothing, "", 1000, 0
  MsgBox "Coverage conversion complete!"
End Sub
```

Part 3 uses the *ConvertFeatureClass* method on *IFeatureDataConverter* to convert *pFCName* to *pOutAcFCName*.

6.4.2 *CoverageToShapefile*

CoverageToShapefile converts a coverage to a shapefile. The macro performs the same function as using the Coverage to Shapefile tool in ArcToolbox or the Data/Export Data command in the context menu of a coverage in ArcMap. ***CoverageToShapefile*** has three parts. Part 1 defines the input coverage's workspace, dataset, and geometry. Part 2 specifies the output workspace and dataset. Part 3 converts the coverage to a shapefile and adds the shapefile as a feature layer in the active map.

Key Interfaces: *IWorkspaceName, IFeatureClassName, IDatasetName, IPropertySet, IName, IFeatureDataConverter.*

Key Members: *PathName, WorkspaceFactoryProgID, name, WorkspaceName, SetProperty, ConnectionProperties, ConvertFeatureClass.*

Usage: Add *breakstrm.arc*, the arc of the coverage *breakstrm*, to an active map. Import ***CoverageToShapefile*** to Visual Basic Editor. Run the macro. The macro creates *breakstrm.shp* and adds the shapefile as a feature layer to the active map.

```
Private Sub CoverageToShapefile()
  ' Part 1: Define the input coverage.
  Dim pInCovWorkspaceName As IWorkspaceName
  Dim pInFCName As IFeatureClassName
  Dim pCovDatasetName As IDatasetName
  ' Define the workspace.
  Set pInCovWorkspaceName = New WorkspaceName
  pInCovWorkspaceName.pathname = "c:\data\chap6"
  pInCovWorkspaceName.WorkspaceFactoryProgID = _
  "esriCore.ArcInfoWorkspaceFactory.1"
  ' Define the dataset.
  Set pInFCName = New FeatureClassName
  Set pCovDatasetName = pInFCName
  pCovDatasetName.Name = "breakstrm:arc"
  Set pCovDatasetName.WorkspaceName = pInCovWorkspaceName
```

Part 1 defines the input coverage. The code first creates *pInCovWorkspaceName* as an instance of the *WorkspaceName* class and defines its *PathName* and *WorkspaceFactoryProgID* properties. Then the code creates *pInFCName* as an instance of the *FeatureClassName* class and switches to the *IDatasetName* interface to define its *Name* and *WorkspaceName* properties.

```
  ' Part 2: Define the output.
  Dim pPropset As IPropertySet
  Dim pOutShWSName As IWorkspaceName
  Dim pOutShFCName As IFeatureClassName
```

```
Dim pOutshDSName As IDatasetName
' Set the connection property.
Set pPropset = New PropertySet
pPropset.SetProperty "DATABASE", "c:\data\chap6"
' Define the workspace.
Set pOutShWSName = New WorkspaceName
pOutShWSName.ConnectionProperties = pPropset
pOutShWSName.WorkspaceFactoryProgID = _
"esriCore.shapefileWorkspaceFactory.1"
' Define the dataset.
Set pOutShFCName = New FeatureClassName
Set pOutshDSName = pOutShFCName
Set pOutshDSName.WorkspaceName = pOutShWSName
pOutshDSName.Name = "breakstrm.shp"
```

Part 2 defines the output. The code first creates *pPropset* as an instance of the *PropertySet* class and uses *pPropset*, along with the shapefile workspace factory, to define an instance of the *WorkspaceName* class referenced by *pOutShWSName*. Next the code creates *pOutShFCName* as an instance of the *FeatureClassName* class and switches to the *IDatasetName* interface to define its *WorkspaceName* and *Name* properties.

```
' Part 3: Convert the coverage to a shapefile and display
' the shapefile.
Dim pName As IName
Dim pCovtoshape As IFeatureDataConverter
Dim pOutShFC As IFeatureClass
Dim pOutSh As IFeatureLayer
Dim pMxDoc As IMxDocument
Dim pMap As IMap
' Convert the coverage to a shapefile.
Set pCovtoshape = New FeatureDataConverter
pCovtoshape.ConvertFeatureClass pInFCName, Nothing, _
Nothing, pOutShFCName, Nothing, Nothing, "", 1000, 0
' Create a feature layer from the output feature class
' name object.
Set pName = pOutShFCName
Set pOutShFC = pName.Open
Set pOutSh = New FeatureLayer
Set pOutSh.FeatureClass = pOutShFC
pOutSh.Name = "breakstrm.shp"
' Add the feature layer to the active map.
```

```
   Set pMxDoc = ThisDocument
   Set pMap = pMxDoc.FocusMap
   pMap.AddLayer pOutSh
   MsgBox "Coverage conversion complete!"
End Sub
```

Part 3 converts *pInFCName* to *pOutShFCName* by using the *ConvertFeature-Class* method on *IFeatureDataConverter*. Next the code opens a feature class referenced by *pOutShFC* from its name object, and creates a feature layer from *pOutShFC*. Finally, the code adds the feature layer referenced by *pOutSh* to the active map.

6.5 PERFORMING RASTERIZATION AND VECTORIZATION

This section covers rasterization and vectorization. Rasterization creates a raster and populates the cells of the raster with values from an attribute. The raster is an ESRI grid, a software-specific raster. The attribute can be a specified field or feature ID by default. Vectorization creates a feature class. The cell values of the input raster are stored in a field called GRIDCODE. Other fields may be added through code.

6.5.1 *FeatureToRaster*

FeatureToRaster converts a feature layer into a permanent raster and adds the raster to the active map. The input feature layer may represent a shapefile, a coverage, or a geodatabase feature class. The output raster contains cell values that correspond to the feature IDs of the input dataset.

The macro performs the same function as using the Convert/Features to Raster command in Spatial Analyst. *FeatureToRaster* has three parts. Part 1 defines the input feature layer. Part 2 performs the conversion and saves the output raster in a specified workspace. Part 3 creates a raster layer from the output and adds the layer to the active map.

> **Key Interfaces:** *IWorkspace, IWorkspaceFactory, IConversionOp, IRasterAnalysis-Environment, IRasterDataset.*
> **Key Members:** *OpenFromFile, SetCellSize, ToRasterDataset, CreateFromDataset, Name.*
> **Usage:** Add *nwroads.shp*, an interstate highway shapefile of the Pacific Northwest, to an active map. Import *FeatureToRaster* to Visual Basic Editor in ArcMap. Run the macro. *FeatureToRaster* adds *roadsgd* as a raster layer to the active map.

```
Private Sub FeatureToRaster()
  ' Part 1: Define the input dataset.
  Dim pMxDoc As IMxDocument
  Dim pMap As IMap
```

```
Dim pInputFL As IFeatureLayer
Dim pInputFC As IFeatureClass
Set pMxDoc = ThisDocument
Set pMap = pMxDoc.FocusMap
Set pInputFL = pMap.Layer(0)
Set pInputFC = pInputFL.FeatureClass
```

Part 1 sets *pInputFC* to be the feature class of the top layer in the active map.

```
' Part 2: Convert feature class to raster.
Dim pWS As IWorkspace
Dim pWSF As IWorkspaceFactory
Dim pConversionOp As IConversionOp
Dim pEnv As IRasterAnalysisEnvironment
Dim pOutRaster As IRasterDataset
' Set the output workspace.
Set pWSF = New RasterWorkspaceFactory
Set pWS = pWSF.OpenFromFile("c:\data\chap6", 0)
' Prepare a raster conversion operator, and perform the
' conversion.
Set pConversionOp = New RasterConversionOp
Set pEnv = pConversionOp
pEnv.SetCellSize esriRasterEnvValue, 5000
Set pOutRaster = pConversionOp.ToRasterDataset _
(pInputFC, "GRID", pWS, "roadsgd")
```

Part 2 performs the conversion. The code first defines the output workspace referenced by *pWS*. Next the code creates *pConversionOp* as an instance of the *RasterConversionOp* class and switches to the *IRasterAnalysisEnvironment* interface to set the output cell size of 5000 (meters). The code then uses the *ToRasterDataset* method on *IConversionOp* to convert *pInputFC* into *pOutRaster* and to save the output in *pWS*.

```
' Part 3: Create a new raster layer and add the layer
' to the active map.
Dim pOutRasterLayer As IRasterLayer
Set pOutRasterLayer = New RasterLayer
pOutRasterLayer.CreateFromDataset pOutRaster
pOutRasterLayer.Name = "roadsgd"
pMap.AddLayer pOutRasterLayer
End Sub
```

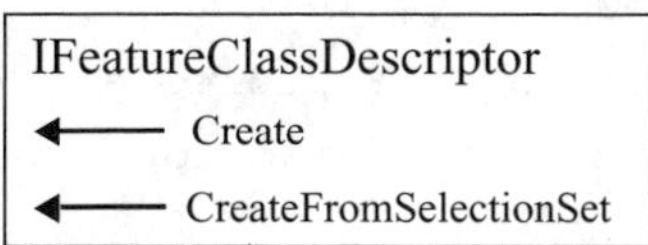

Figure 6.8   Methods on *IFeatureClassDescriptor*.

Part 3 uses the *CreateFromDataset* method on *IRasterLayer* to create a new raster layer from *pOutRaster*, and adds the layer to the active map.

6.5.2   *FCDescriptorToRaster*

FeatureToRaster produces an output raster that has feature IDs as cell values. In some cases, it is more useful to have cell values that correspond to county names or highway numbers instead of feature IDs. *FCDescriptorToRaster*, which uses a feature class descriptor as the input dataset, is designed for those cases. A *FeatureClassDescriptor* object implements *IFeatureClassDescriptor*, which has methods for creating a feature class descriptor based on a specific field of a dataset (Figure 6.8).

FCDescriptorToRaster performs the same function as using the Convert/Features to Raster command in Spatial Analyst. The Convert/Features to Raster dialog allows the user to specify a field that corresponds to the cell value. *FCDescriptorToRaster* has three parts. Part 1 creates a feature class descriptor from a feature layer. Part 2 converts the feature class descriptor into a permanent raster. Part 3 creates a new raster layer from the output raster and adds the layer to the active map.

> **Key Interfaces:** *IFeatureClassDescriptor, IWorkspace, IWorkspaceFactory, IConversionOp, IRasterAnalysisEnvironment, IRasterDataset.*
> **Key Members:** *Create, OpenFromFile, SetCellSize, ToRasterDataset, CreateFromDataset, Name.*
> **Usage:** Add *nwroads.shp* to an active map. Import *FCDescriptorToRaster* to Visual Basic Editor in ArcMap. Run the macro. *FCDescriptorToRaster* adds *rtenumgd* as a raster layer to the active map. The attribute table of *rtenumgd* has an Rte_num1 field that lists the highway numbers.

```
Private Sub FCDescriptorToRaster()
  ' Part 1: Get a handle on the descriptor of a shapefile
  ' to be converted.
  Dim pMxDoc As IMxDocument
  Dim pMap As IMap
  Dim pInputFL As IFeatureLayer
  Dim pInputFC As IFeatureClass
  Dim pFCDescriptor As IFeatureClassDescriptor
  Dim sFieldName As String
  Set pMxDoc = ThisDocument
  Set pMap = pMxDoc.FocusMap
```

```
Set pInputFL = pMap.Layer(0)
Set pInputFC = pInputFL.FeatureClass
' Create a feature class descriptor based on the field
' RTE_NUM1.
Set pFCDescriptor = New FeatureClassDescriptor
sFieldName = "RTE_NUM1"
pFCDescriptor.Create pInputFC, Nothing, sFieldName
```

Part 1 sets *pInputFC* to be the feature class of the top layer in the active map. Next the code creates *pFCDescriptor* as an instance of the *FeatureClassDescriptor* class and assigns RTE_NUM1 to the string variable *sFieldName*. The code then uses the *Create* method on *IFeatureClassDescriptor* and *sFieldName* as an argument to create *pFCDescriptor*.

```
' Part 2: Convert the feature class descriptor to
' a raster.
Dim pWS As IWorkspace
Dim pWSF As IWorkspaceFactory
Dim pConversionOp As IConversionOp
Dim pEnv As IRasterAnalysisEnvironment
Dim pOutRaster As IRasterDataset
Set pWSF = New RasterWorkspaceFactory
Set pWS = pWSF.OpenFromFile("c:\data\chap6", 0)
Set pConversionOp = New RasterConversionOp
Set pEnv = pConversionOp
pEnv.SetCellSize esriRasterEnvValue, 5000
' Use the feature class descriptor as the input for
' conversion.
Set pOutRaster = pConversionOp.ToRasterDataset _
(pFCDescriptor, "GRID", pWS, "rtenumgd")
```

Part 2 sets *pWS* as a reference to the output workspace and *pConversionOp* as an instance of the *RasterConversionOp* class. The code then uses the *ToRaster-Dataset* method and *pFCDescriptor* as an object qualifier to create *pOutRaster*.

```
' Part 3: Create a new raster layer and add the layer
' to the active map.
Dim pOutRasterLayer As IRasterLayer
Set pOutRasterLayer = New RasterLayer
pOutRasterLayer.CreateFromDataset pOutRaster
pOutRasterLayer.Name = "rtenumgd"
pMap.AddLayer pOutRasterLayer
End Sub
```

Part 3 creates a new raster layer from *pOutRaster* and adds the layer to the active map.

6.5.3 *RasterToShapefile*

RasterToShapefile converts a raster to a shapefile. The macro performs the same function as using the Convert/Raster to Features command in Spatial Analyst. The output shapefile contains a default field called GRIDCODE that stores the cell values of the input raster.

RasterToShapefile is organized into three parts. Part 1 defines the input raster. Part 2 performs the conversion and saves the output shapefile in a specified workspace. Part 3 creates a feature layer from the output and adds the layer to the active map.

> **Key Interfaces:** *IWorkspace, IWorkspaceFactory, IConversionOp, IFeatureClass.*
> **Key Members:** *OpenFromFile, RasterDataToLineFeatureData, FeatureClass, Name.*
> **Usage:** Add *nwroads_gd*, a raster showing interstate highways in the Pacific Northwest, to an active map. Import *RasterToShapefile* to Visual Basic Editor in ArcMap. Run the macro. *RasterToShapefile* adds *roads.shp* as a feature layer to the active map.

```
Private Sub RasterToShapefile()
' Part 1: Get a handle on the input raster.
  Dim pMxDoc As IMxDocument
  Dim pMap As IMap
  Dim pInputRL As IRasterLayer
  Dim pInputRaster As IRaster
  Set pMxDoc = ThisDocument
  Set pMap = pMxDoc.FocusMap
  Set pInputRL = pMap.Layer(0)
  Set pInputRaster = pInputRL.Raster
```

Part 1 sets *pInputRaster* to be the raster of the top layer in the active map.

```
' Part 2: Convert the raster to a shapefile.
  Dim pWS As IWorkspace
  Dim pWSF As IWorkspaceFactory
  Dim pConversionOp As IConversionOp
  Dim pOutFClass As IFeatureClass
  ' Set the output workspace.
  Set pWSF = New ShapefileWorkspaceFactory
  Set pWS = pWSF.OpenFromFile("c:\data\chap6", 0)
  ' Prepare a raster conversion operation, and perform the
  ' conversion.
```

```
Set pConversionOp = New RasterConversionOp
Set pOutFClass = pConversionOp. _
RasterDataToLine FeatureData (pInputRaster, pWS, _
"roads.shp", True, True, 5000)
```

Part 2 sets *pWS* to be the workspace of the output shapefile. The code then creates *pConversionOp* as an instance of the *RasterConversionOp* class and uses the *RasterDataToLineFeatureData* method on *IConversionOp* to create *pOutFClass*. The method requires *pInputRaster* and *pWS* as the object qualifiers. Additionally, it has four arguments. The first is the name of the output dataset. The second, if true, means that cells with values of <= 0 or no data belong to the background. The third, if true, means applying line generalization or weeding to the output. The last is the minimum length of dangling arcs on the output.

The *IConversionOp* interface also has a method called *ToFeatureData*, which can convert a raster to a polyline shapefile. But, because *ToFeatureData* does not include the options of line generalization and dangling arcs, the output is typically less desirable than that from *RasterDataToLineFeatureData*.

```
' Part 3: Create a new raster layer and add the layer
' to the active map.
Dim pOutputFeatLayer As IFeatureLayer
Set pOutputFeatLayer = New FeatureLayer
Set pOutputFeatLayer.FeatureClass = pOutFClass
pOutputFeatLayer.Name = pOutFClass.AliasName
pMap.AddLayer pOutputFeatLayer
End Sub
```

Part 3 creates a new feature layer from *pOutFClass* and adds the layer to the active map.

6.5.4 *RasterDescriptorToShapefile*

RasterDescriptorToShapefile uses a raster descriptor as an input and adds a field in addition to GRIDCODE to the output shapefile. Like a feature class descriptor, a raster descriptor is based on a specific field in a raster. A *RasterDescriptor* object implements *IRasterDescriptor*, which has methods for creating raster descriptors (Figure 6.9).

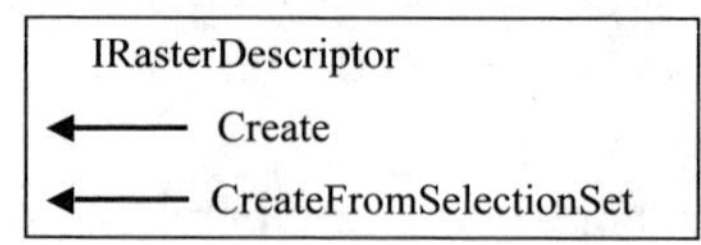

Figure 6.9 Methods on *IRasterDescriptor.*

RasterDescriptorToShapefile has three parts. Part 1 creates a raster descriptor from the input raster. Part 2 performs the conversion and saves the output shapefile in a specified workspace. Part 3 creates a feature layer from the output and adds the layer to the active map.

Key Interfaces: *IRasterDescriptor, IWorkspace, IWorkspaceFactory, IConversionOp, IFeatureClass.*

Key Members: *Create, OpenFromFile, RasterDataToPolygonFeatureData, FeatureClass, Name.*

Usage: Add *nwcounties_gd*, a raster showing counties in the Pacific Northwest, to an active map. Import ***RasterDescriptorToShapefile*** to Visual Basic Editor in ArcMap. Run the macro. ***RasterDescriptorToShapefile*** adds *Fips.shp* as a feature layer to the active map. The attribute table of *Fips* has Fips (federal information processing codes) as a field in addition to the default field of GRIDCODE.

```
Private Sub RasterDescriptorToShapefile()

    ' Part 1: Get a handle on the descriptor of the raster
    ' to be converted.
    Dim pMxDoc As IMxDocument
    Dim pMap As IMap
    Dim pInputRL As IRasterLayer
    Dim pInputRaster As IRaster
    Dim pRasterDescriptor As IRasterDescriptor
    Dim sFieldName As String
    Set pMxDoc = ThisDocument
    Set pMap = pMxDoc.FocusMap
    Set pInputRL = pMap.Layer(0)
    Set pInputRaster = pInputRL.Raster
    ' Create a raster descriptor based on the field Fips.
    Set pRasterDescriptor = New RasterDescriptor
    sFieldName = "Fips"
    pRasterDescriptor.Create pInputRaster, Nothing, sFieldName
```

Part 1 sets *pInputRaster* to be the raster of the top layer in the active map. Next the code creates *pRasterDescriptor* as an instance of the *RasterDescriptor* class and assigns Fips to the string variable *sFieldName*. The code then uses the *Create* method on *IRasterDescriptor* to create *pRasterDescriptor.*

```
    ' Part 2: Convert the raster descriptor to a shapefile.
    Dim pWS As IWorkspace
    Dim pWSF As IWorkspaceFactory
    Dim pConversionOp As IConversionOp
```

```
Dim pOutFClass As IFeatureClass
Set pWSF = New ShapefileWorkspaceFactory
Set pWS = pWSF.OpenFromFile("c:\data\chap6", 0)
Set pConversionOp = New RasterConversionOp
' Use the raster descriptor as the input for conversion.
Set pOutFClass = pConversionOp. _
RasterDataToPolygonFeatureData(pRasterDescriptor, _
pWS, "Fips.shp", True)
```

Part 2 first sets *pWS* as a reference to the workspace for the output shapefile. The code then creates *pConversionOp* as an instance of the *RasterConversionOp* class and uses the *RasterDataToPolygonFeatureData* method on *IConversionOp* to create *pOutFClass*. Notice that the method uses *pRasterDescriptor* as an object qualifier.

```
' Part 3: Create a new raster layer and add the layer
' to the active map.
Dim pOutputFeatLayer As IFeatureLayer
Set pOutputFeatLayer = New FeatureLayer
Set pOutputFeatLayer.FeatureClass = pOutFClass
pOutputFeatLayer.Name = pOutFClass.AliasName
pMap.AddLayer pOutputFeatLayer
End Sub
```

Part 3 creates a feature layer from *pOutFClass* and adds the layer to the active map.

6.6 ADDING XY EVENTS

This section covers conversion of a table with x-, y-coordinates into point features. The coordinates can be either geographic or projected. When converted, each pair of x-, y-coordinates becomes a point.

6.6.1 *XYEvents*

XYEvents adds a point layer from a table with x-, y-coordinates to an active map. The source of the table is a text file. The layer can later be exported to a shapefile based on the coordinate system that has already been defined for x-, y-coordinates. The macro performs the same function as using the Add XY Data command in ArcMap's Tools menu.

XYEvents is organized into three parts. Part 1 defines the text file in terms of its workspace and name. Part 2 defines the event source including its event-to-fields properties and spatial reference. Part 3 creates a new feature layer from the event source and adds the layer to the active map.

> **Key Interfaces:** *IWorkspaceName, ITableName, IDatasetName, IXYEvent2FieldsProperties, ISpatialReferenceFactory, IProjectedCoordinate-System, IXYEventSourceName, IXYEventSource.*
>
> **Key Members:** *PathName, WorkspaceFactoryProgID, Name, WorkspaceName, XFieldName, YFieldName, ZFieldName, CreateProjectedCoordinateSystem, EventProperties, SpatialReference, EventTableName, Open, FeatureClass.*
>
> **Usage:** Import *XYEvents* to ArcMap's Visual Basic Editor. Run the macro. The macro adds a layer named *XYEvents* to the active map. The source for *XYEvents* is *events.txt*, a table with *x*-, *y*-coordinates.

```
Private Sub XYEvents()
  ' Part 1: Define the text file.
  Dim pWorkspaceName As IWorkspaceName
  Dim pTableName As ITableName
  Dim pDatasetName As IDatasetName
  ' Define the input table's workspace.
  Set pWorkspaceName = New WorkspaceName
  pWorkspaceName.PathName = "c:\data\chap6"
  pWorkspaceName.WorkspaceFactoryProgID = _
  "esriCore.TextFileWorkspaceFactory.1"
  ' Define the dataset.
  Set pTableName = New TableName
  Set pDatasetName = pTableName
  pDatasetName.Name = "events.txt"
  Set pDatasetName.WorkspaceName = pWorkspaceName
```

Part 1 defines the input text file. The code creates *pWorkspaceName* as an instance of the *WorkspaceName* class and defines its *PathName* and *WorkspaceFactoryProgID* properties. Notice that the ProgID is esri.TextFileWorkspaceFactory. Next the code creates *pTableName* as an instance of the *TableName* class and switches to the *IDatasetName* interface to specify its properties of name and workspace.

```
  ' Part 2: Define the XY events source name object.
  Dim pXYEvent2FieldsProperties As _
  IXYEvent2FieldsProperties
  Dim pSpatialReferenceFactory As ISpatialReferenceFactory
  Dim pProjectedCoordinateSystem As _
  IProjectedCoordinateSystem
  Dim pXYEventSourceName As IXYEventSourceName
  ' Set the event to fields properties.
```

```
Set pXYEvent2FieldsProperties = _
New XYEvent2FieldsProperties
With pXYEvent2FieldsProperties
  .XFieldName = "easting"
  .YFieldName = "northing"
  .ZFieldName = ""
End With
' Set the spatial reference.
Set pSpatialReferenceFactory = _
New SpatialReferenceEnvironment
Set pProjectedCoordinateSystem = _
pSpatialReferenceFactory. _
CreateProjectedCoordinateSystem _
(esriSRProjCS_NAD1927UTM_11N)
' Specify the event source and its properties.
Set pXYEventSourceName = New XYEventSourceName
With pXYEventSourceName
  Set.EventProperties = pXYEvent2FieldsProperties
  Set.SpatialReference = pProjectedCoordinateSystem
  Set.EventTableName = pTableName
End With
```

Part 2 defines the event source. First, the code creates *pXYEvent2FieldsProperties* as an instance of the *XYEvent2FieldsProperties* class and specifies the *XFieldName*, *YFieldName*, and *ZFieldName* properties on *IXYEvent2FieldsProperties* in a *With* statement. Next the code creates a projected coordinate system referenced by *pProjectedCoordinateSystem*. The code then creates *pXYEventSourceName* as an instance of the *XYEventSourceName* class and defines the properties of *EventProperties*, *SpatialReference*, and *EventTableName* in a *With* block. The referenced objects for these three properties have all been defined previously.

```
' Part 3: Display XY events.
Dim pMxDoc As IMxDocument
Dim pMap As IMap
Dim pName As IName
Dim pXYEventSource As IXYEventSource
Dim pFLayer As IFeatureLayer
Set pMxDoc = ThisDocument
Set pMap = pMxDoc.FocusMap
' Open the events source name object.
Set pName = pXYEventSourceName
Set pXYEventSource = pName.Open
```

```
    ' Create a new events layer and add the layer to the
    ' display.
    Set pFLayer = New FeatureLayer
    Set pFLayer.FeatureClass = pXYEventSource
    pFLayer.Name = "XYEvent"
    pMap.AddLayer pFLayer
End Sub
```

Part 3 opens *pXYEventSource* from *pXYEventSourceName*. An event source object can only be created from an event source name object. Finally, the code creates a new events layer from *pXYEventSource* and adds the layer to the active map.

CHAPTER 7

Coordinate Systems

A coordinate system is a location reference system for spatial features on the Earth's surface. Because a geographic information system (GIS) works with geographically referenced data, the coordinate system plays a key role in a GIS project. There are two types of coordinate systems: geographic and projected. A geographic coordinate system consists of lines of longitude and latitude. A projected coordinate system is a plane coordinate system based on a map projection. For example, the Universal Transverse Mercator (UTM) coordinate system is based on the transverse Mercator projection.

A coordinate system, either geographic or projected, is defined by a set of parameters. The basic parameter of a geographic coordinate system is a datum, which is in turn based on a spheroid. Two commonly used datums in the United States are NAD27 (North American Datum of 1927), based on the Clarke 1866 spheroid, and NAD83 (North American Datum of 1983), based on the GRS80 (Geodetic Reference System 1980) spheroid. The parameters of a projected coordinate system define not only the projection but also the geographic coordinate system on which the projection is based. For example, the NAD27UTM_11N coordinate system is based on NAD27 as well as the projection parameters of UTM_Zone_11 North, which include the central meridian at 117°W, a scale factor of 0.9996, the latitude of projection origin at the equator, a false easting of 500,000 meters, and a false northing of zero.

ArcGIS saves the definition of a coordinate system in a spatial reference file (e.g., a prj file for a shapefile). ArcGIS considers a geographic dataset without the spatial reference information to have an unknown coordinate system. The presence of the spatial reference information is not only required for projecting a dataset from one coordinate system to another but also important for spatial analysis.

This chapter covers coordinate systems. Section 7.1 reviews commands for managing coordinate systems in ArcGIS. Section 7.2 discusses objects that are related to coordinate systems. Section 7.3 includes macros for defining on-the-fly projection. Section 7.4 has macros for defining and importing a coordinate system. Section 7.5 includes macros for datum change. Section 7.6 discusses macros for projecting and reprojecting geographic datasets. All macros start with the listing of key interfaces and key members (i.e., properties and methods) and the usage.

7.1 MANAGING COORDINATE SYSTEMS IN ARCGIS

ArcMap uses on-the-fly projection for data display. On-the-fly projection takes the spatial reference information of each dataset in a data frame and automatically converts these datasets to a common coordinate system. This common coordinate system is the coordinate system of the first layer added to the map. Alternatively, it can be set as a property of the data frame. On-the-fly projection does not actually change the original coordinate system of a dataset; it is designed for data display. If a dataset has an unknown coordinate system, ArcMap may use an assumed coordinate system. For example, NAD27 is the assumed geographic coordinate system.

On-the-fly projection cannot replace the projecting tasks. Projecting a dataset is recommended if the dataset is used frequently in a different coordinate system. Likewise, projecting the datasets to be used in spatial analysis to the same coordinate system is also recommended to obtain the most accurate results.

ArcToolbox has a suite of projection tools for working with coordinate systems. By function, these tools can be classified into three groups: defining coordinate systems, performing geographic transformations, and projecting datasets.

7.1.1 Defining Coordinate Systems

ArcToolbox has the Define Projection tools for defining coordinate systems. There are three basic methods for defining a coordinate system: Select, Import, and New. Select lets the user choose a predefined coordinate system. The parameters of a predefined coordinate system are known and are already coded in the software. Import lets the user copy the spatial reference information from one dataset to another. New is for a custom coordinate system. A custom geographic coordinate system requires a datum, an angular unit, and a prime meridian for its definition. A custom projected coordinate system requires the parameters of a geographic coordinate system as well as the parameters of the projected coordinate system for its definition.

7.1.2 Performing Geographic Transformations

Geographic transformation refers to the transformation or conversion from one geographic coordinate system to another. A common scenario for geographic transformation is datum change such as changing from NAD27 to NAD83. Geographic transformation is included in the Project tools of ArcToolbox. Whenever a projection involves a datum change, a dialog appears with the datum to convert from, the datum to convert to, and a list of transformation methods. In North America, NADCON is a popular method for transformation between NAD27 and NAD83. The Project tools save new coordinates from a geographic transformation into the output dataset.

7.1.3 Projecting Datasets

Projection tools in ArcToolbox produce new datasets based on the defined coordinate systems for the input and the output. There are two common types of projections. The first type projects a dataset from a geographic to a projected coordinate system, such as from NAD27 geographic coordinates to UTM coordinates. The second type projects a dataset from one projected coordinate system to another, such as from UTM coordinates to State Plane coordinates. The second type of projection is sometimes called reprojection.

7.2 ARCOBJECTS FOR COORDINATE SYSTEMS

ArcObjects organizes objects for managing coordinate systems in the Spatial Reference subsystem. A primary component of the subsystem is the *SpatialReference* abstract class. A spatial reference object implements *ISpatialReference*, which provides access to the spatial reference properties of a dataset such as the name of its coordinate system.

The *SpatialReference* class is inherited by three coclasses: *GeographicCoordinateSystem*, *ProjectedCoordinateSystem*, and *UnknownCoordinateSystem* (Figure 7.1). These coclasses manage geographic, projected, and unknown coordinate systems respectively.

A geographic coordinate system object implements *IGeographicCoordinateSystem* and *IGeographicCoordinateSystemEdit* (Figure 7.2). *IGeographicCoordinateSystem* provides access to the properties of coordinate unit, datum,

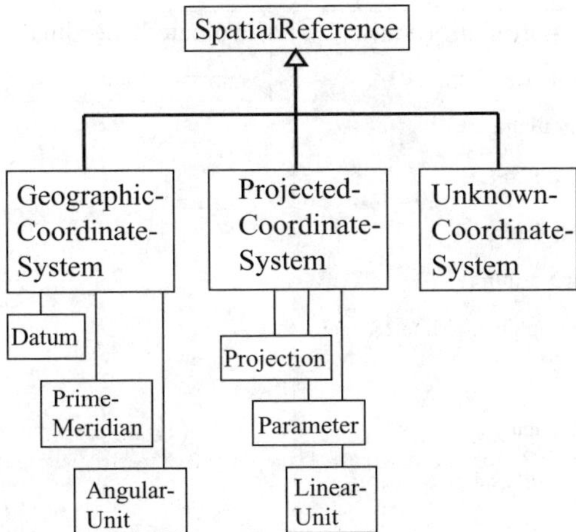

Figure 7.1 *SpatialReference* is an abstract class with three types of coordinate systems: geographic, projected, and unknown. Each geographic and projected coordinate system is associated with a set of parameters.

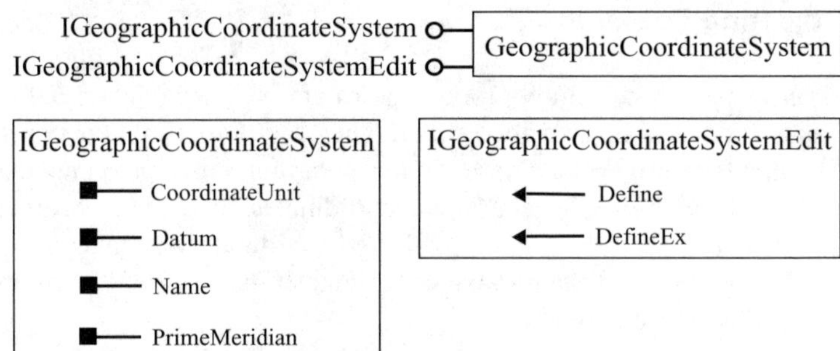

Figure 7.2 A *GeographicCoordinateSystem* object supports *IGeographicCoordinateSystem* and *IGeographicCoordinateSystemEdit*.

name, and prime meridian. *IGeographicCoordinateSystemEdit* has methods for defining the properties of a new geographic coordinate system.

Likewise, a projected coordinate system object implements *IProjected-CoordinateSystem* and *IProjectedCoordinateSystemEdit* (Figure 7.3). *IProjected-CoordinateSystem* provides access to the properties of central meridian, central parallel, coordinate unit, false easting, false northing, geographic coordinate system, name, scale factor, first standard parallel, second standard parallel, and others. *IProjectedCoordinateSystemEdit* has the method to define the properties of a projected coordinate system.

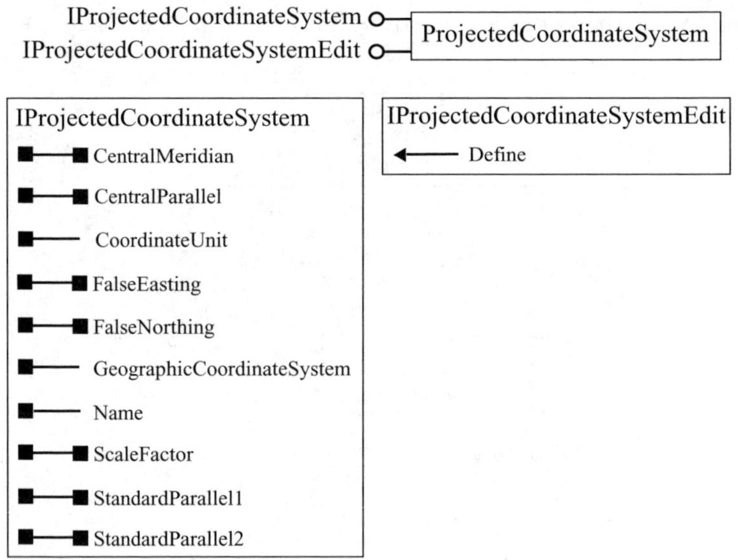

Figure 7.3 A *ProjectedCoordinateSystem* object supports *IProjectedCoordinateSystem* and *IProjectedCoordinateSystemEdit*.

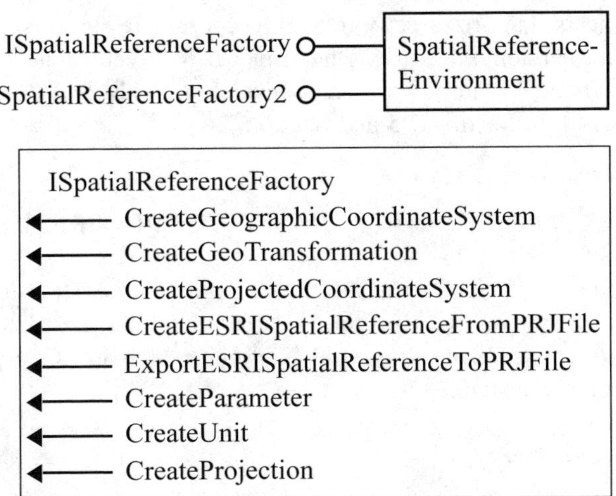

Figure 7.4   Methods on *ISpatialReferenceFactory* that a *SpatialReferenceEnvironment* object supports.

Two other primary components for coordinate systems in ArcObjects are *SpatialReferenceEnvironment* and *GeoTransformation*. The *SpatialReference-Environment* coclass lets the user create coordinate systems by using predefined spatial reference objects. A spatial reference environment object implements *ISpatialReferenceFactory* and *ISpatialReferenceFactory2* (Figure 7.4). Both interfaces have the same methods for using predefined spatial reference objects such as NAD27 and UTM to define a coordinate system. The difference is that *ISpatialReferenceFactory2* has additional methods for working with predefined geographic transformations.

The *GeoTransformation* abstract class manages geographic transformations (Figure 7.5). Of particular interest to ArcGIS users in the United States is *Grid-Transformation*, a subclass of *GeoTransformation*, which uses a grid-based method for converting geographic coordinates between the NAD27 and NAD83 datums.

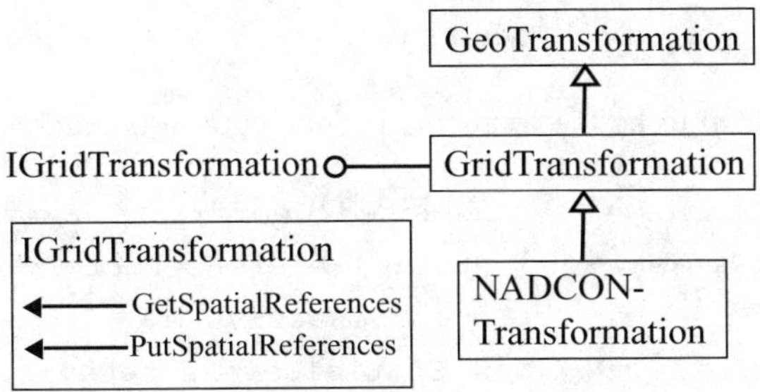

Figure 7.5   *NADCONTransformation* is a type of *GridTransformation*, which is in turn a type of *GeoTransformation*.

Besides objects that are directly related to coordinate systems, macros for projection and reprojection must use other objects to create a new dataset. These additional objects deal mainly with data conversion, fields, and the geometry field. They are discussed in Section 7.5 and Section 7.6.

7.3 MANIPULATING ON-THE-FLY PROJECTION

This section offers two examples of altering the common coordinate system for on-the-fly projection. The first uses the UTM, a predefined coordinate system. The second uses the IDTM (Idaho Statewide Transverse Mercator Coordinate System), a custom coordinate system.

7.3.1   *UTM_OnTheFly*

UTM_OnTheFly adopts a UTM coordinate system for an active map. The macro performs the same function as using the map property to specify a UTM coordinate system for the active map. *UTM_OnTheFly* has two parts. Part 1 defines the active map. Part 2 creates the NAD27UTM_11N coordinate system and assigns it to the active map.

> **Key Interfaces:** *ISpatialReferenceFactory, IProjectedCoordinateSystem.*
> **Key Members:** *CreateProjectedCoordinateSystem, SpatialReference.*
> **Usage:** Import *UTM_OnTheFly* to Visual Basic Editor in ArcMap. Right-click the active map and select Properties. The Coordinate System tab shows no projection for the current coordinate system. Run the macro. NAD_1927_UTM_Zone_11N should appear as the current coordinate system.

```
Private Sub UTM_OnTheFly()
  ' Part 1: Define the active map.
  Dim pMxDoc As IMxDocument
  Dim pMap As IMap
  Set pMxDoc = Application.Document
  Set pMap = pMxDoc.FocusMap
```

Part 1 sets *pMap* to be the focus map of the current document.

```
  ' Part 2: Create a Projected Coordinate System.
  Dim pSpatRefFact As ISpatialReferenceFactory
  Dim pPCS As IProjectedCoordinateSystem
  Set pSpatRefFact = New SpatialReferenceEnvironment
  Set pPCS = pSpatRefFact. _
  CreateProjectedCoordinateSystem _
  (esriSRProjCS_NAD1927UTM_11N)
  Set pMap.SpatialReference = pPCS
```

```
   ' Refresh the map.
     pMxDoc.ActiveView.Refresh
 End Sub
```

Part 2 first creates *pSpatRefFact* as an instance of the *SpatialReference-Environment* class. Next the code uses the *CreateProjectedCoordinateSystem* method on *ISpatialReferenceFactory* to create the NAD27UTM_11N coordinate system, which is referenced by *pPCS*. Then the code sets *pPCS* to be the spatial reference of the active map before refreshing the active view of the map document.

7.3.2 *IDTM_OnTheFly*

IDTM_OnTheFly adopts IDTM as the common coordinate system for an active map. The macro performs the same function as using the map property to specify the IDTM coordinate system for the active map. *IDTM_OnTheFly* has three parts. Part 1 defines the projection, geographic coordinate system, and linear unit of IDTM. Part 2 stores IDTM's projection parameters in an array. Part 3 creates IDTM and sets IDTM to be the active map's coordinate system.

> **Key Interfaces:** *ISpatialReferenceFactory2, IGeographicCoordinateSystem, IProjection, IUnit, ILinearUnit, IParameter, IProjectedCoordinateSystemEdit.*
> **Key Members:** *CreateProjection, CreateGeographicCoordinateSystem, CreateUnit, CreateParameter, Define, SpatialReference.*

> **Usage:** Import *IDTM_OnTheFly* to Visual Basic Editor in ArcMap. Run the macro. To make sure that the macro works, right-click the active map and select Properties. On the Coordinate System tab, UserDefinedPCS with IDTM's parameters should appear as the current coordinate system.

```
Private Sub IDTM_OnTheFly()
  ' Part 1: Define the active map and the IDTM coordinate
  ' system.
  Dim pMxDoc As IMxDocument
  Dim pMap As IMap
  Set pMxDoc = Application.Document
  Set pMap = pMxDoc.FocusMap
  ' Define the coordinate system's projection, geographic
  ' coordinate system (datum), and unit.
  Dim pSpatRefFact As ISpatialReferenceFactory2
  Dim pProjection As IProjection
  Dim pGCS As IGeographicCoordinateSystem
  Dim pUnit As IUnit
  Dim pLinearUnit As ILinearUnit
  Set pSpatRefFact = New SpatialReferenceEnvironment
```

```
Set pProjection = pSpatRefFact.CreateProjection _
(esriSRProjection_TransverseMercator)
Set pGCS = pSpatRefFact. _
CreateGeographicCoordinateSystem _
(esriSRGeoCS_NAD1927)
Set pUnit = pSpatRefFact.CreateUnit(esriSRUnit_Meter)
Set pLinearUnit = pUnit
```

Part 1 creates *pSpatRefFact* as an instance of the *SpatialReferenceEnvironment* class and uses methods on *ISpatialReferenceFactory2* to define the following: transverse Mercator for the projection, NAD27 for the geographic coordinate system, and meter for the linear unit. The *CreateUnit* method initially creates an *IUnit* object. But, because *ILinearUnit* inherits *IUnit*, *pLinearUnit* can be set to be equal to *pUnit*.

```
' Part 2: Store the 5 known parameters of IDTM in an array.
Dim aParamArray(5) As IParameter
Set aParamArray(0) = pSpatRefFact.CreateParameter _
(esriSRParameter_FalseEasting)
aParamArray(0).Value = 500000
Set aParamArray(1) = pSpatRefFact.CreateParameter _
(esriSRParameter_FalseNorthing)
aParamArray(1).Value = 100000
Set aParamArray(2) = pSpatRefFact.CreateParameter _
(esriSRParameter_CentralMeridian)
aParamArray(2).Value = -114
Set aParamArray(3) = pSpatRefFact.CreateParameter _
(esriSRParameter_LatitudeOfOrigin)
aParamArray(3).Value = 42
Set aParamArray(4) = pSpatRefFact.CreateParameter _
(esriSRParameter_ScaleFactor)
aParamArray(4).Value = 0.9996
```

Part 2 uses the *CreateParameter* method on *ISpatialReferenceFactory2* to store the five known parameters of IDTM in an array. Using the index and value properties of *IParameter*, the code stores the false easting of 500,000 as the first element of the array, the false northing of 100,000 as the second element, the central meridian of −114 as the third element, the latitude of origin of 42 as the fourth element, and the scale factor of 0.9996 at the central meridian as the fifth element.

```
' Part 3: Use the Define method of IProjectedCoordinate
' SystemEdit to create IDTM.
Dim pProjCoordSysEdit As IProjectedCoordinateSystemEdit
Set pProjCoordSysEdit = New ProjectedCoordinateSystem
pProjCoordSysEdit.Define "UserDefinedPCS", _
```

```
    "UserDefinedAlias", _
    "UsrDefAbbrv", _
    "Custom IDTM", _
    "Suitable for Idaho", _
    pGCS, _
    pLinearUnit, _
    pProjection, _
    aParamArray
  ' Set the map to use IDTM and refresh the map.
  Set pMap.SpatialReference = pProjCoordSysEdit
  pMxDoc.ActiveView.Refresh
End Sub
```

Part 3 first creates *pProjCoordSysEdit* as an instance of the *Projected-CoordinateSystem* class. The code then uses the *Define* method on *IProjected-CoordinateSystemEdit* to define the coordinate system. The *Define* method requires nine object qualifiers and arguments, each of which is listed in a separate line for clarity. The first five arguments are the name, alias name, abbreviation, remarks, and usage of the coordinate system. The last four object qualifiers relate to the projection parameters from Part 1. After IDTM is created, the macro sets the active map to use IDTM as its spatial reference and refreshes the map.

7.4 DEFINING COORDINATE SYSTEMS

This section covers two scenarios in which the coordinate system of a geographic dataset is defined. The first uses a VBA (Visual Basic for Applications) macro to code the spatial reference information of a dataset. The information is presumably available in the dataset's metadata. The second defines the spatial reference information of a dataset by copying the information from another dataset.

To read or change a dataset's spatial reference requires the use of a *GeoDataset* object. A *GeoDataset* object implements *IGeoDataset* and *IGeoDatasetSchemaEdit* (Figure 7.6). *IGeoDataset* has the read-only properties of *Extent* and *SpatialReference*. *IGeoDatasetSchemaEdit* has the members of *CanAlterSpatialReference* and *AlterSpatialReference*. Both interfaces are used in this section.

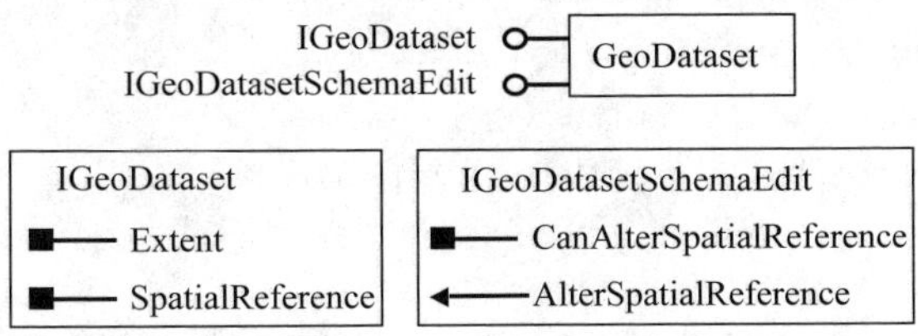

Figure 7.6   A *GeoDataset* object supports *IGeoDataset* and *IGeoDatasetSchemaEdit*.

7.4.1   *DefineGCS*

DefineGCS defines a shapefile's geographic coordinate system as NAD27. The macro performs the same function as using the Define Projection tool in ArcToolbox. *DefineGCS* has three parts. Part 1 creates the NAD27 coordinate system. Part 2 defines the input shapefile. Part 3 alters the shapefile's spatial reference information and exports the information to a prj file.

> **Key Interfaces:** *ISpatialReferenceFactory, IGeographicCoordinateSystem, ISpatial-Reference, IGeoDataset, IGeoDatasetSchemaEdit.*
> **Key Members:** *CreateGeographicCoordinateSystem, ExportESRISpatial-ReferenceToPRJFile, CanAlterSpatialReference, CreateESRISpatialReference-FromPRJFile, AlterSpatialReference, SpatialReference.*
> **Usage:** Add *idll.shp* to the active map in ArcMap. Without a spatial reference file, *idll* has the assumed NAD27 geographic coordinate system. Import *DefineGCS* to Visual Basic Editor in ArcMap. Run the macro. A message box verifies that GCS_North_American_1927 is the new spatial reference for *idll*. The macro also creates a prj file with the prefix of the shapefile name.

```
Private Sub DefineGCS()
  ' Part 1: Create the nad27 coordinate system.
  Dim pSpatRefFact As ISpatialReferenceFactory
  Dim pGeogCS As IGeographicCoordinateSystem
  Dim pSpatialReference As ISpatialReference
  Set pSpatRefFact = New SpatialReferenceEnvironment
  Set pGeogCS = pSpatRefFact. _
  CreateGeographicCoordinateSystem _
  (esriSRGeoCS_NAD1927)
```

Part 1 creates *pSpatRefFact* as an instance of the *SpatialReferenceEnvironment* class and uses the *CreateGeographicCoordinateSystem* method to create a geographic coordinate system based on NAD27 and referenced by *pGeogCS*.

```
  ' Part 2: Define the input shapefile.
  Dim pMxDoc As IMxDocument
  Dim pMap As IMap
  Dim pFeatureLayer As IFeatureLayer
  Dim pFeatureClass As IFeatureClass
  Set pMxDoc = Application.Document
  Set pMap = pMxDoc.FocusMap
  Set pFeatureLayer = pMap.Layer(0)
  Set pFeatureClass = pFeatureLayer.FeatureClass
```

Part 2 defines *pFeatureLayer* to be the top layer in the active map and *pFeature-Class* to be its feature class.

```
' Part 3: Verify and alter the shapefile's spatial
' reference.
Dim pGeoDataset As IGeoDataset
Dim pGeoDatasetEdit As IGeoDatasetSchemaEdit
Set pGeoDatasetEdit = pFeatureClass
' Verify if the shapefile's spatial reference can be
' altered.
If (pGeoDatasetEdit.CanAlterSpatialReference = True) Then
   ' Alter the target layer's spatial reference.
  pGeoDatasetEdit.AlterSpatialReference pGeogCS
Else
  Exit Sub
End If
' Get the spatial reference information and export it
' to a prj file.
Set pGeoDataset = pFeatureClass
Set pSpatialReference = pGeoDataset.SpatialReference
pSpatRefFact.ExportESRISpatialReferenceToPRJFile _
"c:\data\chap7\idll", pSpatialReference
MsgBox pSpatialReference.Name
End Sub
```

Part 3 first performs a QueryInterface (QI) for the *IGeoDatasetSchemaEdit* interface
and uses the *CanAlterSpatialReference* property to test whether the spatial reference of
pFeatureClass can be altered. If it can be, the code uses the *AlterSpatialReference*
method to alter the spatial reference with *pGeogCS*. If it cannot be, exit the sub. Next
the code switches to the *IGeoDataset* interface to assign the spatial reference of
pFeatureClass to *pSpatialReference*. The *ExportESRISpatialReferenceToPRJFile*
method on *ISpatialReferenceFactory* then exports *pSpatialReference* to *idll.prj*. Finally,
a message box displays the name of the updated spatial reference.

7.4.2 *CopySpatialReference*

CopySpatialReference copies the spatial reference from a source layer to a target
layer. The macro performs the same function as using the Project tool in ArcToolbox
to import the spatial reference from one dataset to another. *CopySpatialReference*
assumes that the prj file of the source layer already exists on disk. If it does not
exist, one can use a macro similar to *DefineGCS* to first create the prj file. The
macro has two parts. Part 1 defines the target layer. Part 2 verifies that the target
layer's spatial reference is unknown and can be altered, before copying the prj file
from the source layer.

Key Interfaces: *IGeoDataset, ISpatialReference, ISpatialReferenceFactory,
IGeoDatasetSchemaEdit, IProjectedCoordinateSystem.*

Key Members: *SpatialReference, CanAlterSpatialReference, CreateESRISpatial-ReferenceFromPRJFile, AlterSpatialReference.*

Usage: Add *emidastrm.shp*, a stream shapefile with an unknown coordinate system, to an active map. Import **CopySpatialReference** to Visual Basic Editor in ArcMap. Run the macro. The macro creates *emidastrm.prj* by copying the prj file of *emidalat* on disk. *emidalat* is an elevation grid projected onto the NAD27UTM_11N coordinate system. Its prj file resides in the emidalat folder as *prj.adf*. Two messages appear during the execution of the macro: "unknown," the initial coordinate system for *emidastrm*; and "NAD27UTM_11N," the new spatial reference for *emidastrm*.

```
Private Sub CopySpatialReference()

  ' Part 1: Define the target layer.
  Dim pMxDoc As IMxDocument
  Dim pMap As IMap
  Dim pTargetFL As IFeatureLayer
  Dim pTargetGD As IGeoDataset
  Set pMxDoc = ThisDocument
  Set pMap = pMxDoc.FocusMap
  Set pTargetFL = pMap.Layer(0)
  Set pTargetGD = pTargetFL.FeatureClass
```

Part 1 sets *pTargetFL* to be the top layer in the active map and *pTargetGD* to be its feature class.

```
  ' Part 2: Copy a prj file on disk to be the target layer's.
  Dim pTargetSR As ISpatialReference
  Dim pGeoDatasetEdit As IGeoDatasetSchemaEdit
  Dim pSpatRefFact As ISpatialReferenceFactory
  Dim pProjCoordSys As IProjectedCoordinateSystem
  Set pTargetSR = pTargetGD.SpatialReference
  MsgBox pTargetSR.Name
  ' Verify that the target layer has unknown spatial reference.
  If (pTargetSR.Name = "Unknown") Then
    Set pGeoDatasetEdit = pTargetGD
    ' If the spatial reference can be altered, create a
    ' prj file for the target layer.
    If (pGeoDatasetEdit.CanAlterSpatialReference = True) Then
      Set pSpatRefFact = New SpatialReferenceEnvironment
      Set pProjCoordSys = _
      pSpatRefFact.CreateESRISpatialReferenceFromPRJFile _
      ("c:\data\chap7\emidalat\prj.adf")
```

```
      ' Alter the target layer's spatial reference.
      pGeoDatasetEdit.AlterSpatialReference pProjCoordSys
      ' Get the updated spatial reference and report its
      ' name.
      Set pTargetSR = pTargetGD.SpatialReference
      MsgBox pTargetSR.Name
    Else
      Exit Sub
    End If
  Else
    Exit Sub
  End If
End Sub
```

Part 2 sets *pTargetSR* to be the spatial reference of *pTargetGD*. If the name of *pTargetSR* is unknown, the code switches to the *IGeoDatasetSchemaEdit* interface and verifies that its spatial reference can be altered. If it can be, the code uses the *CreateESRISpatialReferenceFromPRJFile* method on *ISpatialReferenceFactory* to create *pProjCoordSys* from *prj.adf*, the prj file of the source layer. The code then alters the spatial reference of *pTargetGD* with *pProjCoordSys*. A message box follows and reports the name of the target layer's new spatial reference.

7.5 PERFORMING GEOGRAPHIC TRANSFORMATIONS

A datum change may take place between two geographic coordinate systems, between a geographic and a projected coordinate system, or between two projected coordinate systems. Each of the above scenarios requires a geographic transformation. This section deals only with the geographic transformation between two geographic coordinate systems. The other two scenarios are treated as reprojection and are covered in Section 7.6.

7.5.1 *NAD27to83_Map*

NAD27to83_Map transforms the geographic coordinate system of an active map from NAD27 to NAD83. The macro performs the same function as using the map property to change the geographic coordinate system of an active map. Because on-the-fly projection includes datum transformation, NAD83 applies to existing as well as new layers in the active map. *NAD27to83_Map* has two parts. Part 1 performs the geographic transformation from NAD27 to NAD83. Part 2 applies the transformation to the spatial reference property of the active map.

Key Interfaces: *ISpatialReferenceFactory2, IGridTransformation, IGeographicCoordinateSystem.*

Key Members: *CreateGeoTransformation, GetSpatialReferences, SpatialReference.*
Usage: Right-click the active data frame in ArcMap, and select properties. In the next
dialog, click the Clear button to clear the current coordinate system. Add *idll27.shp*
to the map. The map property shows GCS_ North_American_1927 as the current
coordinate system. Import **NAD27to83_Map** to Visual Basic Editor. Run the macro.
The map's current coordinate system should now appear as
GCS_North_American_1983.

```vba
Private Sub Nad27to83_Map()
  ' Part 1: Perform a geographic transformation from NAD27
  ' to NAD83.
  Dim pSpatRefFact As ISpatialReferenceFactory2
  Dim pGeotransNAD27toNAD83 As IGridTransformation
  Dim pFromGCS As IGeographicCoordinateSystem
  Dim pToGCS As IGeographicCoordinateSystem
  Set pSpatRefFact = New SpatialReferenceEnvironment
  Set pGeotransNAD27toNAD83 = _
  pSpatRefFact.CreateGeoTransformation _
  (esriSRGeoTransformation_NAD_1927_TO_NAD_1983_NADCON)
  pGeotransNAD27toNAD83.GetSpatialReferences pFromGCS, _
  pToGCS
  MsgBox "The coordinate system has been changed from " _
  & pFromGCS.Name & "to " & pToGCS.Name
```

Part 1 creates *pSpatRefFact* as an instance of the *SpatialReferenceEnvironment*
class and uses the *CreateGeoTransformation* method on *ISpatialReferenceFactory2*
to create *pGeotransNAD27toNAD83*, a reference to *IGridTransformation*. The transformation is grid based and uses the NADCON method. The code then uses the
GetSpatialReferences method on *IGridTransformation* to verify the geographic coordinate systems involved in the transformation.

```vba
  ' Part 2: Set the spatial reference of the active map.
  Dim pMxDoc As IMxDocument
  Dim pMap As IMap
  Set pMxDoc = Application.Document
  Set pMap = pMxDoc.FocusMap
  Set pMap.SpatialReference = pToGCS
End Sub
```

Part 2 sets the spatial reference of the active map to be that of *pToGCS*, the
output of the geographic transformation from Part 1.

7.5.2 *NAD27to83_Shapefile*

NAD27to83_Shapefile reprojects a shapefile based on NAD27 to a new shapefile based on NAD83. The macro performs the same function as using the Project tool in ArcToolbox to transform the shapefile from NAD27 to NAD83 coordinates. *NAD27to83_Shapefile* has six parts. Part 1 defines the input shapefile. Part 2 performs a geographic transformation from NAD27 to NAD83. Part 3 defines the output shapefile. Part 4 creates fields in the output based on the input fields. Part 5 finds the geometry field and defines the field's spatial reference and spatial index. Part 6 projects the input into the output and reports any processing errors.

Because *NAD27to83_Shapefile* creates a new shapefile from reprojection, the macro requires use of objects that do not belong to the Spatial Reference subsystem. A *FeatureDataConverter* object implements *IFeatureDataConverter* and *IFeatureDataConverter2* (Figure 7.7). Both interfaces have methods for converting feature classes, feature datasets, and tables. *IFeatureDataConverter2* has the additional functionality of working with data subsets. Created by a feature data converter object, an *EnumInvalidObject* enumerator captures objects that have failed the conversion process. Created by an *EnumInvalidObject* enumerator, *InvalidObjectInfo* provides information about failed objects.

A *FieldChecker* object implements *IFieldChecker*, which has methods for creating a new set of fields from another set and for validating the fields (Figure 7.8). A field checker object creates an *EnumFieldError* enumerator and a *FieldError* object for fields that have caused problems in the conversion process.

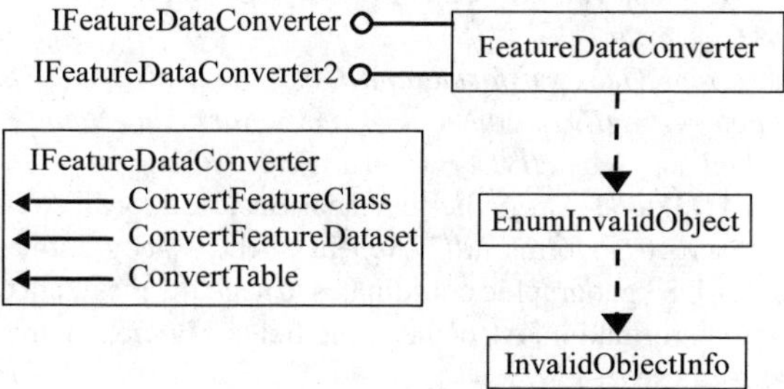

Figure 7.7 A *FeatureDataConverter* object creates an *EnumInvalidObject*, which in turn creates an *InvalidObjectInfo* object.

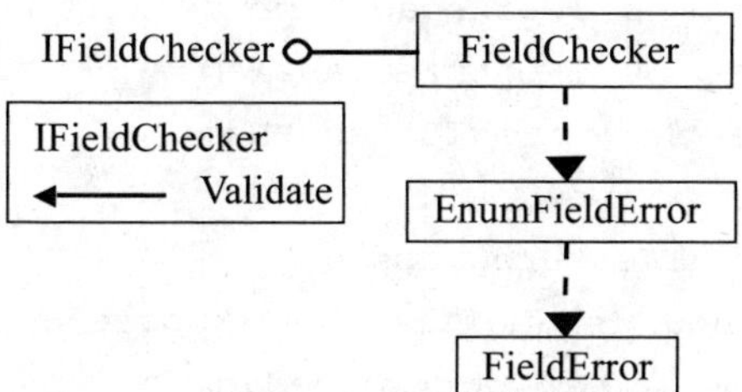

Figure 7.8 A *FieldChecker* object creates an *EnumFieldError* object, which in turn creates a *FieldError* object.

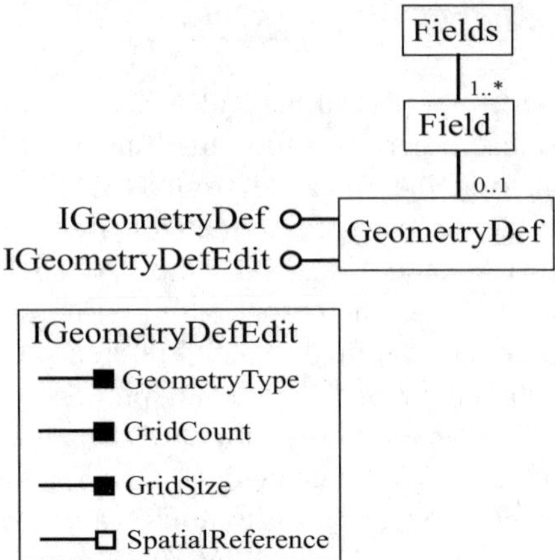

Figure 7.9   The diagram shows the relationship between the *Fields*, *Field*, and *GeometryDef* classes.

Accessed from a field, a *GeometryDef* object defines the spatial properties of a feature class (Figure 7.9). A *GeometryDef* object implements *IGeometryDef* and *IGeometryDefEdit*. *IGeometryDef* has read-only access to the spatial properties, whereas *IGeometryDefEdit* has write-only access to define the spatial properties.

> **Key Interfaces:** *ISpatialReferenceFactory2, IGridTransformation, IGeographic-CoordinateSystem, IWorkspaceName, IFeatureClassName, IDatasetName, IName, IGeoDataset, IFields, IFieldChecker, IField, IGeometryDef, IGeometryDefEdit, IFeatureDataConverter.*
>
> **Key Members:** *CreateGeoTransformation, GetSpatialReferences, SpatialReference, WorkspaceFactoryProgID, PathName, Name, WorkspaceName, Fields, Validate, FieldCount, Field(), GeometryDef, ConvertFeatureClass.*
>
> **Usage:** Import ***NAD27to83_Shapefile*** to Visual Basic Editor in ArcCatalog. Run the macro. The macro transforms *idll27.shp* in NAD27 geographic coordinates into *idll83.shp* in NAD83 geographic coordinates. Check the metadata of both shapefiles to verify the transformation. Also check the fields of *idll83* to make sure that they are identical to those of *idll27*.

```
Private Sub NAD27to83_Shapefile()
  ' Part 1: Define the shapefile.
  Dim pInWSName As IWorkspaceName
  Dim pInFCName As IFeatureClassName
  Dim pInDatasetName As IDatasetName
  Dim pName As IName
  Dim pInFC As IFeatureClass
  ' Set the input shapefile's workspace and path names.
  Set pInWSName = New WorkspaceName
```

```
pInWSName.WorkspaceFactoryProgID = _
"esriCore.ShapefileWorkspaceFactory.1"
pInWSName.PathName = "c:\data\chap7"
' Set the input feature class and dataset names.
Set pInFCName = New FeatureClassName
Set pInDatasetName = pInFCName
pInDatasetName.Name = "idl127"
Set pInDatasetName.WorkspaceName = pInWSName
' Set the input geodataset.
Set pName = pInFCName
Set pInFC = pName.Open
```

Part 1 uses the name objects to define the input shapefile's workspace, feature class, and name. The code creates *pInFCName* as an instance of the *FeatureClass-Name* class and performs a QI for the *IDatasetName* interface to define its name and workspace. Next the code sets *pName*, a reference to *IName*, to be *pInFCName*, and uses the *Open* method on *IName* to open *pInFC*.

```
' Part 2: Perform a geographic transformation from NAD27
' to NAD83.
Dim pSpatRefFact As ISpatialReferenceFactory2
Dim pGeotransNAD27toNAD83 As IGridTransformation
Dim pFromGCS As IGeographicCoordinateSystem
Dim pToGCS As IGeographicCoordinateSystem
Set pSpatRefFact = New SpatialReferenceEnvironment
Set pGeotransNAD27toNAD83 = _
pSpatRefFact.CreateGeoTransformation _
(esriSRGeoTransformation_NAD_1927_TO_NAD_1983_NADCON)
' Verify the From and To geographic coordinate systems.
pGeotransNAD27toNAD83.GetSpatialReferences pFromGCS, _
pToGCS
MsgBox "The geographic transformation is from " & _
pFromGCS.Name & " to " & pToGCS.Name
```

Part 2 performs a geographic transformation from NAD27 to NAD83. The code creates *pSpatRefFact* as an instance of the *SpatialReferenceEnvironment* class and uses the *CreateGeoTransformation* method on *ISpatialReferenceFactory2* to create *pGeotransNAD27toNAD83*. The code then uses the *GetSpatialReferences* method on *IGridTransformation* to return the input and output geographic coordinate systems, which are referenced by *pFromGCS* and *pToGCS* respectively.

```
' Part 3: Define the output shapefile.
Dim pOutWSName As IWorkspaceName
```

```
Dim pOutFCName As IFeatureClassName
Dim pOutDataSetName As IDatasetName
' Set the output shapefile's workspace and path names.
Set pOutWSName = New WorkspaceName
pOutWSName.WorkspaceFactoryProgID = _
"esriCore.ShapeFileWorkspaceFactory.1"
pOutWSName.PathName = "c:\data\chap7"
' Set the output feature class and dataset names.
Set pOutFCName = New FeatureClassName
Set pOutDataSetName = pOutFCName
Set pOutDataSetName.WorkspaceName = pOutWSName
pOutDataSetName.Name = "idl183"
```

Part 3 defines the output shapefile's workspace, feature class, and name in the same way as Part 1. Only the name objects are used.

```
' Part 4: Create the output fields based on the input's
' fields.
Dim pOutFCFields As IFields
Dim pInFCFields As IFields
Dim pFieldCheck As IFieldChecker
Dim i As Long
Set pInFCFields = pInFC.Fields
Set pFieldCheck = New FieldChecker
pFieldCheck.Validate pInFCFields, Nothing, pOutFCFields
```

Part 4 creates *pFieldCheck* as an instance of the *FieldChecker* class and uses the *Validate* method on *IFieldChecker* to create the output fields referenced by *pOutFCFields* from the input fields referenced by *pInFCFields*. Because the input and output are both shapefiles, the code does not perform error checking using the *EnumFieldError* and *FieldError* objects.

```
' Part 5: Locate and define the geometry field.
Dim pGeoField As IField
Dim pOutFCGeoDef As IGeometryDef
Dim pOutFCGeoDefEdit As IGeometryDefEdit
' Find the index of the geometry field.
For i = 0 To pOutFCFields.FieldCount - 1
  If pOutFCFields.Field(i).Type = esriFieldTypeGeometry Then
    Set pGeoField = pOutFCFields.Field(i)
    Exit For
  End If
```

```
    Next i
    ' Get the geometry field's geometry definition.
    Set pOutFCGeoDef = pGeoField.GeometryDef
    ' Define the spatial index and spatial reference.
    Set pOutFCGeoDefEdit = pOutFCGeoDef
    pOutFCGeoDefEdit.GridCount = 1
    pOutFCGeoDefEdit.GridSize(0) = 200
    Set pOutFCGeoDefEdit.SpatialReference = pToGCS
```

Part 5 locates and defines the geometry field. The code loops through *pOut-FCFields*, checks for the geometry type, and saves the geometry field, referenced by *pGeoField*, by its index value. Next the code sets *pOutFCGeoDef* to be the geometry definition of *pGeoField* and switches to the *IGeometryDefEdit* interface to set its properties of *GridCount*, *GridSize*, and *SpatialReference*. Grid count and grid size make up the spatial index of the output shapefile, which is designed to improve the display, spatial query, and feature identification of the shapefile. In this case, the spatial index has one grid and the grid size of 200. The spatial reference assigned to the output shapefile is *pToGCS*, a reference to NAD83.

```
    ' Part 6: Perform data conversion and error checking.
    Dim pFDConverter As IFeatureDataConverter
    Dim pEnumErrors As IEnumInvalidObject
    Dim pErrInfo As InvalidObjectInfo
    ' Perform feature data conversion.
    Set pFDConverter = New FeatureDataConverter
    Set pEnumErrors = pFDConverter.ConvertFeatureClass _
    (pInFCName, Nothing, Nothing, pOutFCName, _
    pOutFCGeoDef, pOutFCFields, "", 1000, 0)
    ' If an error exists, report an error message in the
    ' immediate window.
    Set pErrInfo = pEnumErrors.Next
    If Not pErrInfo Is Nothing Then
      Debug.Print "Conversion completed with errors"
    Else
      Debug.Print "Conversion completed"
    End If
  End Sub
```

Part 6 performs the data conversion and error checking. The code first creates *pFDConverter* as an instance of the *FeatureDataConverter* class and uses the *ConvertFeatureClass* method to import data from the input feature class to the output. The method uses objects that have been previously defined in Parts 4 and 5. Additionally, the code uses a flush interval of 1000, which controls the interval for committing data. (The flush interval is important for loading large amounts of data into a geodatabase.) Finally, the code checks for invalid features during the conversion process. If an invalid feature exists, the Immediate window displays an error message. If not, the window displays the message of "Conversion completed."

7.6 PROJECTING DATASETS

Similar to use of the Project tools in ArcToolbox, projection using VBA code also follows the sequence of defining the coordinate systems of the input and the output. Additionally, the code must be able to handle the creation of the output dataset, which includes the copying of fields from the input dataset and the definition of the geometry field.

7.6.1 *ProjectShapefile*

ProjectShapefile projects a shapefile from NAD27 geographic coordinates to NAD27UTM_11N projected coordinates. The module performs the same function as using the Project tool in ArcToolbox to project a shapefile. *ProjectShapefile* has four parts. Part 1 defines the input shapefile and its spatial reference. Part 2 creates the output fields based on the input fields. Part 3 finds the geometry field and defines the field's spatial reference and spatial index. Part 4 projects the input into the output and reports any processing errors. *ProjectShapefile* is similar to *NAD27to83_Shapefile* in terms of programming except that *ProjectShapefile* uses functions to define the name objects of workspace and dataset, one for the input and the other for the output.

> **Key Interfaces:** *IWorkspaceName, IFeatureClassName, IDatasetName, IName, IGeoDataset, IGeographicCoordinateSystem, ISpatialReferenceFactory, IFields, IField, IFieldChecker, IGeometryDef, IProjectedCoordinateSystem, IGeometryDefEdit, IFeatureDataConverter.*
>
> **Key Members:** *WorkspaceFactoryProgID, PathName, Name, WorkspaceName, CreateGeographicCoordinateSystem, SpatialReference, Fields, Validate, FieldCount, Field( ), GeometryDef, CreateProjectedCoordinateSystem, ConvertFeatureClass.*
>
> **Usage:** Import *ProjectShapefile* to Visual Basic Editor in ArcCatalog. Run the module. The module projects *idll27.shp* in NAD27 geographic coordinates into *idutm27.shp* in NAD27UTM_11N projected coordinates. Check the metadata of the shapefiles to verify the result.

```
Private Sub ProjectShapefile()
    ' Part 1: Define the input shapefile and its spatial
    ' reference.
    Dim pInFCName As IFeatureClassName
    Dim pGeogCS As IGeographicCoordinateSystem
    Dim pSpatRefFact As ISpatialReferenceFactory
```

```
Dim pName As IName
Dim pInFC As IFeatureClass
Dim pInGeoDataset As IGeoDataset
' Use the InputName function to find the input shapefile.
Set pInFCName = InputName("c:\data\chap7", "idl127")
' Set the geographic coordinate system.
Set pSpatRefFact = New SpatialReferenceEnvironment
Set pGeogCS = pSpatRefFact. _
CreateGeographicCoordinateSystem(esriSRGeoCS_NAD1927)
' Assign the spatial reference to the input dataset.
Set pName = pInFCName
Set pInFC = pName.Open
Set pInGeoDataset = pInFC
Set pGeogCS = pInGeoDataset.SpatialReference
```

Part 1 locates the input shapefile, creates a geographic coordinate system, and assigns the coordinate system to the input dataset. The code passes the input shapefile's workspace and name to the ***InputName*** function, runs the function, and gets *pInFCName*, a reference to *IFeatureClassName*, from the function. Next the code creates *pSpatRefFact* as an instance of the *SpatialReferenceEnvironment* class and uses the *CreateGeographicCoordinateSystem* method on *ISpatialReferenceFactory* to create *pGeogCS*, a reference to the NAD27 geographic coordinate system. The code then opens *pInFCName*, sets *pInGeoDataset* to be *pInFC*, and assigns *pGeogCS* to be the spatial reference of *pInGeoDataset*.

```
' Part 2: Create the output fields based on the input's
' fields.
Dim pOutFCFields As IFields
Dim pInFCFields As IFields
Dim pFieldCheck As IFieldChecker
Dim i As Long
Set pInFCFields = pInFC.Fields
Set pFieldCheck = New FieldChecker
pFieldCheck.Validate pInFCFields, Nothing, pOutFCFields
```

Part 2 creates *pFieldCheck* as an instance of the *FieldChecker* class and uses the *Validate* method to create *pOutFCFields* from *pInFCFields*.

```
' Part 3: Locate and define the geometry field.
Dim pGeoField As IField
Dim pOutFCGeoDef As IGeometryDef
Dim pPCS As IProjectedCoordinateSystem
```

```
Dim pOutFCGeoDefEdit As IGeometryDefEdit
' Loop through the fields to locate the geometry field.
For i = 0 To pOutFCFields.FieldCount - 1
  If pOutFCFields.Field(i).Type = esriFieldTypeGeometry _
  Then
    Set pGeoField = pOutFCFields.Field(i)
    Exit For
  End If
Next i
' Set the output's coordinate system.
Set pSpatRefFact = New SpatialReferenceEnvironment
Set pPCS = pSpatRefFact. _
CreateProjectedCoordinateSystem _
(esriSRProjCS_NAD1927UTM_11N)
' Get and edit the geometry field's geometry definition.
Set pOutFCGeoDef = pGeoField.GeometryDef
Set pOutFCGeoDefEdit = pOutFCGeoDef
pOutFCGeoDefEdit.GridCount = 1
pOutFCGeoDefEdit.GridSize(0) = 200
Set pOutFCGeoDefEdit.SpatialReference = pPCS
```

Part 3 locates the geometry field referenced by *pGeoField* by looping through the fields in *pOutFCFields*. Next the code creates *pPCS* as a new projected coordinate system based on NAD1927UTM_11N. The code then sets *pOutFCGeoDef* to be the geometry definition of *pGeoField* and performs a QI for the *IGeometryDefEdit* interface to edit its spatial index and spatial reference.

```
' Part 4: Create the output, and report any errors in
' creating the output.
Dim pFDConverter As IFeatureDataConverter
Dim pOutFCName As IFeatureClassName
Dim pEnumErrors As IEnumInvalidObject
Dim pErrInfo As IInvalidObjectInfo
Set pFDConverter = New FeatureDataConverter
Set pOutFCName = OutputName("c:\data\chap7", "idutm27")
Set pEnumErrors = pFDConverter.ConvertFeatureClass _
(pInFCName, Nothing, Nothing, pOutFCName, _
 pOutFCGeoDef, pOutFCFields, "", 1000, 0)
' If an error exists, show an error message.
Set pErrInfo = pEnumErrors.Next
If Not pErrInfo Is Nothing Then
  Debug.Print "Conversion completed with errors"
```

```
    Else
      Debug.Print "Conversion completed"
    End If
End Sub
```

Part 4 creates the output shapefile and reports any errors in data conversion. The code first runs the ***OutputName*** function to define the output shapefile. Then the code uses the *ConvertFeatureClass* method on *IFeatureDataConverter* to create *pOutFCName*, a reference to *IFeatureClassName*. The Immediate window reports any invalid object during the conversion.

```
Private Function InputName(InWSPath As String, InDataset _
As String) As IFeatureClassName
  Dim pInWSName As IWorkspaceName
  Dim pInFCName As IFeatureClassName
  Dim pInDatasetName As IDatasetName
  ' Set the input shapefile's workspace and path names.
  Set pInWSName = New WorkspaceName
  pInWSName.WorkspaceFactoryProgID = _
  "esriCore.ShapefileWorkspaceFactory.1"
  pInWSName.PathName = InWSPath
  ' Set the input feature class and name.
  Set pInFCName = New FeatureClassName
  Set pInDatasetName = pInFCName
  pInDatasetName.Name = InDataset
  Set pInDatasetName.WorkspaceName = pInWSName
  Set InputName = pInFCName
End Function
```

The ***InputName*** function receives the names of the input shapefile and its workspace path as strings, and returns *pInFCName*, a reference to an *IFeatureClass-Name* object, to ***ProjectShapefile***.

```
Private Function OutputName(OutWSPath As String, _
OutDataset As String) As IFeatureClassName

  Dim pOutWSName As IWorkspaceName
  Dim pOutFCName As IFeatureClassName
  Dim pOutDatasetName As IDatasetName
  ' Set the output shapefile's workspace and path names.
  Set pOutWSName = New WorkspaceName
  pOutWSName.WorkspaceFactoryProgID = _
  "esriCore.ShapeFileWorkspaceFactory.1"
```

```
  pOutWSName.PathName = OutWSPath
  ' Set the output feature class and name.
  Set pOutFCName = New FeatureClassName
  Set pOutDatasetName = pOutFCName
  Set pOutDatasetName.WorkspaceName = pOutWSName
  pOutDatasetName.Name = OutDataset
  Set OutputName = pOutFCName
End Function
```

The ***OutputName*** function receives the names of the output shapefile and its workspace path, and returns *pOutFCName*, a reference to an *IFeatureClassName* object, to ***ProjectShapefile***.

7.6.2   Use of a Different Datum

The only change needed for projecting a shapefile from NAD27 to NAD83UTM_11N coordinates is the line statement for setting the output file's spatial reference in Part 4 of ***ProjectShapefile***:

```
Set pPCS = pSpatRefFact. _
CreateProjectedCoordinateSystem _
(esriSRProjCS_NAD1983UTM_11N)
```

7.6.3   *ReprojectShapefile*

ReprojectShapefile reprojects a shapefile from one projected coordinate system to another. Specifically, it reprojects a shapefile from NAD27UTM_11N to IDTM coordinates. The module performs the same function as using the Project tool in ArcToolbox to project a shapefile.

ReprojectShapefile has five parts. Additionally, the module uses a function to define IDTM. Part 1 defines the input shapefile and its spatial reference. Part 2 defines the output shapefile. Part 3 creates the output fields based on the input fields. Part 4 finds the geometry field and defines the field's spatial reference and spatial index. Part 5 projects the input into the output and reports any processing errors. Many line statements in ***ReprojectShapefile*** have been used in previous sample macros. Therefore, it does not require detailed explanation of its code structure.

Key Interfaces: *IWorkspaceName, IFeatureClassName, IDatasetName, IProjectedCoordinateSystem, ISpatialReferenceFactory, IName, IGeoDataset, IFields, IField, IFieldChecker, IGeometryDef, IGeometryDefEdit, IFeatureDataConverter, IProjection, IGeographicCoordinateSystem, IProjectedCoordinateSystemEdit.*

Key Members: *WorkspaceFactoryProgID, PathName, Name, WorkspaceName, SpatialReference, Fields, Validate, FieldCount, GeometryDef, Field(), CreateProjectedCoordinateSystem, ConvertFeatureClass, CreateProjection, CreateGeographicCoordinateSystem, CreateParameter.*

Usage: Import *ReprojectShapefile* to Visual Basic Editor in ArcCatalog. Run the macro. The macro projects *idutm27.shp* in NAD27 UTM_11N projected coordinates into *idtm.shp* in IDTM projected coordinates. (*idutm27.shp* was created in Section 7.6.1.) Check the metadata of the shapefiles to verify the result.

```
Private Sub ReprojectShapefile()
  ' Part 1: Define the input shapefile and its spatial
  ' reference.
  Dim pInWSName As IWorkspaceName
  Dim pInFCName As IFeatureClassName
  Dim pInDatasetName As IDatasetName
  Dim pInCS As IProjectedCoordinateSystem
  Dim pSpatRefFact As ISpatialReferenceFactory
  Dim pName As IName
  Dim pInFC As IFeatureClass
  Dim pInGeoDataset As IGeoDataset
  ' Define the input shapefile's workspace and path.
  Set pInWSName = New WorkspaceName
  pInWSName.WorkspaceFactoryProgID = _
  "esriCore.ShapefileWorkspaceFactory.1"
  pInWSName.PathName = "c:\data\chap7"
  ' Define the input feature class.
  Set pInFCName = New FeatureClassName
  Set pInDatasetName = pInFCName
  pInDatasetName.Name = "idutm27"
  Set pInDatasetName.WorkspaceName = pInWSName
  ' Define the input shapefile's projected coordinate
  ' system.
  Set pSpatRefFact = New SpatialReferenceEnvironment
  Set pInCS = pSpatRefFact. _
  CreateProjectedCoordinateSystem _
  (esriSRProjCS_NAD1927UTM_11N)
  ' Assign the spatial reference to the input shapefile.
  Set pName = pInFCName
  Set pInFC = pName.Open
  Set pInGeoDataset = pInFC
  Set pInCS = pInGeoDataset.SpatialReference
```

Part 1 sets *pInFC* to be the input shapefile and *pInCS* to be its spatial reference. The code defines *pInCS* as the NAD27 UTM_11N projected coordinate system.

```
  ' Part 2: Define the output shapefile.
  Dim pOutWSName As IWorkspaceName
```

```
Dim pOutFCName As IFeatureClassName
Dim pOutDatasetName As IDatasetName
' Set the output shapefile's workspace and path names.
Set pOutWSName = New WorkspaceName
pOutWSName.WorkspaceFactoryProgID = _
"esriCore.ShapeFileWorkspaceFactory.1"
pOutWSName.PathName = "c:\data\chap7"
' Set the output feature class and dataset names.
Set pOutFCName = New FeatureClassName
Set pOutDatasetName = pOutFCName
Set pOutDatasetName.WorkspaceName = pOutWSName
pOutDatasetName.Name = "idtm"
```

Part 2 defines the workspace and name of the output shapefile.

```
' Part 3: Create the output fields based on the input's
' fields.
Dim pOutFCFields As IFields
Dim pInFCFields As IFields
Dim pFieldCheck As IFieldChecker
Dim i As Long
Set pInFCFields = pInFC.Fields
Set pFieldCheck = New FieldChecker
pFieldCheck.Validate pInFCFields, Nothing, pOutFCFields
```

Part 3 creates the fields for the output based on the fields of the input shapefile.

```
' Part 4: Locate and define the geometry field.
Dim pGeoField As IField
Dim pOutFCGeoDef As IGeometryDef
Dim pPCS As IProjectedCoordinateSystem
Dim pOutFCGeoDefEdit As IGeometryDefEdit
' Loop through the fields to locate the geometry field.
For i = 0 To pOutFCFields.FieldCount - 1
  If pOutFCFields.Field(i).Type = esriFieldTypeGeometry _
  Then
    Set pGeoField = pOutFCFields.Field(i)
    Exit For
  End If
Next i
' Use the OutCS function to define the output's
' coordinate system.
```

```
Set pPCS = OutCS()
' Get and edit the geometry field's geometry definition.
Set pOutFCGeoDef = pGeoField.GeometryDef
Set pOutFCGeoDefEdit = pOutFCGeoDef
pOutFCGeoDefEdit.GridCount = 1
pOutFCGeoDefEdit.GridSize(0) = 200
Set pOutFCGeoDefEdit.SpatialReference = pPCS
```

Part 4 loops through the output fields, locates the geometry field, and defines
the geometry definition of the field. Then the code calls the ***OutCS*** function to get
the projected coordinate system for the output.

```
' Part 5: Create the output, and report any errors in
' creating the output.
Dim pFDConverter As IFeatureDataConverter
Dim pEnumErrors As IEnumInvalidObject
Dim pErrInfo As IInvalidObjectInfo
Set pFDConverter = New FeatureDataConverter
Set pEnumErrors = pFDConverter.ConvertFeatureClass _
(pInFCName, Nothing, Nothing, pOutFCName, _
pOutFCGeoDef, pOutFCFields, "", 1000, 0)
' If an error exists, show an error message.
Set pErrInfo = pEnumErrors.Next
If Not pErrInfo Is Nothing Then
  Debug.Print "Conversion completed with errors"
Else
  Debug.Print "Conversion completed"
End If
End Sub
```

Part 5 creates the output shapefile and reports any errors in creating the output.

```
Private Function OutCS() As IProjectedCoordinateSystem
  Dim pSpatRefFact As ISpatialReferenceFactory2
  Dim pProjection As IProjection
  Dim pGCS As IGeographicCoordinateSystem
  Dim pUnit As IUnit
  Dim pLinearUnit As ILinearUnit
  Dim aParamArray(5) As IParameter
  Dim pProjCoordSysEdit As IProjectedCoordinateSystemEdit
  Dim pProjCoordSys As IProjectedCoordinateSystem
  ' Define the IDTM Coordinate System.
```

```vba
Set pSpatRefFact = New SpatialReferenceEnvironment
Set pProjection = pSpatRefFact.CreateProjection _
(esriSRProjection_TransverseMercator)
Set pGCS = pSpatRefFact. _
CreateGeographicCoordinateSystem(esriSRGeoCS_NAD1927)
Set pUnit = pSpatRefFact.CreateUnit(esriSRUnit_Meter)
Set pLinearUnit = pUnit
' Store the 5 known parameters of IDTM in an array.
Set aParamArray(0) = pSpatRefFact.CreateParameter _
(esriSRParameter_FalseEasting)
aParamArray(0).Value = 500000
Set aParamArray(1) = pSpatRefFact.CreateParameter _
(esriSRParameter_FalseNorthing)
aParamArray(1).Value = 100000
Set aParamArray(2) = pSpatRefFact.CreateParameter _
(esriSRParameter_CentralMeridian)
aParamArray(2).Value = -114
Set aParamArray(3) = pSpatRefFact.CreateParameter _
(esriSRParameter_LatitudeOfOrigin)
aParamArray(3).Value = 42
Set aParamArray(4) = pSpatRefFact.CreateParameter _
(esriSRParameter_ScaleFactor)
aParamArray(4).Value = 0.9996
' Create IDTM by defining its properties.
Set pProjCoordSysEdit = New ProjectedCoordinateSystem
pProjCoordSysEdit.Define "UserDefinedPCS," _
  "UserDefinedAlias," _
  "UsrDefAbbrv," _
  "Custom IDTM," _
  "Suitable for Idaho," _
  pGCS, _
  pLinearUnit, _
  pProjection, _
  aParamArray
Set OutCS = pProjCoordSysEdit
End Function
```

The ***OutCS*** function creates *pProjCoordSysEdit* as an instance of the *ProjectedCoordinateSystem* class and defines its properties including the geographic coordinate system, the linear unit, and five known parameters for the projection. The function then returns the defined projected coordinate system to *ReprojectShapefile.*

CHAPTER **8**

Data Display

Spatial features are characterized by their locations and attributes. Data display involves choice of symbols to show attribute data at the locations of spatial features. Cartographers deal with choice of symbols in two parts: symbol types and visual variables. Symbol types correspond to feature types: point symbols for point features, line symbols for line features, and area symbols for area features. Visual variables distinguish between symbols and communicate data characteristics to the viewer. Choices of visual variables include color, size, texture, shape, and pattern.

Color is a popular visual variable but is often misused. A color has the three visual dimensions of hue, value, and chroma. Hue is the quality that distinguishes one color from another. Value is the lightness or darkness of a color. And chroma refers to the richness of a color. The use of color and its visual dimensions depends on the type of data to be displayed. Cartographic studies have shown that hue is a visual variable better suited for qualitative or categorical data, whereas value and chroma are better suited for quantitative data. Cartographic studies have also recommended a number of color schemes for mapping quantitative data.

Layout design is part of map design. A map requires a title, a legend, a north arrow, a scale bar, and other elements to communicate map information. The task of layout design is to arrange these various elements on a map so that the map would look balanced and organized to the viewer. Cartographers used to use thumbnail sketches to experiment with layout design. Now the experimentation can be easily performed on the computer monitor.

This chapter covers data display with emphasis on symbology, color, and layout. Section 8.1 reviews data display options in ArcGIS. Section 8.2 discusses objects that are related to various aspects of data display. Section 8.3 offers macros for displaying vector data. Section 8.4 includes macros for displaying raster data. Section 8.5 discusses a macro for making a layout page. All macros start with the listing of key intefaces and key members (i.e., properties and methods) and the usage.

8.1 DISPLAYING DATA IN ARCGIS

ArcMap provides many data display options on the Symbology tab of the Layer Properties dialog. These options depend on the data type. The primary data types are vector and raster data. At the secondary level, vector data include point, line, and area features and raster data include categorical and continuous data.

8.1.1 Displaying Vector Data

For vector data, the display options include Features, Categories, Quantities, Charts, and Multiple Attributes. The features option draws all features with a single symbol. The categories option displays unique values from a field or multiple fields. The quantities option offers graduated colors, graduated symbols, proportional symbols, and dot density. Charts include pie, bar/column, and stacked charts. The multiple attributes option uses more than one attribute for data display. ArcMap initially assigns default symbols to a data display option: point symbols for point features, line symbols for line features, and area symbols for area features. The user can alter these default symbols in terms of symbol type, color, size, pattern, and other visual variables.

8.1.2 Displaying Raster Data

For raster data, the display options include Unique Values, Classified, and Stretched. The unique values option displays unique cell values of a raster. The classified option displays classes of cell values. The stretched option stretches cell values to increase the visual contrast in data display. ArcMap initially assigns a default set of area (fill) symbols to a display option. The user can alter these symbols individually or as a group.

8.1.3 Use of Color Ramp and Classification Tools

Color ramp and classification are two tools on the Symbology tab for data display. A color ramp represents a range of distinctive or sequential colors. ArcMap offers a series of predefined color ramps, and the user can choose a color ramp graphically or by description (e.g., yellow to dark red). Each color ramp has a properties dialog that allows the user to select the two end colors and the algorithm for producing the intermediate colors between them. The classification tool lets the user choose number of classes and method for subdividing a dataset into classes. Available classification methods are manual, equal interval, defined interval, quantile, natural breaks, and standard deviation. The natural breaks method, which classifies data values into natural groupings statistically, is the default method.

8.1.4 Designing a Layout

ArcMap offers two options for layout design. The first option is to use a layout template. Current layout templates are grouped into general, industry, USA, and world. Each group has a list of choices. For example, the layout templates for the United States include USA, conterminous USA, and five different regions of the country. The second option is to open a layout page and build on it one map element at a time. These map elements can be graphically manipulated for size change, repositioning, and other modifications on the layout page.

8.2 ARCOBJECTS FOR DATA DISPLAY

A data display macro, especially a layout macro, tends to involve more objects than other types of macros do. This section presents a summary of objects that are important to data display.

8.2.1 Renderer Objects

ArcObjects uses the term *renderer* to describe a set of symbols for displaying data values. A renderer is therefore like a legend. ArcObjects has two general (abstract) classes of renderers: *FeatureRenderer* for vector data and *RasterRenderer* for raster data.

A variety of feature renderers inherit the functionality of the *FeatureRenderer* class (Figure 8.1). Two of these renderers to be covered in the sample macros are *UniqueValueRenderer* and *ClassBreaksRenderer*. A unique value renderer uses a different symbol for each unique value, which may come from a field or a combination of fields. A class breaks renderer uses a different symbol for each class of data values.

ArcObjects organizes feature-based symbols into three general classes: *MarkerSymbol* for point features, *LineSymbol* for line features, and *FillSymbol* for area features (Figure 8.2). Each of these abstract symbol classes is inherited by a number of coclasses. Of these coclasses, *SimpleMarkerSymbol*, *SimpleLineSymbol*, and *SimpleFillSymbol* can generate the commonly used point, line, and area symbols respectively.

The *RasterRenderer* abstract class is inherited by four coclasses: *RasterUniqueValueRenderer*, *RasterClassifyColorRampRenderer*, *RasterStretchColorRampRenderer*, and *RasterRGBRenderer*. As suggested by the name, the first three

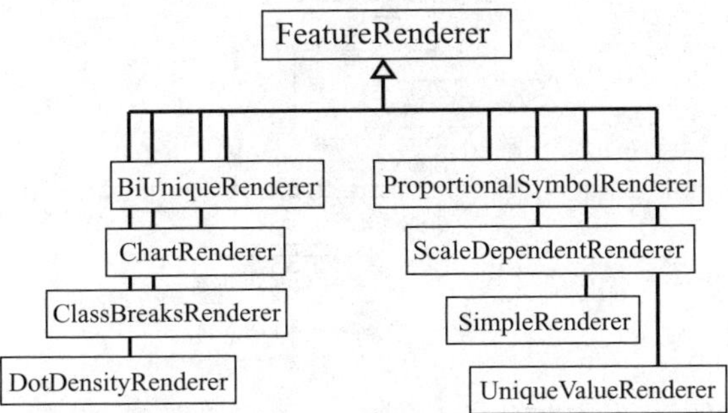

Figure 8.1 *FeatureRenderer* is an abstract class with many feature-based renderer types, each of which is a coclass.

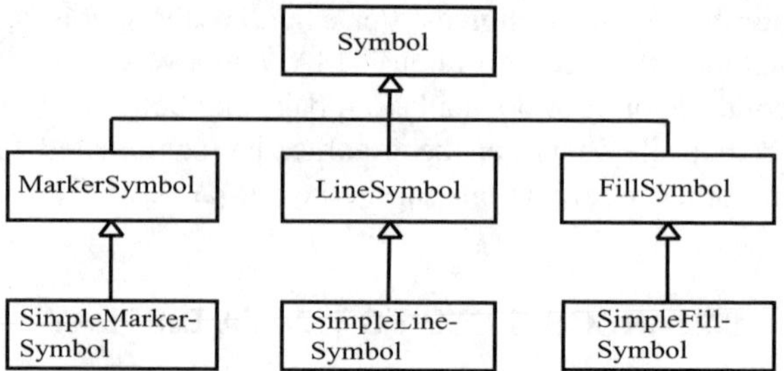

Figure 8.2   *SimpleMarkerSymbol* is a type of *MarkerSymbol*, which is in turn a type of *Symbol*. *SimpleMarkerSymbol* is a coclass, whereas both *MarkerSymbol* and *Symbol* are abstract classes. The same is true for the other two feature-based symbols.

renderers are used for the display options of unique values, classified, and stretched respectively. A raster RGB (red, green, and blue) renderer object is designed for multiband data such as satellite images. The only symbol option for displaying cell-based raster data is the fill symbol.

8.2.2   Classification Objects

A macro for displaying classified data requires a classification object. ArcObjects offers five predefined classification objects: defined interval, equal interval, natural breaks, quantile, and standard deviation (Figure 8.3). A *DefinedInterval* object uses a defined and precise interval such as 100 or 1000. An *EqualInterval* object produces classes with an equal interval. A *NaturalBreaks* object uses a statistical method to create classes with natural breaks between them. A *Quantile* object creates classes with an equal number of values in each class. And a *StandardDeviation* object produces classes that are based on one whole or part of a standard deviation from the mean. In addition to these predefined classification objects, one can also use a user-defined classification.

Use of a predefined classification object typically involves a *TableHistogram* object. A table histogram object implements *IHistogram* and *ITableHistogram*, which

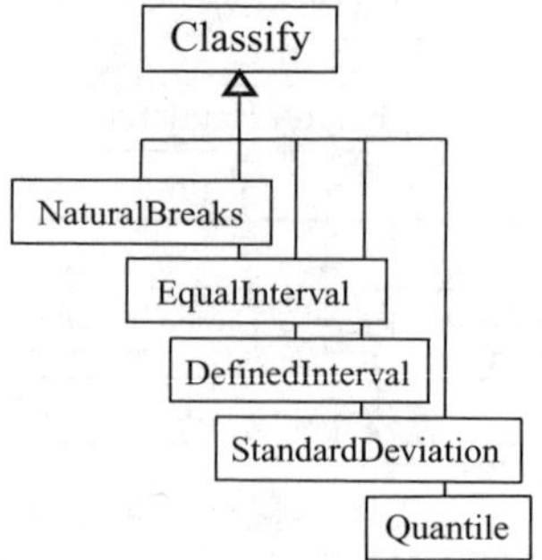

Figure 8.3   *Classify* is an abstract class with five classification types, each of which is a coclass.

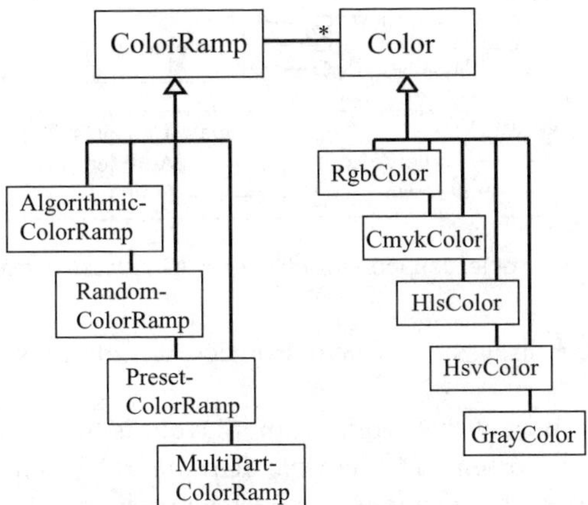

Figure 8.4   The diagram shows the relationship between the *ColorRamp* and *Color* classes and the subtypes of each class.

have members to gather histogram data from a table such as data values and frequencies and to pass the histogram data to a classification object. The classification object can then use the histogram data to compute the class breaks.

8.2.3   Color Ramp and Color Objects

A color ramp is a collection of colors. ArcObjects has four coclasses of color ramps: *RandomColorRamp, AlgorithmicColorRamp, PresetColorRamp,* and *MultiPartColorRamp* (Figure 8.4). A random color ramp creates a series of randomized colors. An algorithmic color ramp produces a sequential series of colors using two end colors and a defined algorithm. A preset color ramp is a series of 13 specific colors. A multipart color ramp is a collection of color ramps.

The algorithms available for generating intermediate colors in an algorithmic color ramp are: esriHSVAlgorithm, esriCIELabAlgorithm, and esriLabLCh-Algorithm. A color ramp produced by either the esriCIELab or esriLabLCh algorithm appears to blend the two end colors, whereas a color ramp produced by the esriHSV algorithm may contain additional hues.

ArcObjects uses a color model to define colors in a color ramp (Figure 8.4). For example, a bright yellow color is represented by (255, 255, 0) in the RGB model. Besides RGB, ArcObjects has CMYK (cyan, magenta, yellow, and black), HLS (hue, lightness, and saturation), HSV (hue, saturation, and value), and Grayscale. Additionally, ArcObjects offers CIELAB as a device-independent color model and uses it to store colors internally.

8.2.4   Layout Objects

A layout consists of various map elements. Each element has its own display style and its own position on a layout page. Because of the complexity of the topic,

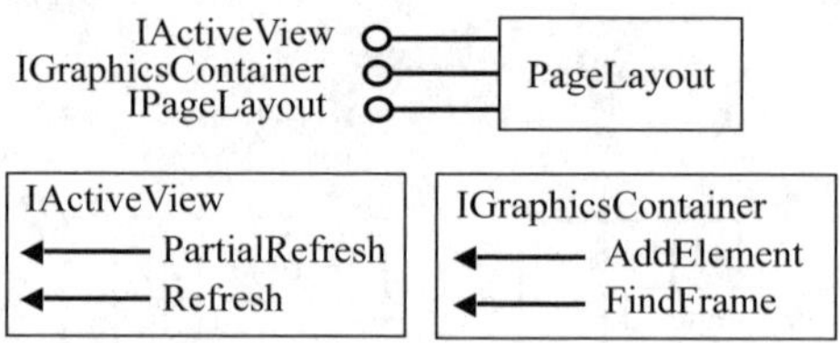

Figure 8.5   *A PageLayout* object supports *IActiveView, IGraphicsContainer,* and *IPageLayout.*

this section limits the discussion of layout objects to only those that are used later in a sample module.

The primary component that works with a layout is the *PageLayout* coclass. A page layout object implements *IActiveView, IGraphicsContainer,* and *IPageLayout* (Figure 8.5). *IActiveView* has methods to refresh, or partially refresh, the layout view. *IGraphicsContainer* has methods for adding and finding elements in a layout. *IPage-Layout* has members that are mainly important to the interaction between the user and a layout.

The *IGraphicsContainer* interface manages two types of elements in a layout: frame elements and text elements (Figure 8.6). Both types of elements implement *IElement,* which has access to the shape and the screen display of the element.

A frame element refers to an object that forms a border around other elements or objects. Two frame elements important to a layout design are *MapFrame* and *MapSurroundFrame* (Figure 8.7). A map frame object implements *IMapFrame,* which has access to a map object (e.g., the focus map in the data view) and can create map surrounds. In ArcObjects, a map surround refers to an element, such as a legend, a north arrow, or a scale bar, that is associated with a map. A map surround frame object supports *IMapSurroundFrame,* which provides access to a map sur-round within the frame. In layout design, each map surround frame must be related to a map frame.

A text element may represent a title or a feature label on a layout page. A *TextElement* object supports *ITextElement,* which can access the text string and symbol. A *TextSymbol* object implements *IFormattedTextSymbol* and *ITextSymbol.* Both interfaces let the user define text symbol properties such as color, font, hori-zontal alignment, size, and others. But *IFormattedTextSymbol* has more options than *ITextSymbol* does, especially in terms of character spacing and display properties. An *StdFont* object implements *IFontDisp,* which allows access to font properties such as name, boldness, and size.

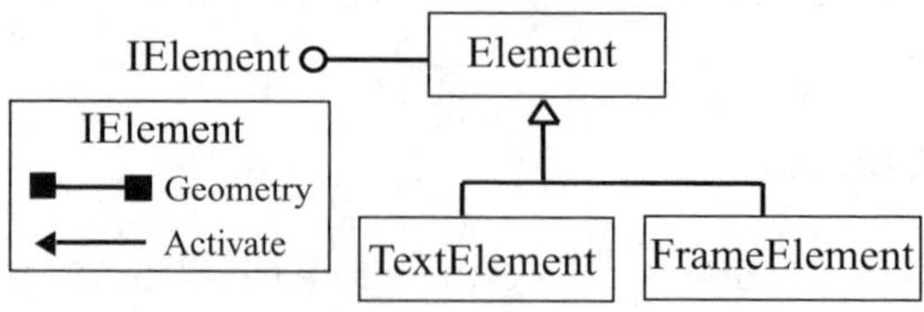

Figure 8.6   *TextElement* and *FrameElement* are types of the *Element* class.

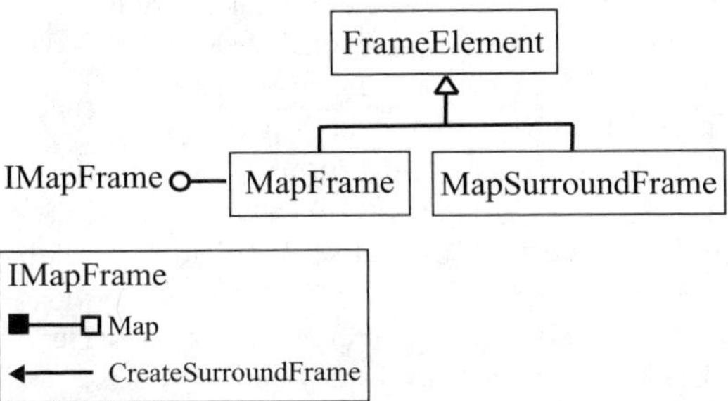

Figure 8.7 *MapFrame* and *MapSurroundFrame* are types of the *FrameElement* class. A map frame object has access to a map and can create map surround objects.

8.3 DISPLAYING VECTOR DATA

This section covers use of VBA (Visual Basic for Applications) macros for displaying vector data. These sample macros cover point, line, and area features; display both numeric and categorical data; and use both predefined and user-defined classification objects. Additionally, one sample macro uses a form to gather the necessary inputs from the user for data display.

8.3.1 *GraduatedColors*

GraduatedColors uses a sequential color ramp and a predefined classification object to display a choropleth map. A choropleth map shows data values based on administrative units. The macro performs the same function as using the Symbology/Quantities/Graduated colors command in the Layer Properties dialog in ArcMap.

GraduatedColors uses a user form to get a field, a classification method, and a number of classes for choropleth mapping (Figure 8.8). The form has two text boxes, a combo box, and two buttons:

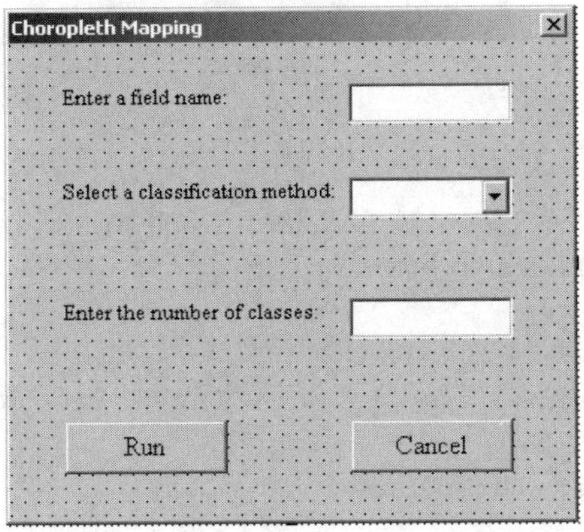

Figure 8.8 *GraduatedColors* uses the form to get a field name, a classification method, and a number of classes for data display.

txtField: a text box for entering the name of a numeric field
cboMethod: a combo box with three predefined classification systems
txtNumber: a text box for entering the number of classes
cmdRun: a command button to execute data display
cmdCancel: a command button to exit the form

Associated with the ***GraduatedColors*** user form are three subs and one function:

UserForm_Initialize: Initialize the user form by populating the method dropdown list.
cmdRun_Click: Use the user inputs to display data.
GetRGBColor: Return a color based on the input RGB values.
cmdCancel_Click: Exit the user form.

Of the above four procedures, ***cmdRun_Click*** involves most code writing and requires explanation. The other three are straightforward.

> **Key Interfaces:** *ITable, ITableHistogram, IHistogram, IClassify, IClassBreaks-Renderer, IAlgorithmicColorRamp, IEnumColors, IFillSymbol, IGeoFeatureLayer.*
> **Key Members:** *Field, Table, GetHistogram, SetHistogramData, Classify, ClassBreaks, BreakCount, MinimumBreak, Algorithm, ToColor, FromColor, Size, CreateRamp, Colors, Color, Break(), Symbol(), Label(), Renderer, PartialRefresh, UpdateContents.*
> **Usage:** Add *idcounty.shp* to an active map. The shapefile has a field called change that shows the rate of population change between 1990 and 2000 in Idaho by county. Import ***GraduatedColors.frm*** to Visual Basic Editor. Right-click UserForm1 and select View Code. Select UserForm from the object list at the upper left of the Code window. Click Run Sub/UserForm. Enter change for the name of the field. Select a classification method from the method dropdown list. Enter a number for the number of classes. Click on the Run button. The module uses a yellow-to-red color ramp to display the rate of population change data.

```vba
Private Sub UserForm_Initialize()

  cboMethod.AddItem "NaturalBreaks"
  cboMethod.AddItem "EqualInterval"
  cboMethod.AddItem "Quantile"

End Sub
```

UserForm_Initialize uses the *AddItem* method to add the three classification methods to the method combo box.

```vba
Private Sub cmdRun_Click()
  ' Part 1: Define the feature layer and derive its
  ' histogram data.
  Dim pMxDoc As IMxDocument
  Dim pMap As IMap
  Dim pLayer As IFeatureLayer
```

```
Dim pTable As ITable
Dim pTableHistogram As ITableHistogram
Dim pHistogram As IHistogram
Dim DataValues As Variant
Dim DataFrequencies As Variant
Set pMxDoc = ThisDocument
Set pMap = pMxDoc.FocusMap
Set pLayer = pMap.Layer(0)
Set pTable = pLayer
' Define the table histogram.
Set pTableHistogram = New TableHistogram
pTableHistogram.Field = txtField.Value
Set pTableHistogram.Table = pTable
' Derive the data values and frequencies from the
' histogram.
Set pHistogram = pTableHistogram
pHistogram.GetHistogram DataValues, DataFrequencies
```

Part 1 of ***cmdRun_Click*** derives from the feature layer the histogram data to be used for classification. The code first sets *pLayer* to be the top layer in the active map and *pTable*, a reference to *ITable*, to be the same as *pLayer*. Next the code creates *pTableHistogram* as an instance of the *TableHistogram* class and defines its *Field* as *txtField.Value*, the field name entered in the user form, and its *Table* as *pTable* (Figure 8.9). The code then performs a QueryInterface (QI) for the *IHistogram* interface and uses the *GetHistogram* method to derive the histogram data from *pTableHistogram*. Of the derived histogram data, *DataValues* contains an array of data values, and *DataFrequencies* contains an array of frequencies corresponding to the data values.

```
' Part 2: Create a class breaks renderer.
Dim pClassify As IClassify
Dim Classes() As Double
```

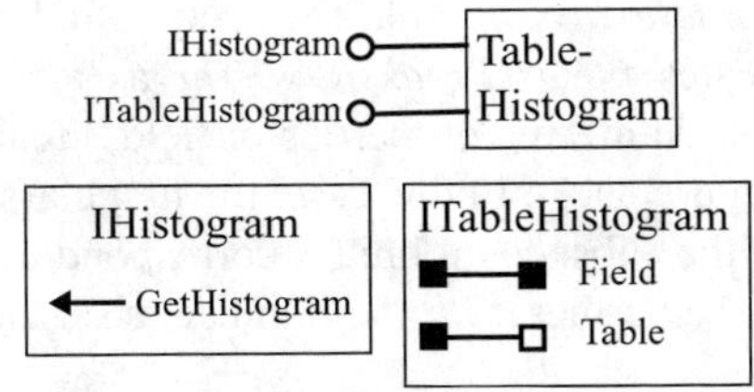

Figure 8.9 A *TableHistogram* object supports *IHistogram* and *ITableHistogram*. *ITableHistogram* has the properties to identify the table and the field. *IHistogram* has a method to get the histogram data.

```
Dim pClassBreaksRenderer As IClassBreaksRenderer
' Set up the classification method.
Select Case cboMethod.ListIndex
  Case 0
    Set pClassify = New NaturalBreaks
  Case 1
    Set pClassify = New EqualInterval
  Case 2
    Set pClassify = New Quantile
End Select
' Prepare a classify object.
pClassify.SetHistogramData DataValues, DataFrequencies
pClassify.Classify Val(txtNumber.Value)
' Create an array of class breaks.
Classes = pClassify.ClassBreaks
' Prepare a class breaks renderer.
Set pClassBreaksRenderer = New ClassBreaksRenderer
With pClassBreaksRenderer
  .Field = txtField.Value
  .BreakCount = Val(txtNumber.Value)
  .MinimumBreak = Classes(0)
End With
```

Part 2 of ***cmdRun_Click*** performs three tasks: using the selected classification method and the histogram data to prepare a classification object, deriving class breaks from the classification object, and linking the class breaks to a class breaks renderer. For the first task, the code creates *pClassify* as an instance of the *NaturalBreaks*, *EqualInterval*, or *Quantile* class, depending on the user's choice. Next the code uses methods on *IClassify* to set the histogram data from Part 1 and to classify the histogram data into the number of classes entered by the user (i.e., *txtNumber.Value*) (Figure 8.10). For the second task, the code stores the class breaks into the array variable *Classes*. By default, the array is indexed from zero. For the third task, the code creates *pClassBreaksRenderer* as an instance of the *ClassBreaksRenderer* class and defines its properties of field, break count, and minimum break in a *With* block (Figure 8.11). In this case, the first class break is the minimum value in the dataset, and the subsequent breaks correspond to the upper class limits. The number of class breaks is therefore the number of classes plus one.

```
' Part 3: Create a color ramp.
Dim pAlgoRamp As IAlgorithmicColorRamp
Dim pColors As IEnumColors
```

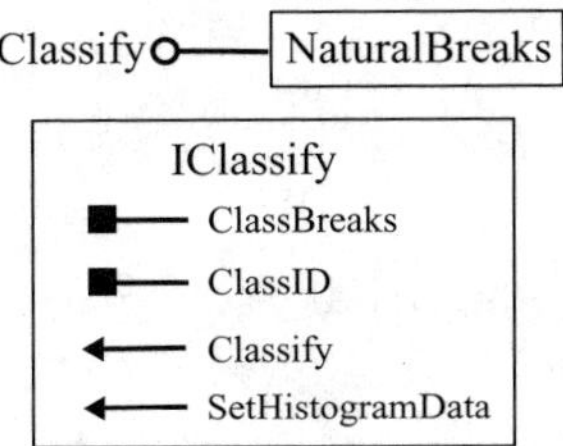

Figure 8.10   Properties and methods on the *IClassify* interface. All predefined classification classes including *NaturalBreaks* implement the *IClassify* interface.

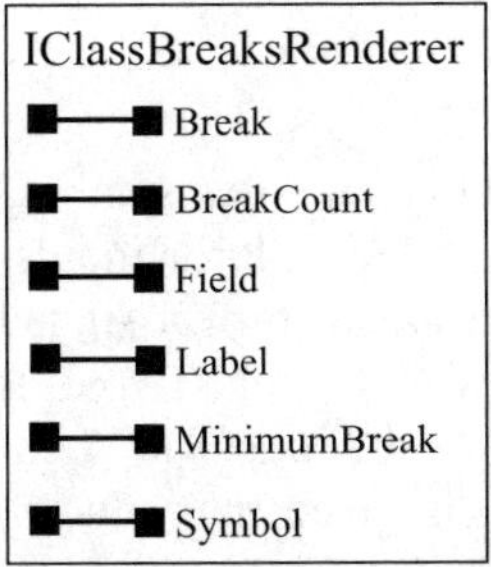

Figure 8.11   Properties on the *IClassBreaksRenderer* interface for defining graduated color symbols.

```
' Prepare a color ramp.
Set pAlgoRamp = New AlgorithmicColorRamp
With pAlgoRamp
  .Algorithm = esriCIELabAlgorithm
  .ToColor = GetRGBColor(255, 0, 0)
  .FromColor = GetRGBColor(255, 255, 0)
  .Size = Val(txtNumber.Value)
  .CreateRamp (True)
End With
' Store the colors.
Set pColors = pAlgoRamp.Colors
```

Part 3 of **cmdRun_Click** creates a color ramp. The code creates *pAlgoRamp* as an instance of the *AlgorithmicColorRamp* class, and defines its properties of algorithm, end colors, and size before creating the ramp. The color ramp starts from a yellow color and ends with a red color. The code uses the **GetRGBColor** function to define the end colors and the esriCIELabAlgorithm to generate the intermediate colors. After they are generated, these colors are stored as a sequence of colors referenced by *pColors*.

```
' Part 4: Assign a color symbol, break, and label
' to each of the classes.
  Dim pFillSymbol As IFillSymbol
  Dim I As Integer
  For I = 0 To pClassBreaksRenderer.BreakCount - 1
    Set pFillSymbol = New SimpleFillSymbol
    pFillSymbol.Color = pColors.Next
    pClassBreaksRenderer.Symbol(I) = pFillSymbol
    pClassBreaksRenderer.Break(I) = Classes(I + 1)
    pClassBreaksRenderer.Label(I) = CSng(Classes(I)) _
    & " - " & CSng(Classes(I + 1))
Next I
```

Part 4 of **cmdRun_Click** defines a color symbol, a break, and a label for each class in the class breaks renderer. The color symbols stored in *pColors* are initially assigned to *pFillSymbol*, which are in turn assigned as symbols to each class in *pClassBreaksRenderer* in a *For...Next* loop. Two other properties of *pClassBreaks-Renderer* are assigned within the same loop. The *Break(I)* property, which represents the upper bound of each class, is given the value of *Breaks(I + 1)*. The *Label(I)* property is given the value of *CSng(Classes(I)) & " - " & CSng(Classes(I + 1))*. The *CSng* (Conversion to Single) function truncates the long fractional part of numeric values.

```
' Part 5: Draw the graduated color map.
Dim pGeoFeatureLayer As IGeoFeatureLayer
' Assign the renderer to the feature layer.
Set pGeoFeatureLayer = pLayer
Set pGeoFeatureLayer.Renderer = pClassBreaksRenderer
' Refresh the map and its table of contents.
pMxDoc.ActiveView.PartialRefresh esriViewGeography, _
pLayer, Nothing
pMxDoc.UpdateContents
End Sub
```

Part 5 of **cmdRun_Click** performs a QI for the *IGeoFeatureLayer* interface and assigns *pClassBreaksRenderer* to be the renderer. *IGeoFeatureLayer* controls the display of a feature layer. The code then refreshes the view and updates the document's table of contents.

```
Private Function GetRGBColor(R As Long, G As Long, B _
As Long)
  Dim pColor As IRgbColor
  Set pColor = New RgbColor
```

```
  pColor.Red = R
  pColor.Green = G
  pColor.Blue = B
  GetRGBColor = pColor
End Function
```

GetRGBColor uses the input values of R, G, and B from **cmdRun_Click** in Part 3 to create a new RGB color referenced by *pColor*. **GetRGBColor** then returns *pColor* to **cmdRun_Click.**

```
Private Sub cmdCancel_Click()
  End
End Sub
```

cmdCancel_Click exits the user form.

8.3.2   *GraduatedSymbols*

GraduatedSymbols uses different-sized circles to display different ranges of a field's values. The classification of the field values is user-defined rather than predefined. The macro performs the same function as using the Symbology/Quantities/ Graduated symbols command in the Layer Properties dialog in ArcMap.

GraduatedSymbols has three parts. Part 1 creates a class breaks renderer. Part 2 prepares the symbol, break, and label for each class in the renderer. Part 3 assigns the renderer to the feature layer and refreshes the view.

> **Key Interfaces:** *IClassBreaksRenderer, ISimpleMarkerSymbol, IGeoFeatureLayer.*
> **Key Members:** *Field, BreakCount, Color, Outline, OutlineColor, Size, Style, Symbol(),*
>   *Break(), Label(), Renderer, PartialRefresh, UpdateContents.*
> **Usage:** Add *idlcity.shp*, a shapefile that contains ten largest cities in Idaho, to an active
>   map. Import **GraduatedSymbols** to Visual Basic Editor. Run the macro. The macro
>   produces a map showing the cities in graduated circles and the city names.

```
Private Sub GraduatedSymbols()
  ' Part 1: Create a class break renderer.
  Dim pClassBreaksRenderer As IClassBreaksRenderer
  Set pClassBreaksRenderer = New ClassBreaksRenderer
  With pClassBreaksRenderer
    .Field = "Population"
    .BreakCount = 3
  End With
```

Part 1 creates *pClassBreaksRenderer* as an instance of the *ClassBreaksRenderer* class and defines its properties of field and break count.

```vba
' Part 2: Set the symbol, break, and label for each class.
Dim pMarkerSymbol As ISimpleMarkerSymbol
' Set the first class's symbol, break, and label.
Set pMarkerSymbol = New SimpleMarkerSymbol
With pMarkerSymbol
   .Color = GetRGBColor(255, 255, 0)
   .Outline = True
   .OutlineColor = GetRGBColor(0, 0, 0)
   .Size = 12
   .Style = esriSMSCircle
End With
pClassBreaksRenderer.Symbol(0) = pMarkerSymbol
pClassBreaksRenderer.Break(0) = 20000
pClassBreaksRenderer.Label(0) = "14300 - 20000"
' Set the second class's symbol, break, and label.
Set pMarkerSymbol = New SimpleMarkerSymbol
With pMarkerSymbol
   .Color = GetRGBColor(255, 125, 0)
   .Outline = True
   .OutlineColor = GetRGBColor(0, 0, 0)
   .Size = 18
   .Style = esriSMSCircle
End With
pClassBreaksRenderer.Symbol(1) = pMarkerSymbol
pClassBreaksRenderer.Break(1) = 30000
pClassBreaksRenderer.Label(1) = "20001 - 30000"
' Set the third class's symbol, break, and label.
Set pMarkerSymbol = New SimpleMarkerSymbol
With pMarkerSymbol
   .Color = GetRGBColor(255, 0, 0)
   .Outline = True
   .OutlineColor = GetRGBColor(0, 0, 0)
   .Size = 24
   .Style = esriSMSCircle
End With
pClassBreaksRenderer.Symbol(2) = pMarkerSymbol
pClassBreaksRenderer.Break(2) = 125660
pClassBreaksRenderer.Label(2) = "30001 - 125660"
```

Part 2 sets the symbol, break, and label for each class of the renderer. The code creates *pMarkerSymbol* as an instance of the *SimpleMarkerSymbol* class and defines the following five symbol properties: *Color* is the symbol color, *Outline* indicates whether or not to draw the outline, *OutlineColor* is the symbol outline color, *Size* is the symbol size in points, and *Style* is the symbol style (e.g., circle, square, or diamond). The **GetRGBColor** function provides the colors for the symbol and the symbol outline. Then the code assigns *pMarkerSymbol* to be the symbol of the first class, followed by assigning the class's upper limit and label. The same procedure for the first class applies to the other two classes.

```
' Part 3: Draw the feature layer and refresh the
' table of contents.
Dim pMxDoc As IMxDocument
Dim pMap As IMap
Dim pLayer As ILayer
Dim pGeoFeatureLayer As IGeoFeatureLayer
Set pMxDoc = ThisDocument
Set pMap = pMxDoc.FocusMap
Set pLayer = pMap.Layer(0)
' Set the renderer and annotation for drawing.
Set pGeoFeatureLayer = pLayer
Set pGeoFeatureLayer.Renderer = pClassBreaksRenderer
pGeoFeatureLayer.DisplayField = "City_Name"
pGeoFeatureLayer.DisplayAnnotation = True
pMxDoc.ActiveView.PartialRefresh esriViewGeography, _
pLayer, Nothing
pMxDoc.UpdateContents
End Sub
```

Part 3 performs a QI for the *IGeoFeatureLayer* interface to assign the renderer and to display the field City_Name as annotation. The code then refreshes the map and updates the table of contents.

8.3.3 *UniqueSymbols*

UniqueSymbols uses a set of symbols to display each unique value of a field. These unique values represent categorical data such as different road types. The macro performs the same function as using the Symbology/Categories/Unique values command in the Layer Properties dialog in ArcMap.

UniqueSymbols has two parts. Part 1 creates a unique value renderer and populates the renderer with symbols for each unique value of a specified field. Part 2 assigns the renderer to the feature layer and refreshes the view.

Key Interfaces: *IUniqueValueRenderer, ILineSymbol, IGeoFeatureLayer.*
Key Members: *FieldCount, Field( ), Color, Width, AddValue, Renderer, PartialRefresh, UpdateContents.*
Usage: Add *idroads.shp* to an active map. Import ***UniqueSymbols*** to Visual Basic Editor. Run the macro. The macro produces a map showing the interstate, U.S., and state highways in three different line symbols.

```
Private Sub UniqueSymbols()
  ' Part 1: Prepare a unique value renderer.
  Dim pUniqueValueRenderer As IUniqueValueRenderer
  Dim pSym1 As ILineSymbol
  Dim pSym2 As ILineSymbol
  Dim pSym3 As ILineSymbol
  ' Define the renderer.
  Set pUniqueValueRenderer = New UniqueValueRenderer
  pUniqueValueRenderer.FieldCount = 1
  pUniqueValueRenderer.Field(0) = "Route_Desc"
  ' Add the first symbol to the renderer.
  Set pSym1 = New SimpleLineSymbol
  pSym1.Color = GetRGBColor(255, 0, 0)
  pSym1.Width = 3
  pUniqueValueRenderer.AddValue "Interstate", "", pSym1
  ' Add the second symbol to the renderer.
  Set pSym2 = New SimpleLineSymbol
  pSym2.Color = GetRGBColor(255, 100, 0)
  pSym2.Width = 2
  pUniqueValueRenderer.AddValue "U.S.", "", pSym2
  ' Add the third symbol to the renderer.
  Set pSym3 = New SimpleLineSymbol
  pSym3.Color = GetRGBColor(255, 150, 0)
  pSym3.Width = 1
  pUniqueValueRenderer.AddValue "State", "", pSym3
```

Part 1 creates a unique value renderer with three symbols. The code creates *pUniqueValueRenderer* as an instance of the *UniqueValueRenderer* class and defines its field count as one and its field as Route_Desc (Figure 8.12). These properties are necessary because a unique value renderer can apply to two or more fields. Next the code creates *pSym1* as an instance of the *SimpleLineSymbol* class and defines its color and width properties. The color is generated by the ***GetRGBColor*** function, and the width is specified in points. Then the code uses the *AddValue* method on *IUniqueValueRenderer* to add the unique value (i.e., Interstate) and corresponding symbol (i.e., *pSym1*) to the renderer. The same procedure is repeated for the second and third unique values.

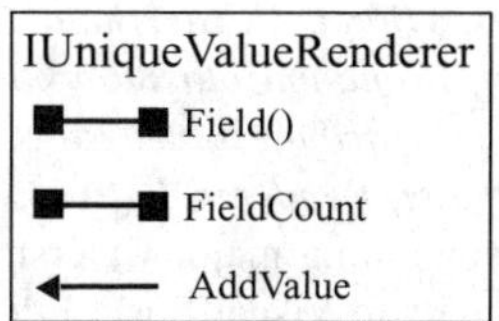

Figure 8.12   Members on the *IUniqueValueRenderer* interface for defining unique symbols.

```
' Part 2: Assign the renderer to the feature layer
' and refresh the map and its table of contents.
Dim pMxDoc As IMxDocument
Dim pMap As IMap
Dim pLayer As IFeatureLayer
Dim pGeoFeatureLayer As IGeoFeatureLayer
Set pMxDoc = ThisDocument
Set pMap = pMxDoc.FocusMap
Set pLayer = pMap.Layer(0)
Set pGeoFeatureLayer = pLayer
Set pGeoFeatureLayer.Renderer = pUniqueValueRenderer
pMxDoc.ActiveView.PartialRefresh esriViewGeography, _
pLayer, Nothing
pMxDoc.UpdateContents
End Sub
```

Part 2 sets *pLayer* to be the top layer in the active map. Next the code switches to the *IGeoFeatureLayer* interface to assign the renderer. Finally, the code refreshes the view and updates the table of contents.

8.4 DISPLAYING RASTER DATA

Unlike vector data, which may be represented by point, line, or area (fill) symbols, raster data are represented only by fill symbols. Raster data can be categorical or numeric, however. Therefore, fill symbols for displaying raster data can be unique symbols or graduated color symbols.

8.4.1 *RasterUniqueSymbols*

RasterUniqueSymbols uses a symbol for each unique cell value to draw a raster layer. The macro performs the same function as using the Symbology/Unique Values command in the Layer Properties dialog in ArcMap.

RasterUniqueSymbols has four parts. Part 1 defines the raster dataset to draw. Part 2 creates a raster unique value renderer and connects the renderer to the raster dataset. Part 3 assigns colors generated from a random color ramp and labels to each unique value. Part 4 applies the renderer to the raster layer and refreshes the view.

Key Interfaces: *ITable, IRasterBand, IRasterBandCollection, IRasterUniqueValue-Renderer, IRasterRenderer, IRandomColorRamp, ISimpleFillSymbol.*

Key Members: *Raster, Item(), AttributeTable, RowCount, FindField, Size, Create-Ramp, GetRow, Value, Update, Renderer, Refresh, UpdateContents.*

Usage: Add *hucgd*, a raster containing major watersheds in Idaho, to an active map. Import **RasterUniqueSymbols** to Visual Basic Editor. Run the macro. The macro displays each watershed in *hucgd* with a unique symbol. These unique symbols differ from those initial symbols used by ArcMap.

```
Private Sub RasterUniqueSymbols()
  ' Part 1: Define the raster dataset to draw.
  Dim pMxDoc As IMxDocument
  Dim pMap As IMap
  Dim pRLayer As IRasterLayer
  Dim pRaster As IRaster
  Dim pTable As ITable
  Dim pBand As IRasterBand
  Dim pBandCol As IRasterBandCollection
  Dim TableExist As Boolean
  Dim NumOfValues As Integer
  Dim FieldIndex As Integer
  Dim FieldName As String
  Set pMxDoc = ThisDocument
  Set pMap = pMxDoc.FocusMap
  Set pRLayer = pMap.Layer(0)
  ' Work with the first band of the raster.
  Set pRaster = pRLayer.Raster
  Set pBandCol = pRaster
  Set pBand = pBandCol.Item(0)
  ' Make sure the band has an attribute table. If not,
  ' exit sub.
  pBand.HasTable TableExist
  If Not TableExist Then Exit Sub
  Set pTable = pBand.AttributeTable
  ' Get the row count.
  NumOfValues = pTable.RowCount(Nothing)
  ' Find the field Value.
  FieldName = "Value"
  FieldIndex = pTable.FindField(FieldName)
```

Part 1 defines the raster dataset and the field to draw. The code sets *pRaster* to be the raster of the top layer in the active map. Next the code switches to the *IRaster-BandCollection* interface to set *pBand* as the first band of *pRaster*. If *pBand* has a table, then the code assigns the attribute table to *pTable*. If not, the sub stops. Using *pTable* as the source, the code assigns the number of rows to the *NumOfValues* variable and the index of the field Value to the *FieldIndex* variable.

```
' Part 2: Define a raster unique value renderer.
Dim pUVRen As IRasterUniqueValueRenderer
Dim pRasRen As IRasterRenderer
Set pUVRen = New RasterUniqueValueRenderer
Set pRasRen = pUVRen
Set pRasRen.Raster = pRaster
pRasRen.Update
```

Part 2 defines the renderer for displaying *pRaster*. The code creates *pUVRen* as an instance of the *RasterUniqueValueRenderer* class. A raster unique value renderer object supports *IRasterRenderer* and *IRasterUniqueValueRenderer* (Figure 8.13). Next the code switches to the *IRasterRenderer* interface to define *pRaster* as the raster and to update the renderer. A raster renderer must be updated for any changes that have been made.

```
' Part 3: Assign symbol and label to each unique
' value in the renderer.
Dim pRamp As IRandomColorRamp
Dim I As Long
Dim pRow As IRow
Dim UniqValue As Variant
Dim pFSymbol As ISimpleFillSymbol
' Create a random color ramp.
Set pRamp = New RandomColorRamp
```

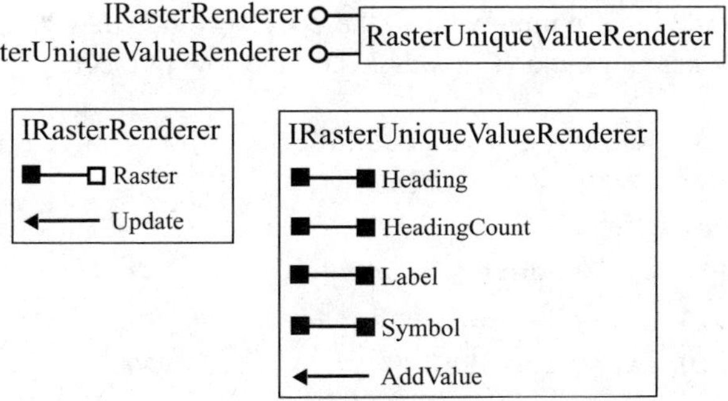

Figure 8.13 A *RasterUniqueValueRenderer* object supports *IRasterRenderer* and *IRaster-UniqueValueRenderer*. These two interfaces provide access to the unique symbols and raster for data display.

```
With pRamp
   .Size = NumOfValues
   .CreateRamp (True)
End With
' Add unique values, labels, and symbols to the renderer.
For I = 0 To NumOfValues - 1
   Set pRow = pTable.GetRow(I)
   ' Add a unique value.
   UniqValue = pRow.Value(FieldIndex)
   pUVRen.AddValue 0, I, UniqValue
   ' Add corresponding label.
   pUVRen.Label(0, I) = CStr(UniqValue)
   ' Add corresponding symbol.
   Set pFSymbol = New SimpleFillSymbol
   pFSymbol.Color = pRamp.Color(I)
   pUVRen.Symbol(0, I) = pFSymbol
Next I
```

Part 3 populates the symbol and label in the renderer. The code first creates *pRamp* as an instance of the *RandomColorRamp* class and sets the number of colors to be the same as the number of unique values before creating the ramp. Then the code uses a *For...Next* loop to add the unique values and their corresponding labels and symbols to *pUVRen*. The unique value is the value at the row defined by *I* and the column defined by *FieldIndex*. The label is the string converted from the unique value. And the symbol is a simple fill symbol with a color generated from *pRamp*. The members of *AddValue*, *Label*, and *Symbol* on *IRasterUniqueValueRenderer* all use two indices: the first represents the *iHeading* and the second represents the *IClass*. Headings are used for organizing unique values. One can create a new heading, for example, by combining two or more unique values. Because this macro works with individual unique values, the code specifies *iHeading* as zero (i.e., the first and only heading) and *IClass* as *I* (i.e., the row number).

```
' Part 4: Update the renderer, draw the layer,
' and refresh the view.
pRasRen.Update
Set pRLayer.Renderer = pUVRen
pMxDoc.ActiveView.Refresh
pMxDoc.UpdateContents
End Sub
```

Part 4 updates *pRasRen* before assigning it to be the renderer for *pRLayer*. Finally, the code refreshes the view and updates the table of contents.

8.4.2 *RasterClassifyColorRamp*

RasterClassifyColorRamp displays a raster layer by using a classification object, a raster classify renderer, and a user-defined color ramp. The macro performs the same function as the Symbology/Classified command in the Layer Properties dialog in ArcMap. *RasterClassifyColorRamp* has three parts. Part 1 derives the histogram data from the raster for display. Part 2 prepares a raster renderer. Part 3 defines the properties of the renderer and uses the renderer to draw the raster layer.

> **Key Interfaces:** *IRaster, ITable, IRasterBandCollection, IRasterBand, ITable-Histogram, IHistogram, IClassify, IRasterClassifyColorRampRenderer, IRasterClassifyUIProperties, IRasterRenderer, IAlgorithmicColorRamp, IEnumColors, IFillSymbol.*
>
> **Key Members:** *Raster, Item(), AttributeTable, Field, Table, GetHistogram, SetHistogramData, Classify, ClassBreaks, ClassID, ClassCount, ClassField, ClassificationMethod, Update, Algorithm, FromColor, ToColor, Size, Create-Ramp, Color, Symbol(), Break(), Label(), Refresh, UpdateContents.*
>
> **Usage:** Add *intemida*, an integer elevation raster, to an active map. Import *RasterClassifyColorRamp* to Visual Basic Editor. Run the macro. The macro displays *intemida* using ten equal interval classes and a color ramp from red to blue.

```
Private Sub RasterClassifyColorRamp()
  ' Part 1: Prepare the raster dataset for display.
  Dim pMxDoc As IMxDocument
  Dim pMap As IMap
  Dim pRLayer As IRasterLayer
  Dim pRaster As IRaster
  Dim pTable As ITable
  Dim pBandCol As IRasterBandCollection
  Dim pRasBand As IRasterBand
  Dim TestTable As Boolean
  Dim pTableHist As ITableHistogram
  Dim pHist As IHistogram
  Dim vValues As Variant
  Dim vFrequencies As Variant
  ' Define the raster dataset and verify it has a table.
  Set pMxDoc = ThisDocument
  Set pMap = pMxDoc.FocusMap
  Set pRLayer = pMap.Layer(0)
  Set pRaster = pRLayer.Raster
  Set pBandCol = pRaster
  Set pRasBand = pBandCol.Item(0)
  pRasBand.HasTable TestTable
```

```vba
If TestTable = False Then Exit Sub
Set pTable = pRasBand.AttributeTable
' Derive the histogram data from the table.
Set pTableHist = New TableHistogram
pTableHist.Field = "Value"
Set pTableHist.Table = pTable
Set pHist = pTableHist
pHist.GetHistogram vValues, vFrequencies
```

Part 1 verifies that the raster has a table and derives the histogram data from the table. The code first defines the following objects: *pRaster* to be the raster of the first layer in the active map, *pBandCol* to be the same as *pRaster*, and *pRasBand* to be the first band of the band collection. Next the code verifies that *pRasBand* has a table and sets *pTable* to be the attribute table of *pRasBand*. The rest of Part 1 derives the histogram data from *pTable*. The code creates *pTableHist* as an instance of the *TableHistogram* class and defines its field and table properties. Then the code switches to the *IHistogram* interface and uses the *GetHistogram* method to derive the data values and frequencies from *pTable*.

```vba
' Part 2: Prepare the raster renderer.
Dim pClassify As IClassify
Dim ClassBreak As Variant
Dim pUID As UID
Dim pClassRen As IRasterClassifyColorRampRenderer
Dim pClassProp As IRasterClassifyUIProperties
Dim pRasRen As IRasterRenderer
' Prepare an equal interval classification object
' with 10 classes.
Set pClassify = New EqualInterval
pClassify.SetHistogramData vValues, vFrequencies
pClassify.Classify 10
' Store the class breaks.
ClassBreak = pClassify.ClassBreaks
' Obtain the classification UID.
Set pUID = pClassify.ClassID
' Prepare a raster classify renderer.
Set pClassRen = New RasterClassifyColorRampRenderer
pClassRen.ClassCount = 10
pClassRen.ClassField = "Value"
' Define the classification method.
Set pClassProp = pClassRen
```

```
Set pClassProp.ClassificationMethod = pUID
' Define the raster and update the renderer.
Set pRasRen = pClassRen
Set pRasRen.Raster = pRaster
pRasRen.Update
```

Part 2 prepares a classification object and a raster renderer. This preparation requires several steps. First, the code creates *pClassify* as an instance of the *EqualInterval* class. The code then uses members on *IClassify* to get the histogram data for a classification with ten classes, to save the class breaks into an array referenced by *ClassBreak*, and to obtain the classification's unique identifier (UID). Second, the code creates *pClassRen* as an instance of the *RasterClassifyColorRampRenderer* class and defines its properties of class count and class field. Third, the code performs a QI for *IRasterClassifyUI-Properties* to specify *pUID* for the classification method. Finally, the code switches to the *IRasterRenderer* interface to define the raster and to update the raster. Figure 8.14 shows the interfaces that a raster classify color ramp renderer object supports.

```
' Part 3: Define the properties of the renderer,
' and use the renderer to draw the map.
Dim pAlgoRamp As IAlgorithmicColorRamp
Dim pColors As IEnumColors
Dim pFillSymbol As IFillSymbol
Dim I As Integer
' Prepare a color ramp.
Set pAlgoRamp = New AlgorithmicColorRamp
With pAlgoRamp
  .Algorithm = esriHSVAlgorithm
  .FromColor = GetRGBColor(255, 0, 0)
  .ToColor = GetRGBColor(0, 0, 255)
  .Size = 10
  .CreateRamp (True)
```

Figure 8.14 A *RasterClassifyColorRampRenderer* object supports *IRasterClassifyColor-RampRenderer*, *IRasterClassifyUIProperties*, and *IRasterRenderer*. The first two interfaces can define the classification of raster data and the third, as shown in Figure 8.13, connects the raster to its renderer.

```
End With
Set pColors = pAlgoRamp.Colors
' Define the symbol, break, and label for each class.
For I = 0 To pClassRen.ClassCount - 1
  Set pFillSymbol = New SimpleFillSymbol
  pFillSymbol.Color = pColors.Next
  pClassRen.Symbol(I) = pFillSymbol
  pClassRen.Break(I) = ClassBreak(I)
  pClassRen.Label(I) = ClassBreak(I) & " - " & _
  ClassBreak(I + 1)
Next I
' Assign the raster renderer to the layer and
' refresh the map.
pRasRen.Update
Set pRLayer.Renderer = pRasRen
pMxDoc.ActiveView.Refresh
pMxDoc.UpdateContents
End Sub
```

Part 3 provides the symbol, break, and label to the renderer and uses the renderer to draw the raster layer. The code first creates *pAlgoRamp* as an instance of the *AlgorithmicColorRamp* class, and defines its properties. The ***GetRGBColor*** function is again used to define the two end colors of the color ramp. The colors from *pAlgoRamp* are stored in an enumerator referenced by *pColors*. Next the code uses a *For...Next* loop to assign the symbol, break, and label to each class in *pClassRen*. Finally, the code updates *pRasRen* with the changes made through *pClassRen* and uses *pRasRen* as the renderer to draw the raster layer.

8.4.3   *RasterUserDefinedColorRamp*

RasterUserDefinedColorRamp displays a classified raster layer by using user-defined class breaks and color ramp. The macro performs the same function as using the Symbology/Classified command in the Layer Properties dialog in ArcMap. ***RasterUserDefinedColorRamp*** has three parts. Part 1 defines the raster and prepares a raster renderer. Part 2 creates a color ramp, followed by specifying the label, break, and symbol of each class for the renderer. Part 3 assigns the renderer to the raster and refreshes the map.

Key Interfaces: *IRaster, IRasterClassifyColorRampRenderer, IRasterRenderer, IAlgorithmicColorRamp, IFillSymbol.*

Key Members: *Raster, ClassCount, Update, Algorithm, FromColor, ToColor, Size; CreateRamp, Break(), label(), Color, Symbol(), Renderer, Refresh, Update-Contents.*

Usage: Add *emidalat*, an elevation raster, to an active map. Import ***RasterUser-DefinedColorRamp*** to Visual Basic Editor. Run the macro. The macro redraws *emidalat* in three classes using symbols from a color ramp.

```
Private Sub RasterUserDefinedColorRamp()
  ' Part 1: Define the raster dataset and a raster renderer.
  Dim pMxDoc As IMxDocument
  Dim pMap As IMap
  Dim pRLayer As IRasterLayer
  Dim pRaster As IRaster
  Dim pClassRen As IRasterClassifyColorRampRenderer
  Dim pRasRen As IRasterRenderer
  ' Define the raster dataset.
  Set pMxDoc = ThisDocument
  Set pMap = pMxDoc.FocusMap
  Set pRLayer = pMap.Layer(0)
  Set pRaster = pRLayer.Raster
  ' Define a raster classify color ramp renderer.
  Set pClassRen = New RasterClassifyColorRampRenderer
  pClassRen.ClassCount = 3
  Set pRasRen = pClassRen
  Set pRasRen.Raster = pRaster
  pRasRen.Update
```

Part 1 defines the raster and a raster renderer. The code sets *pRaster* to be the raster of the layer to draw. Next the code creates *pClassRen* as an instance of the *RasterClassifyColorRampRenderer* class and specifies three for the class count. Then the code switches to the *IRasterRenderer* interface to define the raster and to update the renderer.

```
  ' Part 2: Specify the properties for the renderer.
  Dim pRamp As IAlgorithmicColorRamp
  Dim pFSymbol As IFillSymbol
  ' Create a color ramp.
  Set pRamp = New AlgorithmicColorRamp
  With pRamp
    .Algorithm = esriCIELabAlgorithm
    .FromColor = GetRGBColor(0, 255, 255)
    .ToColor = GetRGBColor(0, 0, 255)
    .Size = 3
    .CreateRamp True
```

```
End With
' Specify the label, break, and symbol for each
' class in the renderer.
Set pFSymbol = New SimpleFillSymbol
pFSymbol.Color = pRamp.Color(0)
pClassRen.Symbol(0) = pFSymbol
pClassRen.Break(0) = 855
pClassRen.Label(0) = "855 - 1000"
pFSymbol.Color = pRamp.Color(1)
pClassRen.Symbol(1) = pFSymbol
pClassRen.Break(1) = 1000
pClassRen.Label(1) = "1000 - 1200"
pFSymbol.Color = pRamp.Color(2)
pClassRen.Symbol(2) = pFSymbol
pClassRen.Break(2) = 1200
pClassRen.Label(2) = "1200 - 1350"
```

Part 2 specifies the properties of the renderer. The code first creates *pRamp* as an instance of the *AlgorithmicColorRamp* class and defines its properties. Then the code uses colors from *pRamp* and hard-coded class breaks and labels to define the properties for each class in *pClassRen*.

```
' Part 3: Assign the renderer to the layer and
' refresh the map.
pRasRen.Update
Set pRLayer.Renderer = pRasRen
pMxDoc.ActiveView.Refresh
pMxDoc.UpdateContents
End Sub
```

After the renderer is set up in Part 2, Part 3 assigns the updated *pRasRen* to the raster layer, refreshes the map, and updates the document's table of contents.

8.5 MAKING A PAGE LAYOUT

A page layout may contain one or more maps and various map elements. A module for making a page layout typically involves more objects and code lines than other types of applications.

8.5.1 *Layout*

Layout prepares a layout of a thematic map showing the rate of population change by county in Idaho between 1990 and 2000. The layout includes the map body, a title, a subtitle, a legend, a north arrow, and a scale bar. The module performs the same function as using the Insert menu in the Layout View to add the different map elements to the layout. *Layout* is organized into five subs. *Start* is the startup sub, which calls the other subs. *AddTitle* adds the title and subtitle, *AddLegend* adds the legend, *AddNorthArrow* adds the north arrow, and *AddScaleBar* adds the scale bar to the layout.

Key Interfaces: *IGraphicsContainer, IMapFrame, IElement, IActiveView, IPage-Layout, ITextElement, IFormattedTextSymbol, IFontDisp, IPoint, IEnvelope, IMapSurroundFrame, IMapSurround, IMarkerNorthArrow, ICharacter-MarkerSymbol.*

Key Members: *PageLayout, FindFrame, Text, Name, Size, Bold, Font, Case, HorizontalAlignment, Symbol, Geometry, AddElement, Value, PutCoords, Create-SurroundFrame, Activate, MarkerSymbol, CharacterIndex.*

Usage: Add *idcounty.lyr* to a new map. Make sure that there are no other maps in the map document. The layer file shows the rate of population change by county between 1990 and 2000 in Idaho. In ArcMap's table of contents, double-click and delete *idcounty* (the layer name) and CHANGE (the field name). (Using the layer name and the field name on the layout would be confusing to the viewer). Change from Data View to Layout View. Use the handle of the map frame to resize the map so that it fills the page. Import *Layout* to Visual Basic Editor. Run the module. The module adds the title, subtitle, legend, north arrow, and scale bar to the layout. Because the map and the map elements are graphic elements, they can be reduced, enlarged, and moved on the page layout if necessary.

```
Public Sub Start()
  ' Set the variables and run the subs.
  Dim pMxDoc As IMxDocument
  Dim pPageLayout As IPageLayout
  Dim pGraphicsContainer As IGraphicsContainer
  Dim pActiveView As IActiveView
  Dim pMapFrame As IMapFrame
  Dim pElement As IElement
  ' Set the layout view.
  Set pMxDoc = Application.Document
  Set pPageLayout = pMxDoc.PageLayout
  Set pGraphicsContainer = pPageLayout
  Set pActiveView = pPageLayout
  ' Set the map frame.
  Set pMapFrame = pGraphicsContainer.FindFrame _
  (pMxDoc.FocusMap)
  Set pElement = pMapFrame
  ' Call subs to add map elements.
  Call AddTitleSubtitle
  Call AddLegend(pElement)
```

```
    Call AddNorthArrow(pElement)
    Call AddScalebar(pElement)
    ' Refresh the layout.
    pActiveView.PartialRefresh esriViewGraphics, _
    Nothing, Nothing
End Sub
```

Start defines the page layout and runs the subs to add the title, subtitle, legend, north arrow, and scale bar to the layout. The code sets *pPageLayout* to be the page layout of the map document, and sets both *pGraphicsContainer* and *pActiveView* to be the same as *pPageLayout*. Next, **Start** sets *pMapFrame* to be the map frame for the focus map of the map document. *IMapFrame* provides access to the map within the frame and has methods to create map surrounds, such as the legend, north arrow, and scale bar that are associated with the map. Therefore, *pElement*, which is set to be *pMapFrame*, must be passed as an argument to the subs that add the legend, north arrow, and scale bar to the layout. After all elements are added to the layout, the code refreshes the layout view.

```
    Private Sub AddTitleSubtitle()
      ' Part 1: Set the page layout.
      Dim pMxDoc As IMxDocument
      Dim pPageLayout As IPageLayout
      Dim pGraphicsContainer As IGraphicsContainer
      Set pMxDoc = Application.Document
      Set pPageLayout = pMxDoc.PageLayout
      Set pGraphicsContainer = pPageLayout
```

Part 1 of **AddTitleSubtitle** defines the page layout.

```
      ' Part 2: Add the title.
      Dim pTextElement As ITextElement
      Dim pTextSymbol As IFormattedTextSymbol
      Dim pTextFont As IFontDisp
      Dim pElement As IElement
      Dim pPoint As IPoint
      ' Define the text font.
      Set pTextFont = New StdFont
      With pTextFont
        .Name = "Times New Roman"
        .Size = 24
        .Bold = True
      End With
```

```
' Define the text symbol.
Set pTextSymbol = New TextSymbol
With pTextSymbol
  .Font = pTextFont
  .Case = esriTCAllCaps
  .HorizontalAlignment = esriTHACenter
End With
' Define the title as a text element.
Set pTextElement = New TextElement
pTextElement.Text = "Idaho 1990-2000"
pTextElement.Symbol = pTextSymbol
' Define the position to plot the title.
Set pElement = pTextElement
Set pPoint = New Point
pPoint.X = 5#
pPoint.Y = 10#
pElement.Geometry = pPoint
' Add the title to the graphics container.
pGraphicsContainer.AddElement pTextElement, 0
```

Part 2 of ***AddTitleSubtitle*** adds a title to the layout. The code performs two tasks: define the title as a text element and locate the title on the layout. The definition of a text element includes a text symbol, the definition of which in turn includes a text font (Figure 8.15). Therefore the code first creates *pTextFont* as an instance of the *StdFont* class and defines its properties of name, size, and boldness in a *With* block. Next the code creates *pTextSymbol* as an instance of the *TextSymbol* class and defines its properties of font, case, and horizontal alignment. To complete the first task of defining the title, the code creates *pTextElement* as an instance of the *TextElement* class and defines its properties of text and symbol. The code starts the second task by creating *pPoint* as an instance of the *Point* class and defines the point's *X* and *Y* properties. Both *X* and *Y* values are measured in inches, with the origin at the lower left corner of the page layout. The code then switches to the *IElement* interface and

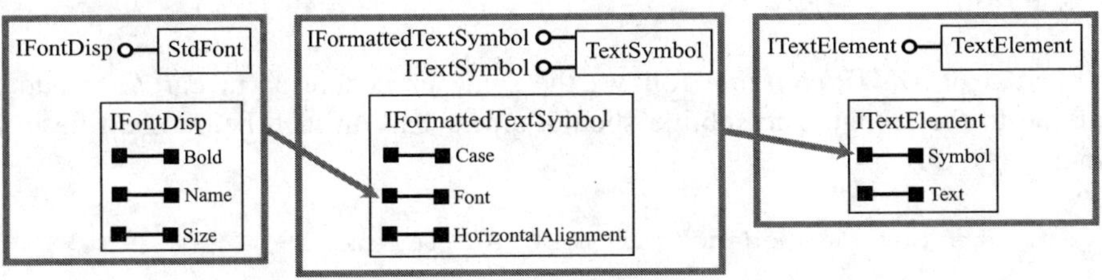

Figure 8.15   The diagram shows text-related objects and their properties for adding the title and subtitle to the layout. *IFontDisp* defines the font for the text symbol, and *IFormattedTextSymbol* defines the symbol for the text element.

assigns *pPoint* to be the geometry of *pTextElement*. Because the horizontal alignment of the text symbol is center justified (i.e., esriTHACenter), *pPoint* is at the center point of the title. After the definition is complete, the code uses the *AddElement* method on *IGraphicsContainer* to add *pTextElement* to the graphics container.

```vba
' Part 3: Add the subtitle.
' Define the text font.
Set pTextFont = New StdFont
With pTextFont
  .Name = "Times New Roman"
  .Size = 14
End With
' Define the text symbol.
Set pTextSymbol = New TextSymbol
With pTextSymbol
  .Font = pTextFont
  .Case = esriTCAllCaps
  .HorizontalAlignment = esriTHACenter
End With
' Define the subtitle as a text element.
Set pTextElement = New TextElement
pTextElement.Text = "Rate of Population Change by County"
pTextElement.Symbol = pTextSymbol
' Define the position to plot the subtitle.
Set pElement = pTextElement
Set pPoint = New Point
pPoint.X = 5#
pPoint.Y = 9.5
pElement.Geometry = pPoint
' Add the subtitle to the graphic container.
pGraphicsContainer.AddElement pTextElement, 0
End Sub
```

Part 3 of **AddTitleSubtitle** follows the same procedure as in Part 2 to add a subtitle to the layout. The subtitle should appear half an inch below the title in a smaller text size.

```vba
Private Sub AddLegend(pElement As IElement)
' Part 1: Define the page layout.
Dim pMxDoc As IMxDocument
Dim pPageLayout As IPageLayout
```

```
Dim pGraphicsContainer As IGraphicsContainer
Dim pActiveView As IActiveView
Set pMxDoc = Application.Document
Set pPageLayout = pMxDoc.PageLayout
Set pGraphicsContainer = pPageLayout
Set pActiveView = pPageLayout
```

Part 1 of *AddLegend* defines the page layout.

```
' Part 2: Create a legend map surround frame.
Dim pMapFrame As IMapFrame
Dim pID As New UID
Dim pMapSurroundFrame As IMapSurroundFrame
Dim pMapSurround As IMapSurround
' Get the map frame.
Set pMapFrame = pElement
' Create a legend map surround frame.
pID.Value = "esriCore.Legend"
Set pMapSurroundFrame = pMapFrame. _
CreateSurroundFrame (pID, pMapSurround)
```

Part 2 of *AddLegend* creates a map surround frame for the legend. The code
first sets *pMapFrame* to be *pElement*, an argument passed from *Layout*. Next the
code sets the value of *pID*, a new unique identifier (*UID*) object, to be "esriCore.Leg-
end." Then the code uses the *CreateSurroundFrame* method on *IMapFrame* to create
a legend within a frame. The legend is referenced by *pMapSurround*, and the frame
is referenced by *pMapSurroundFrame*.

```
' Part 3: Create the legend and add the legend to
' the graphics container.
Dim pFrameElement As IElement
Dim pEnvelope As IEnvelope
' Define the geometry of the legend.
Set pEnvelope = New Envelope
pEnvelope.PutCoords 5.5, 7, 6.5, 8.4
Set pFrameElement = pMapSurroundFrame
pFrameElement.Geometry = pEnvelope
' Activate the screen display of the legend.
pFrameElement.Activate pActiveView.ScreenDisplay
' Add the legend to the graphics container.
pGraphicsContainer.AddElement pFrameElement, 0
End Sub
```

Part 3 of **AddLegend** creates the legend on the layout. The code first creates *pEnvelope* as an instance of the *Envelope* class and uses the *PutCoords* method to define the envelope's xmin, ymin, xmax, and ymax values. An envelope object represents a rectangular shape, and in this case, it controls the size and position of the legend. Next the code switches to the *IElement* interface to assign *pEnvelope* to be the geometry of the map surround frame. Finally, the code activates the screen display of the legend frame and adds the frame to the graphics container.

```
Private Sub AddNorthArrow(pElement As IElement)
  ' Part 1: Define the page layout.
  Dim pMxDoc As IMxDocument
  Dim pPageLayout As IPageLayout
  Dim pGraphicsContainer As IGraphicsContainer
  Dim pActiveView As IActiveView
  Set pMxDoc = Application.Document
  Set pPageLayout = pMxDoc.PageLayout
  Set pGraphicsContainer = pPageLayout
  Set pActiveView = pPageLayout
```

Part 1 of **AddNorthArrow** defines the page layout.

```
  ' Part 2: Create a north arrow map surround frame.
  Dim pMapFrame As IMapFrame
  Dim pID As New UID
  Dim pMapSurroundFrame As IMapSurroundFrame
  Dim pFrameElement As IElement
  Dim pMapSurround As IMapSurround
  Dim pMarkerNorthArrow As IMarkerNorthArrow
  Dim pCharacterMarkerSymbol As ICharacterMarkerSymbol
  ' Get the map frame.
  Set pMapFrame = pElement
  ' Create a north arrow map surround frame.
  pID.Value = "esriCore.MarkerNorthArrow"
  ' Choose a north arrow design other than the default.
  Set pMarkerNorthArrow = New MarkerNorthArrow
  Set pCharacterMarkerSymbol = pMarkerNorthArrow. _
  MarkerSymbol
  pCharacterMarkerSymbol.CharacterIndex = 176
  pMarkerNorthArrow.MarkerSymbol = pCharacterMarkerSymbol
  Set pMapSurround = pMarkerNorthArrow
  Set pMapSurroundFrame = pMapFrame. _
  CreateSurroundFrame(pID, pMapSurround)
```

Part 2 of ***AddNorthArrow*** creates a map surround frame for the north arrow. The code first sets *pMapFrame* to be *pElement,* an argument passed from ***Layout***. Next the code sets the value of *pID* to be "esriCore.MarkerNorthArrow." ArcObjects offers a wide variety of north arrow objects. The default option is a rather fancy north arrow. But one can use additional line statements to specify for a simpler north arrow. The code creates *pMarkerNorthArrow* as an instance of the *MarkerNorth-Arrow* class, and initially sets *pCharacterMarkerSymbol* to be its marker symbol. Then the code specifies *pCharacterMarkerSymbol* to be the character at index 176 and assigns the symbol to be the marker symbol for *pMarkerNorthArrow.* In this case, the north arrow, a simpler symbol than the default, is a character marker symbol at the index of 176. Finally, the code uses the chosen symbol as an object qualifier to create a north arrow map surround frame. (To see other character marker symbols for the north arrow, do the following in the Layout View of ArcMap: select North Arrow from the Insert menu, click Properties in the North Arrow Selector dialog, and then click the Character dropdown arrow in the North Arrow dialog. The index value of the character symbol shows up in the ToolTip message box.)

```
' Part 3: Create the north arrow and add it to
' the graphics container.
Dim pEnvelope As IEnvelope
' Create a envelope for the north arrow.
Set pEnvelope = New Envelope
pEnvelope.PutCoords 5.7, 6.2, 5.9, 6.4
Set pFrameElement = pMapSurroundFrame
pFrameElement.Geometry = pEnvelope
pFrameElement.Activate pActiveView.ScreenDisplay
' Add the north arrow to the graphics container.
pGraphicsContainer.AddElement pFrameElement, 0
End Sub
```

Part 3 of ***AddNorthArrow*** creates a new envelope and assigns the envelope to be the geometry of the map surround frame for the north arrow. Then the code activates the screen display of the frame element and adds the element to the graphics container.

```
Private Sub AddScalebar(pElement As IElement)
  ' Part 1: Define the page layout.
  Dim pMxDoc As IMxDocument
  Dim pPageLayout As IPageLayout
  Dim pGraphicsContainer As IGraphicsContainer
  Dim pActiveView As IActiveView
  Set pMxDoc = Application.Document
  Set pPageLayout = pMxDoc.PageLayout
```

```
      Set pGraphicsContainer = pPageLayout
      Set pActiveView = pPageLayout
```

Part 1 of ***AddScaleBar*** defines the page layout.

```
      ' Part 2: Create a scalebar map surround frame.
      Dim pMapFrame As IMapFrame
      Dim pID As New UID
      Dim pMapSurroundFrame As IMapSurroundFrame
      Dim pFrameElement As IElement
      Dim pMapSurround As IMapSurround
      ' Get the map frame.
      Set pMapFrame = pElement
      ' Create a scale bar map surround frame.
      pID.Value = "esriCore.Scalebar"
      Set pMapSurround = New AlternatingScaleBar
      Set pMapSurroundFrame = pMapFrame. _
      CreateSurroundFrame(pID, pMapSurround)
```

Part 2 of ***AddScalebar*** creates a scale bar map surround frame. The code first sets *pMapFrame* to be *pElement*, an argument passed from ***Layout***. Next the code defines the value of *pID* to be "esriCore.Scalebar." Then the code creates *pMap-Surround* as an instance of the *AlternatingScaleBar* class and uses this scale bar option to create a map surround frame. An *AlternatingScaleBar* object uses two symbols such as black and white to create a scale bar.

```
      ' Part 3: Create the scalebar and add it to the
      ' graphics container.
      Dim pEnvelope As IEnvelope
      ' Create a envelope for the scale bar.
      Set pEnvelope = New Envelope
      pEnvelope.PutCoords 5.2, 5.4, 7.2, 6#
      Set pFrameElement = pMapSurroundFrame
      pFrameElement.Geometry = pEnvelope
      pFrameElement.Activate pActiveView.ScreenDisplay
      ' Add the scale bar to the graphics container.
      pGraphicsContainer.AddElement pFrameElement, 0
   End Sub
```

Part 3 of *AddScalebar* creates a new envelope and assigns the envelope to be the geometry of the map surround frame for the scale bar. The code then activates the screen display of the frame element and adds the element to the graphics container.

CHAPTER 9

Data Exploration

Data exploration refers to activities that involve data-centered query and analysis. For some users of geographic information systems (GIS), data exploration serves as a starting point in formulating research questions and hypotheses. For others, data exploration allows them to examine the general trends in the data, to take a close look at data subsets, and to focus on possible relationships between datasets.

Data exploration in ArcGIS can be based on an attribute query or a spatial query. An attribute query uses an SQL (Structured Query Language) statement and attribute data to select records from a table. An attribute query can be performed on a feature class or an attribute table. A spatial query uses a cursor, a graphic object, or a spatial relationship to select features. The use of a cursor or a graphic object (e.g., a circle) for selecting features is straightforward. The use of a spatial relationship, on the other hand, requires a statement that can link features to select and features to be selected spatially. Common spatial relationships are containment, intersect, and proximity. The result from an attribute or spatial query is a data subset, which can be highlighted in a table, a feature layer, or both.

A data exploration task may use both attribute and spatial queries. This happens when the selection of a data subset involves both attribute and spatial conditions. One of the sample macros in this chapter selects thermal wells and springs that are within a certain distance of an interstate highway (a spatial condition) and have water temperatures above a certain reading (an attribute condition). The macro must therefore perform both attribute and spatial queries.

Data exploration in GIS may also involve derivation of descriptive statistics such as minimum, maximum, range, mean, and standard deviation from numeric data. These descriptive statistics are derived on a field using either all records or a data subset in a table.

This chapter covers data exploration. Section 9.1 reviews data exploration options in ArcGIS. Section 9.2 discusses objects that are important to data exploration. Section 9.3 offers macros for attribute query. Section 9.4 offers macros for spatial query. Section

9.5 includes macros for combining attribute and spatial queries. Section 9.6 has macros for deriving and reporting descriptive statistics. All macros start with the listing of key interfaces and key members (i.e., properties and methods) and the usage.

9.1 EXPLORING DATA IN ARCGIS

ArcMap has the Select by Attributes command for attribute queries. SQL is the query language. The basic syntax of an SQL statement is:

select <attribute list>
from <table>
where <condition>

The *select* keyword selects field(s) from a database, the *from* keyword selects table(s) from a database, and the *where* keyword specifies the condition for data query. SQL is integrated with the user interface in ArcMap; the table to be selected from is the activated table, and fields are available in a dropdown list. The user only has to prepare the *where* clause (i.e., the query expression) in the Select by Attributes dialog.

ArcMap has Select Features, Select by Graphics, and Select by Location for spatial queries. These three commands correspond to selecting features by using a cursor, a graphic, and a spatial relationship, respectively. Select by Location is the only command that uses a dialog. The dialog requires the user to specify one or more layers whose features will be selected and a layer whose features will be used for selection. The dialog offers 11 spatial relationships to connect features to be selected and features used for selection. These relationships are: "are completely within," "completely contain," "have their center in," "contain," "are contained by," "intersect," "are crossed by the outline of," "are within a distance of," "share a line segment with," "touch the boundary of," and "are identical to." The spatial concepts of containment, intersect, and proximity are the basis for these relationships.

Both Select by Attributes and Select by Location offer the selection methods of "create a new selection," "add to current selection," "remove from current selection," and "select from current selection." Therefore, after selecting a feature subset, we may add features to, remove features from, or select features from the subset or we may select a new subset.

There are two ways to derive descriptive statistics on a field in ArcMap. The context menu of a field in a feature attribute table has the Statistics command. When selected, the command displays the field's statistics using all or selected records. The Statistics command is also available in the Selection menu. This command derives statistics for selected features.

9.2 ARCOBJECTS FOR DATA EXPLORATION

The primary components for data query in ArcObjects are *QueryFilter* and *SpatialFilter* (Figure 9.1). A query filter object can retrieve a data subset from a

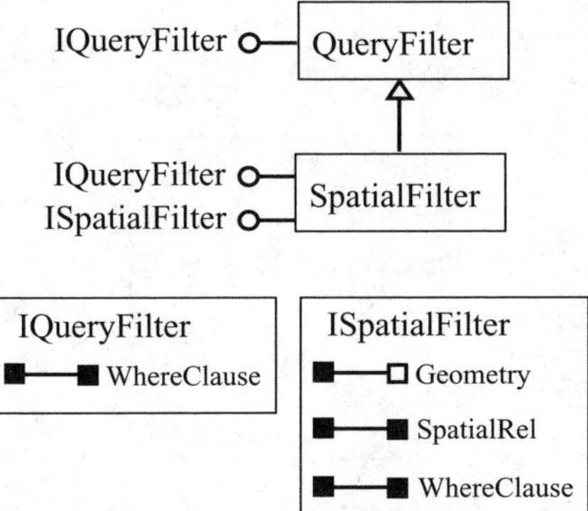

Figure 9.1   *SpatialFilter* is a type of *QueryFilter*. Notice that a spatial filter object has its own properties in addition to the *WhereClause* property that it inherits from *QueryFilter*.

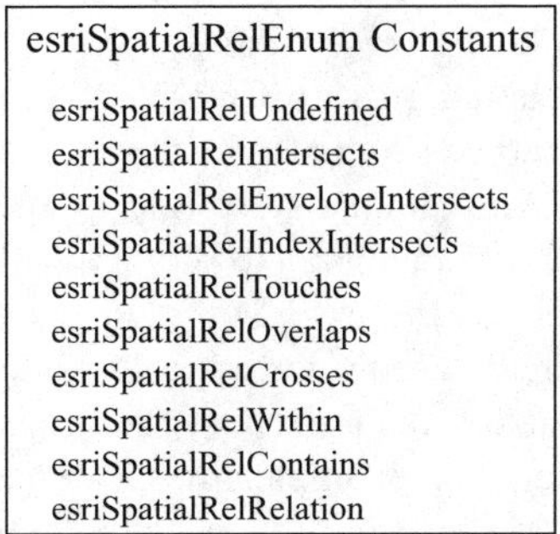

Figure 9.2   Predefined spatial relationships in ArcObjects

feature layer, a feature class, or a table by using the condition expressed in the *WhereClause* property.

The *SpatialFilter* class is a type of the *QueryFilter* class. This means that a spatial filter object can include both spatial and attribute constraints in performing a data query. Two important properties of a spatial filter object are *Geometry* and *SpatialRel*. The *Geometry* property defines the query geometry to filter results. The query geometry may be the shape of a selected polygon in a feature layer, a rectangle based on the extent drawn by the user, or a buffer zone around a highway segment. The *SpatialRel* property defines the spatial relationship for filtering. ArcObjects has 10 predefined spatial relationships (Figure 9.2).

9.2.1   Use of a Query Filter

A query filter object can be used on a feature layer or a feature class. When used on a feature layer, a query filter produces a set of selected features for visual inspection. A *FeatureLayer* object implements *IFeatureSelection*, and the *Select-Features* method on *IFeatureSelection* can select features by using a query filter

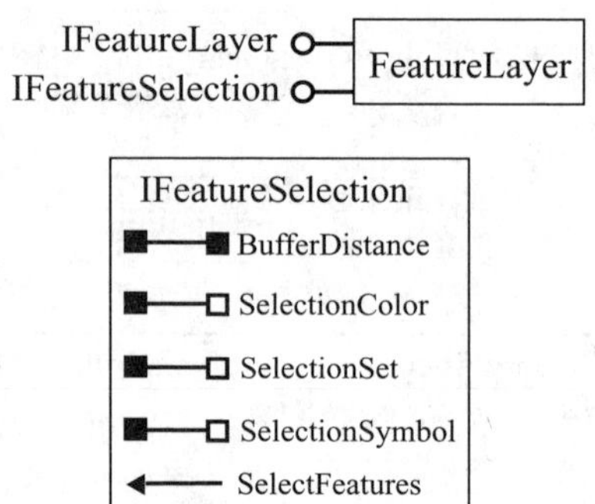

Figure 9.3   Properties and methods on *IFeatureSelection*.

(Figure 9.3). *IFeatureSelection* also has properties to set the color and symbol for displaying the selected features.

When used on a feature class, a query filter produces a data subset for data manipulation or presentation. Figure 9.4 shows that a *SelectionSet* object can be created from a table object and a query filter object. A selection set object supports *ISelectonSet*, which has methods to manage the selection set such as making it permanent or combining it with another selection set. A *TableWindow* object implements *ITableWindow*, which has members to set up a table for data presentation and to highlight the selected subset (Figure 9.5).

Because the relationship between selected features and a selection set is like that between a feature layer and its feature class, there are ways to connect selected features and a selection set. For example, the *SelectionSet* method on *IFeature-Selection* produces a selection set from the selected features.

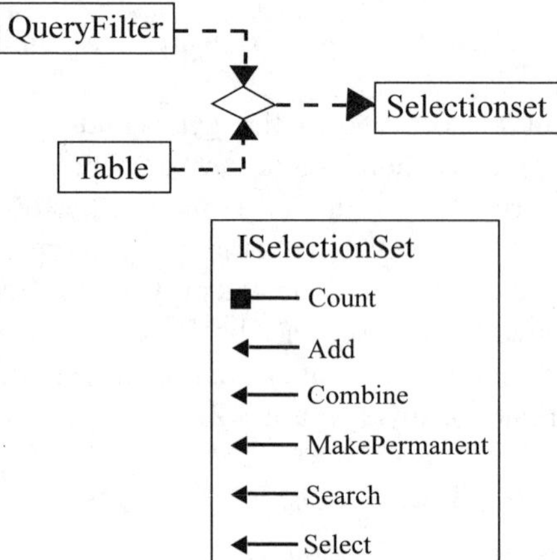

Figure 9.4   A *QueryFilter* object and a *Table* object together can create a *SelectionSet* object. *ISelectionSet* has members to manage the selection set.

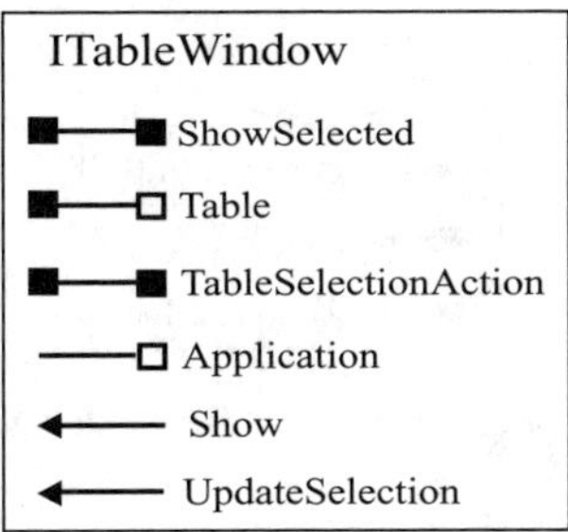

Figure 9.5 *ITableWindow* has members to set up a table for view.

9.2.2 Cursor

A *cursor* is a data-access object that allows the programmer to step through each row for such purposes as counting and editing. Figure 9.6 shows two ways for creating a cursor: combining a query filter and a table (or a feature class), or combining a query filter and a selection set.

A cursor object created from a feature class or a feature layer is a feature cursor. A feature cursor object supports *IFeatureCursor* (Figure 9.7). Two important methods on *IFeatureCursor* are *NextFeature* and *UpdateFeature*. The *NextFeature* method advances the position of the cursor by one and returns the feature at that position. The method is therefore useful for setting up a loop to step through each feature in the cursor. The *UpdateFeature* method can update the feature corresponding to the current position of the cursor.

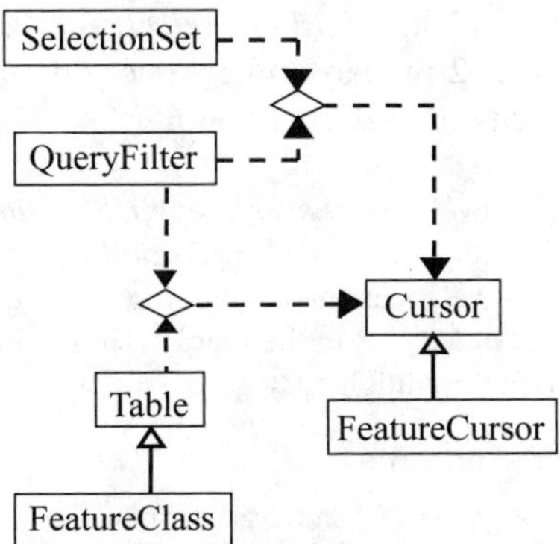

Figure 9.6 The diagram shows two ways for creating a *Cursor* object.

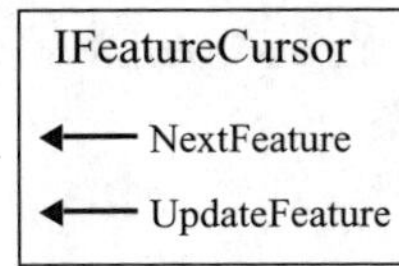

Figure 9.7 Methods on *IFeatureCursor*.

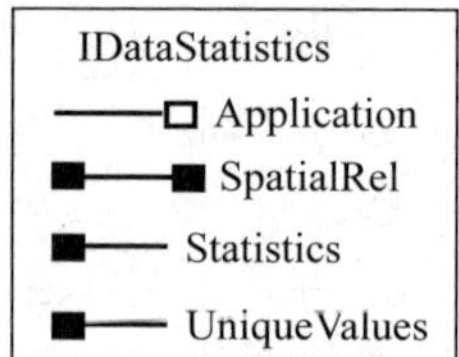

Figure 9.8   *IDataStatistics* has properties for accessing statistics and unique values on a field.

9.2.3 Data Statistics

The primary component for deriving descriptive statistics in ArcObjects is *DataStatistics*. A data statistics object implements *IDataStatistics*, which has properties that let the user set a field to gather statistics on and a cursor to use either all records or just the selected records (Figure 9.8).

9.3 PERFORMING ATTRIBUTE QUERY

This section includes three sample macros for an attribute query from a feature layer and from a table.

9.3.1 *SelectFeatures*

SelectFeatures selects from a feature layer features that meet a query expression and highlights the selected features. The macro performs the same function as the Select by Attributes command in ArcMap. *SelectFeatures* has two parts. Part 1 defines the feature layer. Part 2 prepares a new query filter, uses the query filter to select features, and highlights the selected features.

> **Key Interfaces:** *IActiveView, IFeatureLayer, IFeatureSelection, IQueryFilter.*
> **Key Members:** *WhereClause, SelectFeatures, PartialRefresh.*
> **Usage:** Add *idcities.shp* to the focus map as the top layer. Import *SelectFeatures* to Visual Basic Editor in ArcMap. Run the macro. Those cities that have population > 10,000 are selected and highlighted.

```vba
Private Sub SelectFeatures()
  ' Part 1: Define the feature layer.
  Dim pDoc As IMxDocument
  Dim pMap As IMap
  Dim pActiveView As IActiveView
  Dim pFeatureLayer As IFeatureLayer
  Set pDoc = ThisDocument
  Set pMap = pDoc.FocusMap
  Set pActiveView = pMap
  Set pFeatureLayer = pMap.Layer(0)
```

Part 1 sets *pFeatureLayer* to be the top layer in the active map.

```
 ' Part 2: Select features.
 Dim pQueryFilter As IQueryFilter
 Dim pFeatureSelection As IFeatureSelection
 ' Prepare a query filter.
 Set pQueryFilter = New QueryFilter
 pQueryFilter.WhereClause = "Population > 10000"
 ' Refresh the old selection if any to erase it.
 pActiveView.PartialRefresh esriViewGeoSelection, _
 Nothing, Nothing
 ' Select features.
 Set pFeatureSelection = pFeatureLayer
 pFeatureSelection.SelectFeatures pQueryFilter, _
 esriSelectionResultNew, False
 ' Refresh again to draw the new selection.
 pActiveView.PartialRefresh esriViewGeoSelection, _
 Nothing, Nothing
End Sub
```

Part 2 first creates *pQueryFilter* as an instance of the *QueryFilter* class and defines its *WhereClause* property. Next the code refreshes the active view to erase any previous selection. Then the code performs a QueryInterface (QI) for *IFeature-Selection* and uses the *SelectFeatures* method to create a new selection of features. Finally, the code refreshes the active view and draws the new selection. The *esriView-GeoSelection* option limits the refresh to the selected features only.

9.3.2 *SelectRecords*

SelectRecords selects records from the attribute table of a shapefile (i.e., a feature class), displays the selected records in a table window, and displays the feature layer with selected features highlighted. *SelectRecords* has four parts. Part 1 defines the feature class. Part 2 prepares a query filter and uses it to select records that meet the query expression. Part 3 prepares a table window, shows the table, and highlights the selected records. Part 4 displays a new feature layer based on the feature class and highlights features of the selected records.

> **Key Interfaces:** *IWorkspaceName, IDatasetName, IName, ITable, IQueryFilter, IScratchWorkspaceFactory, ISelectionSet, ITableWindow, IFeatureLayer, IFeature-Selection, IActiveView.*
>
> **Key Members:** *WorkspaceFactoryProgID, PathName, WorkspaceName, Name, Open, WhereClause, Select, Table, TableSelectionAction, UpdateSelection, Application, Show, FeatureClass, PartialRefresh, SelectionSet.*

Usage: Import *SelectRecords* to Visual Basic Editor in ArcMap. Run the macro. The macro opens the attributes of *idcounty* table, highlights the selected records, adds *idcounty* to the active map, and highlights features of the selected records.

```
Private Sub SelectRecords()
  ' Part 1: Define the table.
  Dim pWSName As IWorkspaceName
  Dim pDatasetName As IDatasetName
  Dim pName As IName
  Dim pTable As ITable
  ' Get the dbf file.
  Set pWSName = New WorkspaceName
  pWSName.WorkspaceFactoryProgID = _
  "esriCore.Shapefile-WorkspaceFactory.1"
  pWSName.PathName = "c:\data\chap9"
  Set pDatasetName = New TableName
  pDatasetName.Name = "idcounty.dbf"
  Set pDatasetName.WorkspaceName = pWSName
  Set pName = pDatasetName
  ' Open the dbf table.
  Set pTable = pName.Open
```

Part 1 creates *pDatasetName* as an instance of the *TableName* class and defines its workspace and name. Next the code switches to the *IName* interface and uses the *Open* method to open *pTable*, a reference to the actual table.

```
  ' Part 2: Perform selection.
  Dim pQFilter As IQueryFilter
  Dim pScratchWS As IWorkspace
  Dim pScratchWSFactory As IScratchWorkspaceFactory
  Dim pSelectionSet As ISelectionSet
  ' Prepare a query filter.
  Set pQFilter = New QueryFilter
  pQFilter.WhereClause = "Change > 30"
  ' Select records and save the selection in a
  ' scratch workspace.
  Set pScratchWSFactory = New ScratchWorkspaceFactory
  Set pScratchWS = pScratchWSFactory. _
  DefaultScratchWorkspace
  Set pSelectionSet = pTable.Select(pQFilter, _
  esriSelectionTypeHybrid, esriSelectionOptionNormal, _
  pScratchWS)
```

Part 2 creates *pQFilter* as an instance of the *QueryFilter* class and defines its *WhereClause* property. Next the code creates a scratch workspace referenced by *pScratchWS*. The code then uses the *Select* method on *ITable* to create a selection set referenced by *pSelectionSet*. The *Select* method requires two predefined selection options in addition to the object qualifiers of *pQFilter* and *pScratchWS*.

```
' Part 3: Show the selection in a table window.
Dim pTableWindow As ITableWindow
' Create a table window and specify its properties
' and methods.
Set pTableWindow = New TableWindow
With pTableWindow
  Set.Table = pTable
  .TableSelectionAction = esriSelectFeatures
  .ShowSelected = False
  Set.Application = Application
  .Show True
  .UpdateSelection pSelectionSet
End With
```

Part 3 creates *pTableWindow* as an instance of the *TableWindow* class and defines its display properties as follows: *pTable* for *Table*, esriSelectFeatures for *Table-SelectionAction*, False for *ShowSelected*, Application for *Application*, True for *Show*, and *pSelectionSet* for *UpdateSelection*. False for *ShowSelected* means that all records are displayed. True for *ShowSelected*, on the other hand, means that only selected records are displayed. Application for *Application* is an important property because it provides *pTableWindow* with the reference to the application (i.e., ArcMap). If the property is not specified, the macro will crash.

```
' Part 4: Show selected features in the feature layer.
  Dim pMxDoc As IMxDocument
  Dim pMap As IMap
  Dim pFeatureLayer As IFeatureLayer
  Dim pFeatSelection As IFeatureSelection
  Dim pActiveView As IActiveView
  Set pMxDoc = ThisDocument
  Set pMap = pMxDoc.FocusMap
  ' Use the dBASE table to create a new feature layer.
  Set pFeatureLayer = New FeatureLayer
  Set pFeatureLayer.FeatureClass = pTable
  pFeatureLayer.Name = "idcounty"
```

```
' Add the new layer to the active map.
pMap.AddLayer pFeatureLayer
' Refresh the view to draw selected features.
Set pActiveView = pMap
Set pFeatSelection = pFeatureLayer
Set pFeatSelection.SelectionSet = pSelectionSet
pActiveView.PartialRefresh esriViewGeoSelection, _
Nothing, Nothing
End Sub
```

Part 4 creates *pFeatureLayer* as an instance of the *FeatureLayer* class, sets *pTable* to be its feature class, and sets idcounty to be its name. Next the code adds *pFeatureLayer* to the active map. Finally, the code refreshes the view to draw the selected features in *pSelectionSet.*

9.4 PERFORMING SPATIAL QUERY

A VBA (Visual Basic for Applications) macro for a spatial query requires the use of a spatial filter object. A spatial filter in turn requires a query object whose geometry will be used for selecting features and a spatial relationship by which features of a target layer will be selected. Because a spatial filter object can have both spatial and attribute constraints, it is more versatile than a query filter object for custom applications.

9.4.1 *SpatialQuery*

SpatialQuery uses the shape of a preselected county and the spatial relationship of containment to select cities that fall within the county. The macro performs the same function as using the Select By Location command in ArcMap to select features from the city layer that are completely within the selected feature in the county layer.

SpatialQuery has three parts. Part 1 defines the selected feature on the layer for selection (i.e., the county layer). Part 2 sets up a spatial filter. Part 3 selects features and uses a cursor to count the number of selected features (i.e., selected cities).

> **Key Interfaces:** *IFeatureSelection, ISelectionSet, IFeatureCursor, IFeature, ISpatial-Filter, IFeatureClass.*
> **Key Members:** *SelectionSet, Search, NextFeature, Geometry, Shape, SpatialRel, FeatureClass.*
> **Usage:** Add *idcounty.shp* and *idcities.shp* to the active map. *idcities* must be on top of *idcounty.* Use the Select Features tool to select a county. Import **SpatialQuery** to Visual Basic Editor in ArcMap. Run the macro. A message box reports the number of cities selected.

```
Private Sub SpatialQuery()
' Part 1: Define the selected feature on the layer
' for selection.
```

```
Dim pMxDoc As IMxDocument
Dim pCountyLayer As IFeatureSelection
Dim pCountySelection As ISelectionSet
Dim pCountyCursor As IFeatureCursor
Dim pCounty As IFeature
Set pMxDoc = ThisDocument
Set pCountyLayer = pMxDoc.FocusMap.Layer(1)
Set pCountySelection = pCountyLayer.SelectionSet
' Create a cursor from the selected feature in Layer (1).
pCountySelection.Search Nothing, True, pCountyCursor
' Return the selected feature from the cursor.
Set pCounty = pCountyCursor.NextFeature
' Exit the sub if no feature has been selected.
If pCounty Is Nothing Then
  MsgBox "Please select a county"
  Exit Sub
End If
```

Part 1 locates and verifies the selected feature on the layer for selection. The code sets *pCountyLayer* to be the second layer in the active map and *pCounty-Selection* to be the selection set of the layer. Next the code uses the *Search* method on *ISelectionSet* to create a feature cursor referenced by *pCountyCursor*. Because the macro is designed to work with a selected county, the feature cursor should contain a single county. The *NextFeature* method on *IFeatureCursor* returns the county and assigns it to *pCounty*. If *pCountyCursor* contains no selected feature, the macro terminates.

```
' Part 2: Prepare a spatial filter.
Dim pSpatialFilter As ISpatialFilter
Set pSpatialFilter = New SpatialFilter
Set pSpatialFilter.Geometry = pCounty.Shape
pSpatialFilter.SpatialRel = esriSpatialRelContains
```

Part 2 creates *pSpatialFilter* as an instance of the *SpatialFilter* class and defines its query geometry and spatial relationship. In this case, the query geometry is the shape of *pCounty* and the spatial relationship is esriSpatialRelContains, which stipulates that the query geometry contains the target geometry (i.e., cities).

```
' Part 3: Select features and report number of
' selected features.
Dim pCityLayer As IFeatureLayer
Dim pCityFClass As IFeatureClass
```

```
    Dim pCityCursor As IFeatureCursor
    Dim pCity As IFeature
    Dim intCount As Integer
    Set pCityLayer = pMxDoc.FocusMap.Layer(0)
    ' Select features from Layer (0) and save them to
    ' a cursor.
    Set pCityFClass = pCityLayer.FeatureClass
    Set pCityCursor = pCityFClass.Search _
    (pSpatialFilter, False)
    ' Loop through the cursor and count number of
    ' selected features.
    Set pCity = pCityCursor.NextFeature
    Do Until pCity Is Nothing
      intCount = intCount + 1
      Set pCity = pCityCursor.NextFeature
    Loop
    MsgBox "This county has " & intCount & " cities"
  End Sub
```

Part 3 sets *pCityLayer* to be the top layer in the active map and *pCityFClass* to be its feature class. Next the code uses the *Search* method on *IFeatureClass* to create a feature cursor referenced by *pCityCursor*. (An alternative is to use the *Search* method on *IFeatureLayer* to create the cursor.) The code then uses a *Do…Loop* to step through each feature in *pCityCursor* and counts the number of features with the *intCount* variable. Finally, the code reports *intCount* in a message box.

9.4.2 *SpatialQueryByName*

Instead of using a preselected county, **SpatialQueryByName** uses an input box to get the name of a county from the user. The code then selects and highlights the county before selecting cities that are contained within the county. The only difference between **SpatialQuery** and **SpatialQueryByName** is Part 1. Parts 2 and 3 are the same and are not listed below. (SpatialQueryByName.txt on the companion CD has all three parts.)

To use **SpatialQueryByName**, first add *idcounty.shp* and *idcities.shp* to the active map. *idcities* must be on top of *idcounty*. Import **SpatialQueryByName** to Visual Basic Editor in ArcMap. Run the macro. Enter the name of a county such as Ada. (The evaluation of the county name is case sensitive.) A message box reports the number of cities selected within Ada County.

```
  Private Sub SpatialQueryByName()
    ' Part 1: Define the selected feature on the layer
    ' for selection.
    Dim pMxDoc As IMxDocument
```

```
Dim pActiveview As IActiveview
Dim pCountyLayer As IFeatureLayer
Dim pFeatureSelection As IFeatureSelection
Dim name As String
Dim pQueryFilter As IQueryFilter
Dim pCountySelection As ISelectionSet
Dim pCountyCursor As IFeatureCursor
Dim pCounty As IFeature
Set pMxDoc = ThisDocument
Set pActiveview = pMxDoc.FocusMap
Set pCountyLayer = pMxDoc.FocusMap.Layer(1)
' Get a county name from the user.
name = InputBox("Enter a county name: ", "")
' Prepare a new query filter.
Set pQueryFilter = New QueryFilter
pQueryFilter.WhereClause = "CO_NAME = '" & name & "'"
' Refresh or erase any previous selection.
pActiveview.PartialRefresh esriviewGeoSelection, _
Nothing, Nothing
' Select features.
Set pFeatureSelection = pCountyLayer
' Refresh again to draw the new selection.
pFeatureSelection.SelectFeatures pQueryFilter, _
esriSelectionResultNew, False
pActiveview.PartialRefresh esriviewGeoSelection, _
Nothing, Nothing
' Create a selection set from the selected feature.
Set pCountySelection = pFeatureSelection.SelectionSet
' Create a cursor from the selection set.
pCountySelection.Search Nothing, True, pCountyCursor
' Return the selected feature.
Set pCounty = pCountyCursor.NextFeature
```

Part 1 uses an input box to get a county name from the user and assigns it to the string variable *name*. Next the code switches to the *IFeatureSelection* interface and uses the *SelectFeatures* method to select the county. The code then highlights the selected county in the map. The rest of Part 1 processes the selected county before using it for selecting cities. The code sets *pCountySelection* to be the selection set of *pFeatureSelection*. Then the code creates a feature cursor and uses the *Next-Feature* method to return the selected county.

9.4.3 *MultipleSpatialQueries*

MultipleSpatialQueries selects cities that are within two or more selected counties. The code uses a *Do...Loop* statement to iterate through each county and adds selected cities to those that are already in a cursor. *MultipleSpatialQueries* has three parts. Part 1 defines the layer for selection (i.e., county layer) and the selected features on the layer. Part 2 steps through each selected county, uses a spatial filter to select cities that are within the county, and adds the selected cities to a cursor. Part 3 draws all selected features including counties and cities and reports the number of selected cities.

> **Key Interfaces:** *IFeatureSelection, ISelectionSet, IFeatureCursor, IFeature, ISpatial-Filter, IFeatureSelection, IActiveView.*
> **Key Members:** *SelectionSet, Search, Geometry, SpatialRel, SelectFeatures, NextFeature.*
> **Usage:** Add *idcounty.shp* and *idcities.shp* to the active map. *idcities* must be on top of *idcounty.* Use the Select Features tool to select two or more counties. Import *MultipleSpatialQueries* to Visual Basic Editor in ArcMap. Run the macro. The macro draws all selected counties and cities, and a message box reports the number of cities selected.

```
Private Sub MultipleSpatialQueries()
  ' Part 1: Define the selected counties.
  Dim pMxDoc As IMxDocument
  Dim pCountyLayer As IFeatureSelection
  Dim pCountySelection As ISelectionSet
  Dim pCountyCursor As IFeatureCursor
  Dim pCounty As IFeature
  Set pMxDoc = ThisDocument
  Set pCountyLayer = pMxDoc.FocusMap.Layer(1)
  Set pCountySelection = pCountyLayer.SelectionSet
  ' Create a cursor from the selected counties.
  pCountySelection.Search Nothing, True, pCountyCursor
```

Part 1 defines *pCountyLayer* to be the second layer in the active map and creates *pCountyCursor* as a feature cursor that contains the selection set in *pCountyLayer.*

```
  ' Part 2: Loop through each selected county to
  ' select cities.
  Dim intCount As Integer
  Dim pCityLayer As IFeatureLayer
  Dim pSpatialFilter As ISpatialFilter
  Dim pCitySelection As IFeatureSelection
  intCount = 0
```

```
Set pCityLayer = pMxDoc.FocusMap.Layer(0)
Set pCitySelection = pCityLayer
' Step through each county to select cities.
Set pCounty = pCountyCursor.NextFeature
Do Until pCounty Is Nothing
  ' Prepare a spatial filter.
  Set pSpatialFilter = New SpatialFilter
  Set pSpatialFilter.Geometry = pCounty.Shape
  pSpatialFilter.SpatialRel = esriSpatialRelContains
  ' Select cities and add them to those already selected.
  pCitySelection.SelectFeatures pSpatialFilter, _
  esriSelectionResultAdd, False
  Set pCounty = pCountyCursor.NextFeature
Loop
```

Part 2 first sets *pCityLayer* to be the top layer in the active map. Next the code uses a *Do...Loop* to step through each selected feature in *pCountyCursor.* Within the loop, the code prepares a spatial filter and defines its properties of geometry and spatial relationship. The code then uses the *SelectFeatures* method on *IFeature-Selection* to select cities that are within the county. The argument for the selection method is esriSelectionResultAdd, which adds selected features to those already selected.

```
' Part 3: Draw all selected features and report number
' of selected cities.
Dim pActiveView As IActiveView
Dim pSelCity As ISelectionSet
Dim pCityCursor As IFeatureCursor
Dim pCity As IFeature
Set pActiveView = pMxDoc.FocusMap
' Refresh selected features.
pActiveView.PartialRefresh esriviewGeoSelection, _
Nothing, Nothing
' Prepare a cursor for selected cities.
Set pSelCity = pCitySelection.SelectionSet
pSelCity.Search Nothing, True, pCityCursor
' Count number of cities in the cursor.
Set pCity = pCityCursor.NextFeature
Do Until pCity Is Nothing
  intCount = intCount + 1
  Set pCity = pCityCursor.NextFeature
```

```
Loop
 MsgBox "These counties have " & intCount & " cities."
End Sub
```

Part 3 draws all selected features by using the *PartialRefresh* method on *IActive-View*. The code then creates a cursor for the selection set of *pCitySelection* from Part 2 and uses a *Do...Loop* to count the number of selected cities in the cursor.

9.4.4 *SelectByShape*

SelectByShape uses a rectangle element drawn by the user as the query geometry to select cities that meet an attribute query expression. The macro performs the same function as using the Select Features tool, followed by Select by Attributes, in ArcMap to select features. *SelectByShape* is organized into three parts. Part 1 uses the extent drawn by the user to prepare a rectangle element. Part 2 prepares a fill symbol, shown only with an outline in red, for the rectangle. Part 3 prepares a spatial filter, selects cities within the rectangle, and counts the number of cities selected.

Key Interfaces: *IGraphicsContainer, IEnvelope, IRubberBand, IElement, IFillShapeElement, IFillSymbol, IColor, ILineSymbol, ISpatialFilter, IFeature-Cursor, IFeature.*

Key Members: *TrackNew, Geometry, Symbol, Color, Outline, Transparency, Width, AddElement, WhereClause, SpatialRel, Search, NextFeature.*

Usage: Add *idcities.shp* to the active map. Follow the steps below to attach *Select-ByShape* to a tool so that the user can use the tool to draw a rectangle and to run the macro.

   a. Double-click the empty space to the right of a toolbar in ArcMap to open the Customize dialog.
   b. Click New in the Customize dialog. In the New Toolbar dialog, use the dropdown menu to select Untitled for Save in. Click OK. Custom Toolbar 1 is added as a new toolbar to ArcMap.
   c. Click the Commands tab. Select UIControls from the list of categories. Click New UIControl to open its dialog. Check the option for UIToolControl and then Create. Drag Project UIToolControl1 from the Commands list to the new toolbar.
   d. Right-click UIToolControl1 and, if desired, select Change Button Image to change the tool icon. Right-click UIToolControl1 again and select View Source to open Visual Basic Editor. Make sure that the object list on the upper left of the Code window shows UIToolControl1 and the procedure list on the upper right shows MouseDown rather than the default option of Select. Copy and paste *SelectByShape* to the Code window. Close Visual Basic Editor.
   e. Click on UIToolControl1 and draw a rectangle on *idcities*. The rectangle is shown in a red outline symbol, and a message box reports how many cities meet the selection criteria.
   f. The rectangle is a graphic element, which can be deleted from the display by first using the Select Elements tool to select it.

```
Private Sub UIToolControl1_MouseDown(ByVal button As _
Long, ByVal shift As Long, ByVal x As Long, _
ByVal y As Long)
  ' Part 1: Get a rectangle element drawn by the user.
 Dim pEnv As IEnvelope
 Dim pRubberEnv As IRubberBand
 Dim pMxDoc As IMxDocument
 Dim pElem As IElement
 ' Create a new rubber envelope.
 Set pRubberEnv = New RubberEnvelope
 ' Return a new envelope from the tracker object.
 Set pMxDoc = ThisDocument
 Set pEnv = pRubberEnv.TrackNew _
 (pMxDoc.ActiveView.ScreenDisplay, Nothing)
 ' Create a new envelope element.
 Set pElem = New RectangleElement
 pElem.Geometry = pEnv
```

Part 1 makes a rectangle element based on the extent drawn by the user. The code first creates *pRubberEnv* as an instance of the *RubberEnvelope* class, which implements *IRubberBand* (Figure 9.9). Next the code uses the *TrackNew* method on *IRubberBand* to track (rubberband) the shape on the screen and assigns the shape to an *IEnvelope* object referenced by *pEnv*. The code then creates *pElem* as an instance of the *RectangleElement* class and assigns *pEnv* to be its geometry (Figure 9.10).

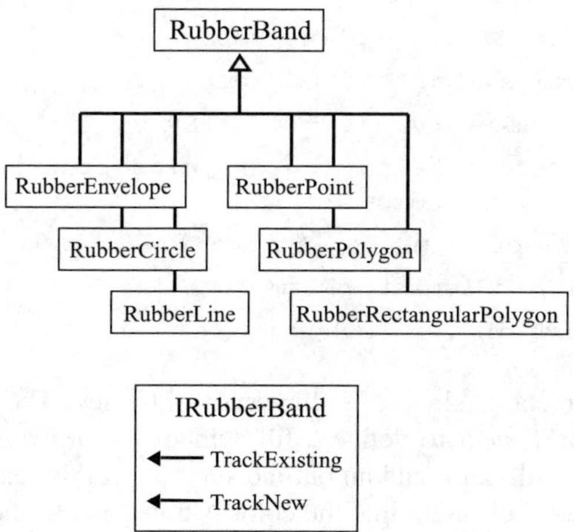

Figure 9.9 *RubberEnvelope* is one of six types of *RubberBand*. A rubber envelope object shares the methods on *IRubberBand*.

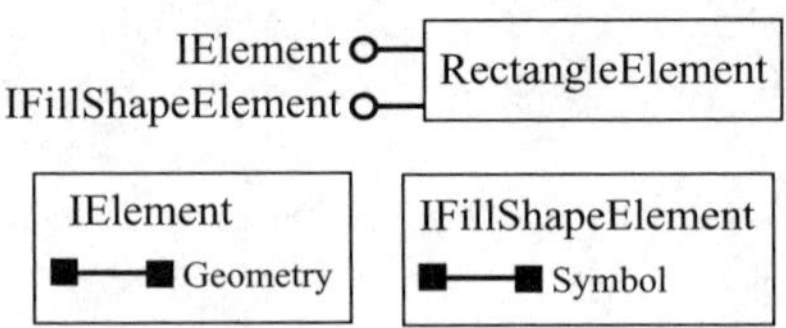

Figure 9.10   A *RectangleElement* object supports *IElement* and *IFillShapeElement*. These two interfaces provide access to the geometry and symbol of the object.

```
' Part 2: Add the rectangle with a red outline to view.
Dim pFillShapeElement As IFillShapeElement
Dim pFillSymbol As IFillSymbol
Dim pColor As IColor
Dim pLineSymbol As ILineSymbol
Dim pGContainer As IGraphicsContainer
Set pFillShapeElement = pElem
' Set the fill symbol.
Set pFillSymbol = pFillShapeElement.Symbol
Set pColor = pFillSymbol.Color
pColor.Transparency = 0
pFillSymbol.Color = pColor
' Set the outline symbol.
Set pLineSymbol = pFillSymbol.Outline
pColor.Transparency = 255
pColor.RGB = RGB(255, 0, 0)
pLineSymbol.Width = 0.1
pFillSymbol.Outline = pLineSymbol
' Assign the symbol to the rectangle.
pFillShapeElement.Symbol = pFillSymbol
' Add the rectangle to the graphics container,
' and refresh the display.
Set pGContainer = pMxDoc.ActiveView
pGContainer.AddElement pElem, 0
pMxDoc.ActiveView.Refresh
```

Part 2 adds the rectangle in a red outline symbol to view. The code first performs a QI for *IFillShapeElement* to define a fill symbol for *pElem* (Figure 9.10). The symbol consists of a fill color and an outline symbol. In this case, the fill color has a transparency of zero, meaning that the color is transparent. The outline symbol is a thin line symbol in solid red. The *RGB* property of *pColor* defines the color of the outline symbol, which is derived by using Visual Basic's RGB function for defining red, green, and blue. The *Width* property of *pLineSymbol* defines the width of the outline symbol. After the fill symbol is defined, the code assigns *pFillSymbol* to be the symbol of *pElem*. Finally, the code adds *pElem* to the graphics container, which is set to be the active view of the document, and refreshes the view.

```
' Part 3: Use the rectangle to search features in
' the top layer.
Dim pSpatialFilter As ISpatialFilter
Dim pCityLayer As IFeatureLayer
Dim pCityCursor As IFeatureCursor
Dim pCity As IFeature
Dim intCount As Integer
' Prepare a spatial filter.
Set pSpatialFilter = New SpatialFilter
pSpatialFilter.WhereClause = "Population > 5000"
Set pSpatialFilter.Geometry = pEnv
pSpatialFilter.SpatialRel = esriSpatialRelContains
' Create a cursor by searching the top layer's
' feature class.
Set pCityLayer = pMxDoc.FocusMap.Layer(0)
Set pCityCursor = pCityLayer.Search _
(pSpatialFilter, False)
' Count number of features in the cursor.
Set pCity = pCityCursor.NextFeature
Do Until pCity Is Nothing
  intCount = intCount + 1
  Set pCity = pCityCursor.NextFeature
Loop
MsgBox "There are " & intCount & _
" cities over 5000 within the rectangular area."
End Sub
```

Part 3 creates *pSpatialFilter* as an instance of the *SpatialFilter* class and defines its properties of *WhereClause*, geometry, and spatial relationship. Next the code sets *pCityLayer* to be the top layer in the active map and uses the *Search* method on *IFeatureLayer* to create a feature cursor. Finally, the code steps through each feature in the cursor and reports the number of selected cities in a message box.

9.5 COMBINING SPATIAL AND ATTRIBUTE QUERIES

Because a spatial filter object can use both spatial and attribute constraints, it seems redundant to have a separate section on combining spatial and attribute queries. In many scenarios, however, a macro needs to perform an attribute query or a spatial query or both more than once; this creates a more challenging task to solve than the sample macros covered so far in this chapter.

9.5.1 *BufferSelect*

BufferSelect selects thermal springs and wells that have water temperatures higher than 60°C and are within 8000 meters of an interstate in Idaho. The macro has three parts. Part 1 defines a road layer, selects interstates by attributes, and saves the interstates into a feature cursor. Part 2 loops through each interstate in the cursor, buffers the interstate with a distance of 8000 meters, and uses a spatial filter object to select thermal springs and wells. Part 3 draws all selected features and reports the total number of thermal springs and wells selected.

BufferSelect uses *ITopologicalOperator* to create the buffer zone around the interstates. Implemented by a point, polyline, or polygon object, *ITopological-Operator* has methods such as buffer, clip, intersect, and union to create new geometries (Figure 9.11).

Key Interfaces: *IFeatureSelection, IQueryFilter, ISelectionSet, IFeatureCursor, ISpatialFilter, IElement, IFeature, ITopologicalOperator, IActiveView.*

Key Members: *WhereClause, SelectFeatures, BufferDistance, SelectionSet, Search, NextFeature, Shape, Buffer, Geometry, SpatialRel, PartialRefresh.*

Usage: Add *idroads.shp* and *thermal.shp* to the active map. *idroads* must be on top of *thermal*. Import *BufferSelect* to Visual Basic Editor in ArcMap. Run the macro. A message box reports the number of thermal springs and wells that meet the selection criteria. The interstates and the selected thermal springs and wells are highlighted on the map.

```vba
Private Sub BufferSelect()
  ' Part 1: Create a cursor of interstates.
  Dim pMxDoc As IMxDocument
  Dim pMap As IMap
  Dim pRoadLayer As IFeatureLayer
  Dim pFeatSelection As IFeatureSelection
  Dim pQueryFilter As IQueryFilter
  Dim pRoadSelSet As ISelectionSet
  Dim pRoadCursor As IFeatureCursor
```

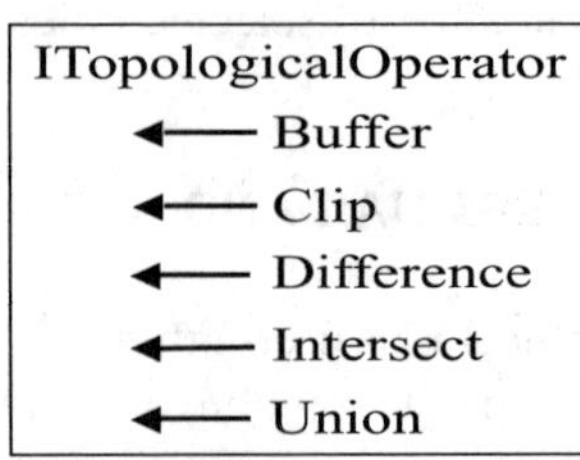

Figure 9.11 Methods on *ITopologicalOperator.*

```
Set pMxDoc = ThisDocument
Set pMap = pMxDoc.FocusMap
Set pRoadLayer = pMap.Layer(0)
Set pFeatSelection = pRoadLayer
' Select interstates.
Set pQueryFilter = New QueryFilter
pQueryFilter.WhereClause = "Route_Desc = 'Interstate'"
pFeatSelection.SelectFeatures pQueryFilter, _
esriSelectionResultNew, False
' Display an 8000-meter buffer around the interstates.
pFeatSelection.BufferDistance = 8000
' Create a feature cursor of selected interstates.
Set pRoadSelSet = pFeatSelection.SelectionSet
pRoadSelSet.Search Nothing, False, pRoadCursor
```

Part 1 first defines *pRoadLayer* as the top layer in the active map. To select the
interstates, the code creates a query filter object, switches to the *IFeatureSelection*
interface, and uses the *SelectFeatures* method to create a feature selection referenced
by *pFeatSelection*. The *BufferDistance* property of *pFeatSelection* is set to display
a buffer distance of 8000 meters. Then the code assigns the selection set of *pFeat-
Selection* to *pRoadSelSet* and creates from *pRoadSelSet* a feature cursor referenced
by *pRoadCursor*. Because the creation of the cursor uses no query filter, all selected
interstates are included in the cursor.

```
' Part 2: Buffer each interstate, and select thermals.
Dim pSpatialFilter As ISpatialFilter
Dim pThermalLayer As IFeatureLayer
Dim pElement As IElement
Dim pThermalSelection As IFeatureSelection
Dim pRoad As IFeature
Dim pTopoOperator As ITopologicalOperator
' Prepare a spatial filter.
Set pSpatialFilter = New SpatialFilter
pSpatialFilter.WhereClause = "temp > 60"
pSpatialFilter.SpatialRel = esriSpatialRelContains
' Define the thermal layer whose features will
' be selected.
Set pThermalLayer = pMxDoc.FocusMap.Layer(1)
Set pThermalSelection = pThermalLayer
' Step through each interstate and select thermals.
Set pRoad = pRoadCursor.NextFeature
```

```
Do Until pRoad Is Nothing
  ' Create an 8000-meter buffer from the road.
  Set pTopoOperator = pRoad.Shape
  Set pElement = New PolygonElement
  pElement.Geometry = pTopoOperator.Buffer(8000)
  ' Define the geometry of the spatial filter.
  Set pSpatialFilter.Geometry = pElement.Geometry
  ' Select thermals and add them to the selection set.
  pThermalSelection.SelectFeatures pSpatialFilter, _
  esriSelectionResultAdd, False
  Set pRoad = pRoadCursor.NextFeature
Loop
```

Part 2 selects thermal springs and wells that are within 8000 meters of an interstate and have water temperatures above 60°C. The code initially creates a spatial filter object referenced by *pSpatialFilter* and defines its attribute constraint and spatial relationship. Next the code sets *pThermalLayer* to be the second layer in the active map and switches to the *IFeatureSelection* interface. Then the code uses a *Do...Loop* to step through each selected interstate. Within the loop, two tasks are performed. First, the code uses the *Buffer* method on *ITopologicalOperator* to create an 8000-meter buffer polygon around the shape of each interstate. This buffer polygon is then assigned to be the geometry of *pSpatialFilter*. Second, the code uses the *SelectFeatures* method on *IFeatureSelection* to create a feature selection. The argument used for the selection method is esriSelectionResultAdd, which adds selected thermal springs and wells to the current selection. The *Do...Loop* continues until all interstates are exhausted.

```
  ' Part 3: Draw all selected features, and report
  ' number of thermals selected.
  Dim pActiveView As IActiveView
  Dim pThermalSelSet As ISelectionSet
  Set pActiveView = pMxDoc.FocusMap
  pActiveView.PartialRefresh esriViewGeoSelection, _
  Nothing, Nothing
  Set pThermalSelSet = pThermalSelection.SelectionSet
  MsgBox "There are " & pThermalSelSet.Count & _
  " thermals selected"
End Sub
```

Part 3 draws all selected interstates and thermal springs and wells. Finally, a message box reports the number of thermal springs and wells that meet the selection criteria.

9.5.2 *IntersectSelect*

IntersectSelect selects counties in Idaho that intersect the interstate highways and have a rate of population increase over 15% during the past decade. Like **BufferSelect**, *IntersectSelect* uses both attribute and spatial queries. One major difference is that *IntersectSelect* applies the intersect relationship in selecting counties.

IntersectSelect has three parts. Part 1 selects interstates from the road layer and creates a feature cursor of the selected features. Part 2 uses a spatial filter object to select counties that intersect an interstate and have a high rate of population growth. Part 3 shows the selected features and reports the number of counties selected.

Key Interfaces: *IFeatureSelection, IQueryFilter, ISelectionSet, IFeatureCursor, ISpatialFilter, IElement, IFeature, IActiveView.*

Key Members: *WhereClause, SelectFeatures, SelectionSet, Search, NextFeature, Shape, Geometry, SpatialRel, PartialRefresh.*

Usage: Add *idroads.shp* and *idcounty.shp* to the active map. *idroads* must be on top of *idcounty.* (If necessary, clear selected features of *idroads.*) Import **IntersectSelect** to Visual Basic Editor in ArcMap. Run the macro. A message box reports the number of counties that meet the selection criteria. The interstates and the selected counties are highlighted on the map.

```
Private Sub IntersectSelect()
  ' Part 1: Create a cursor of interstates.
  Dim pMxDoc As IMxDocument
  Dim pMap As IMap
  Dim pRoadLayer As IFeatureLayer
  Dim pFeatSelection As IFeatureSelection
  Dim pQueryFilter As IQueryFilter
  Dim pRoadSelSet As ISelectionSet
  Dim pRoadCursor As IFeatureCursor
  Set pMxDoc = ThisDocument
  Set pMap = pMxDoc.FocusMap
  Set pRoadLayer = pMap.Layer(0)
  Set pFeatSelection = pRoadLayer
  ' Select interstates.
  Set pQueryFilter = New QueryFilter
  pQueryFilter.WhereClause = "Route_Desc = 'Interstate'"
  pFeatSelection.SelectFeatures pQueryFilter, _
  esriSelectionResultNew, False
  ' Create a feature cursor of selected interstates.
  Set pRoadSelSet = pFeatSelection.SelectionSet
  pRoadSelSet.Search Nothing, False, pRoadCursor
```

Part 1 uses a query filter to select interstates from the road layer and creates a feature cursor from the selected interstates.

```
' Part 2: Select high-growth counties that intersect
' an interstate.
Dim pCountyLayer As IFeatureLayer
Dim pElement As IElement
Dim pCountySelection As IFeatureSelection
Dim pRoad As IFeature
Dim pSpatialFilter As ISpatialFilter
Set pCountyLayer = pMxDoc.FocusMap.Layer(1)
Set pCountySelection = pCountyLayer
' Prepare a spatial filter.
Set pSpatialFilter = New SpatialFilter
pSpatialFilter.WhereClause = "change > 15"
pSpatialFilter.SpatialRel = esriSpatialRelIntersects
' Step through each interstate and select counties.
Set pRoad = pRoadCursor.NextFeature
Do Until pRoad Is Nothing
   ' Define the geometry of the spatial filter.
   Set pSpatialFilter.Geometry = pRoad.Shape
   ' Select counties and add them to the selection set.
   pCountySelection.SelectFeatures pSpatialFilter, _
   esriSelectionResultAdd, False
   Set pRoad = pRoadCursor.NextFeature
Loop
```

Part 2 first creates a spatial filter object and defines its attribute constraint and spatial relationship. Next the code steps through each interstate in a *Do...Loop*. Within the loop, the code defines the query geometry as the shape of the interstate and selects counties that intersect the interstate. As the *Do...Loop* continues, selected counties are added to the current selection set.

```
' Part 3: Draw all selected features and report
' number of counties selected.
Dim pActiveView As IActiveView
Dim pCountySelSet As ISelectionSet
Set pActiveView = pMxDoc.FocusMap
' Draw all selected features.
pActiveView.PartialRefresh esriViewGeoSelection, _
Nothing, Nothing
Set pCountySelSet = pCountySelection.SelectionSet
```

```
    MsgBox "There are " & pCountySelSet.Count & _
    "counties selected"
  End Sub
```

Part 3 draws all selected interstates and counties. Then, a message box reports the number of selected counties.

9.6 DERIVING DESCRIPTIVE STATISTICS

This section discusses how to program ArcObjects to derive descriptive statistics on a field of a feature layer. These statistics can be based on all records or a subset of records.

9.6.1 *DataStatistics*

DataStatistics derives and reports descriptive statistics on a field using all records of a feature layer. The macro performs the same function as using the Statistics command in a field's context menu in ArcMap. *DataStatistics* has three parts. Part 1 defines the feature layer and creates a cursor from the layer. Part 2 creates a data statistics object and uses the object to derive a field's descriptive statistics. Part 3 displays the statistics in a message box.

Key Interfaces: *ICursor, IDataStatistics, IStatisticsResults.*
Key Members: *Search, Field, Cursor, Statistics, Maximum, Minimum, Mean.*
Usage: Add *idcounty.shp* to an active map. Import *DataStatistics* to Visual Basic Editor. Run the macro. The macro uses a message box to display the maximum, minimum, and mean values of the field change.

```
Private Sub DataStatistics()
  ' Part 1: Define the feature layer and cursor.
  Dim pMxDoc As IMxDocument
  Dim pFLayer As IFeatureLayer
  Dim pCursor As ICursor
  Set pMxDoc = ThisDocument
  Set pFLayer = pMxDoc.FocusMap.Layer(0)
  Set pCursor = pFLayer.Search(Nothing, False)
```

Part 1 Sets *pFLayer* to be the top layer in the active map and creates *pCursor* that includes every feature in *pFLayer.*

```
  ' Part 2: Derive statistics on the field change.
  Dim pData As IDataStatistics
  Dim pStatResults As IStatisticsResults
  Dim pChangeMax As Double
```

```
Dim pChangeMin As Double
Dim pChangeMean As Double
' Create a data statistics object.
Set pData = New DataStatistics
pData.Field = "change"
Set pData.Cursor = pCursor
Set pStatResults = pData.Statistics
' Get the maximum, minimum, and mean values.
pChangeMax = pStatResults.Maximum
pChangeMin = pStatResults.Minimum
pChangeMean = pStatResults.Mean
```

Part 2 creates *pData* as an instance of the *DataStatistics* class and defines the following properties for the object: *Field* to be change, *Cursor* to be *pCursor*, and *Statistics* to be *pStatResults*. The code then uses members of *IStatisticsResults* to save the maximum, minimum, and mean statistics into the proper variables.

```
' Part 3: Display the statistics in a message box.
Dim sMsg As String
sMsg = "Statistics of the field change are:" & vbCrLf
sMsg = sMsg & "====================" & vbCrLf
sMsg = sMsg & "Maximum: " & pChangeMax & vbCrLf
sMsg = sMsg & "Minimum: " & pChangeMin & vbCrLf
sMsg = sMsg & "Mean: " & pChangeMean & vbCrLf
sMsg = sMsg & "===================="
MsgBox sMsg
End Sub
```

Part 3 uses a message box to display the statistics derived from Part 2. The constant vbCrLf creates a new line.

9.6.2 *DataSubsetStatistics*

DataSubsetStatistics reports statistics on a set of selected records. The macro performs the same function as the Statistics command on a field of selected records in ArcMap. *DataSubsetStatistics* has three parts. Part 1 defines the feature layer and creates a cursor from the layer. Part 2 creates a data statistics object and uses the object to derive a field's descriptive statistics. Part 3 displays the statistics in a message box.

Key Interfaces: *IQueryFilter, ICursor, IDataStatistics, IStatisticsResults.*
Key Members: *WhereClause, Search, Field, Cursor, Statistics, Maximum, Minimum, Mean.*

Usage: Add *idcounty.shp* to an active map. Import ***DataSubsetStatistics*** to Visual Basic Editor. Run the macro. The macro uses a message box to display the maximum, minimum, and mean values of the field change for those counties that have change > 20.

```
Private Sub DataSubsetStatistics()
  ' Part 1: Get a handle on the feature layer and cursor.
  Dim pMxDoc As IMxDocument
  Dim pFLayer As IFeatureLayer
  Dim pQueryFilter As IQueryFilter
  Dim pCursor As ICursor
  Set pMxDoc = ThisDocument
  Set pFLayer = pMxDoc.FocusMap.Layer(0)
  Set pQueryFilter = New QueryFilter
  pQueryFilter.WhereClause = "change > 20"
  Set pCursor = pFLayer.Search(pQueryFilter, False)
```

Part 1 creates a query filter object and defines its *WhereClause* property. The code then creates a cursor that contains only those counties that have change > 20.

```
  ' Part 2: Derive statistics on the field change.
  Dim pData As IDataStatistics
  Dim pStatResults As IStatisticsResults
  Dim pChangeMax As Double
  Dim pChangeMin As Double
  Dim pChangeMean As Double
  ' Define an IDataStatistics object.
  Set pData = New DataStatistics
  pData.Field = "change"
  Set pData.Cursor = pCursor
  Set pStatResults = pData.Statistics
  ' Get the maximum, minimum, and mean values.
  pChangeMax = pStatResults.Maximum
  pChangeMin = pStatResults.Minimum
  pChangeMean = pStatResults.Mean
```

Part 2 creates a data statistics object and derives its maximum, minimum, and mean.

```
  ' Part 3: Display the statistics in a message box.
  Dim sMsg As String
  sMsg = "Statistics of the field change are:" & vbCrLf
```

```
    sMsg = sMsg & "====================" & vbCrLf
    sMsg = sMsg & "Maximum: " & pChangeMax & vbCrLf
    sMsg = sMsg & "Minimum: " & pChangeMin & vbCrLf
    sMsg = sMsg & "Mean: " & pChangeMean & vbCrLf
    sMsg = sMsg & "===================="
    MsgBox sMsg
End Sub
```

Part 3 reports the statistics in a message box.

Vector Data Operations

Vector data analysis is based on the geometric objects of point, line, and polygon. Vector-based operations typically involve the shape of spatial features. Some operations also involve feature attributes. Four common types of vector data analyses are buffering, overlay, spatial join, and feature manipulation.

Buffering creates a buffer zone that is within a specified distance of each spatial feature in the input layer. If an input layer has 20 line segments, the first step in buffering is to create 20 buffer zones. These buffer zones are often dissolved to remove the overlapped areas between them. An important requirement for buffering is that the input layer must have the spatial reference information that is needed for distance measurements.

Overlay combines the shapes and attributes of two layers to create the output. One of the two layers is the input layer and the other, the overlay layer. Each feature in the output contains a combination of attributes from both layers, and this combination separates the feature from its neighbors. Two common overlay methods are union and intersect. Union preserves all features from both layers, whereas intersect preserves only features that fall within the overlapped area between the layers.

Spatial join joins attribute data from two layers by using a spatial relationship such as nearest neighbor, containment, or intersection. If nearest neighbor is applied, distance measures between features of the two layers can also be included in the output.

Feature manipulation refers to various methods for manipulating features in a layer or between two layers. Unlike overlay, feature manipulation does not combine the shapes and attributes of the two layers into the output. Common manipulations include clip, dissolve, and merge.

This chapter covers vector data operations. Section 10.1 reviews vector data analysis using ArcGIS. Section 10.2 discusses objects that are related to vector data analysis. Section 10.3 offers a macro for buffering. Section 10.4 offers a macro for an overlay operation. Section 10.5 has a macro for a spatial join operation. Section

10.6 includes macros for feature manipulation operations. All macros start with the listing of key interfaces and key members (i.e., properties and methods) and the usage.

10.1 ANALYZING VECTOR DATA IN ARCGIS

Depending on the type of input data, vector data analysis in ArcGIS varies in terms of the variety of operations. ArcToolbox has many more feature-based analysis tools than ArcMap does. However, those tools in ArcToolbox are reserved for coverages rather than shapefiles. Because ArcObjects is the development platform for ArcMap, this section focuses on vector data analysis using ArcMap.

The Buffer Wizard in ArcMap is a tool for buffering. The wizard uses dialogs to get the input layer, buffer distance, and buffering method from the user and to save the output on disk. The Buffer Wizard requires the presence of the spatial reference information of the input layer.

The GeoProcessing Wizard in ArcMap can perform overlay and feature manipulation. The overlay choices are union and intersect, and the feature manipulation tools include dissolve, merge, and clip.

Spatial join in ArcMap joins attributes from two feature attribute tables based on a spatial relationship between features. The spatial relationship can be proximity, so that the joining of attributes takes place between features of a layer and their closest features in another layer. Alternatively, the spatial relationship can be containment, so that the joining of attributes takes place between features of a layer and the features of another layer that contains them. Spatial join can use a maximum search distance as a constraint, and the result of spatial join can include distance measures or aggregate statistics.

10.2 ARCOBJECTS FOR VECTOR DATA ANALYSIS

The primary components for vector data analysis are *FeatureCursorBuffer*, *BasicGeoprocessor*, and *SpatialJoin*. A feature cursor buffer object supports *IFeatureCursorBuffer*, which has different methods and properties for the buffering operation (Figure 10.1). As examples, the *Dissolve* property determines whether overlapping buffered features should be dissolved, the *RingDistance* property sets the number of buffer rings, and the *ValueDistance* property specifies the constant buffer distance. *IFeatureCursorBuffer2* has additional properties that specify the spatial reference systems of the data frame, the source data, the target, and the buffering. A feature cursor buffer object also implements *IBufferProcessingParameter*, which has access to members that set and retrieve parameters for the buffering process.

A basic geoprocessor object implements *IBasicGeoprocessor. IBasicGeoprocessor* offers methods for intersect, union, clip, dissolve, and merge, the same choices as in the GeoProcessing Wizard dialog in ArcMap (Figure 10.2). All methods except for merge allow use of selected features as inputs.

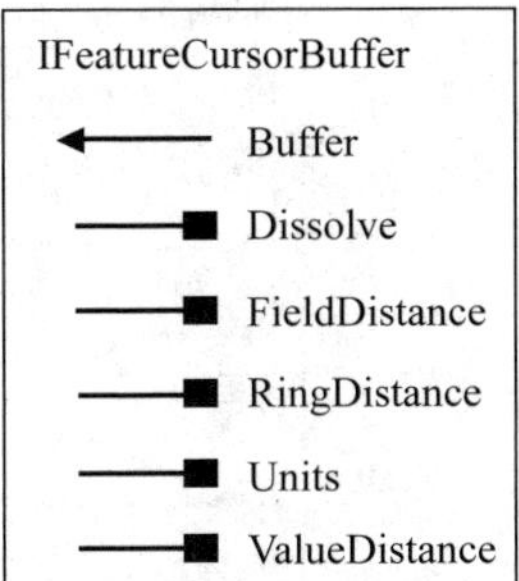

Figure 10.1 *FeatureCursorBuffer* has properties that can define the buffering parameters.

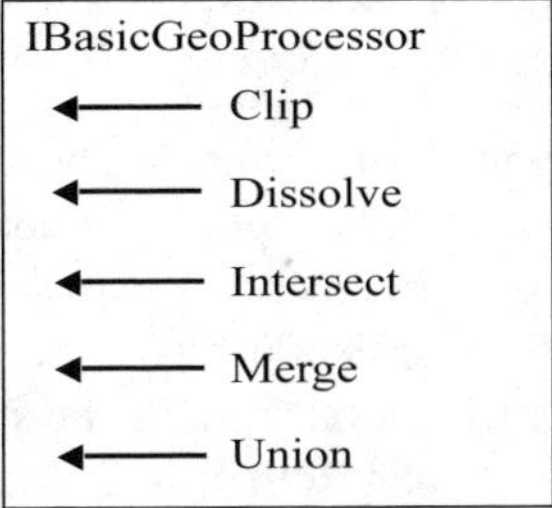

Figure 10.2 *IBasicGeoProcessor* has methods that are the same as in the GeoProcessing Wizard.

A spatial join object supports *ISpatialJoin*, which provides methods for joining attributes of features based on a spatial relationship between features (Figure 10.3). The *JoinAggregate* method joins features using aggregate statistics. The *JoinNearest* method joins features using the nearest relationship. And the *JoinWithin* method joins features using the containment relationship.

Because vector data analysis deals with discrete features of points, lines, and areas, the properties and methods of these discrete features may become important for some applications. One example is the updating of the area and perimeter values of polygons on the output shapefile of an overlay operation. Another example is the derivation of centroids of a polygon feature class. Figure 10.4 summarizes properties and methods on *IPoint*, *ICurve*, and *IArea* that are important for vector data analysis.

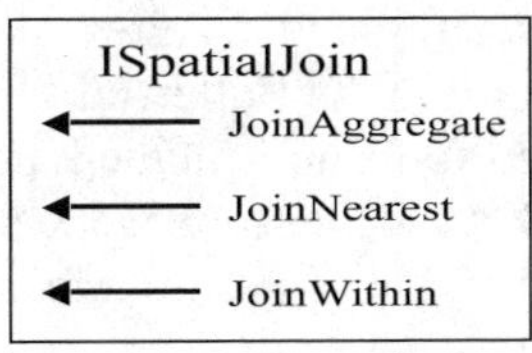

Figure 10.3 Methods on *ISpatialJoin.*

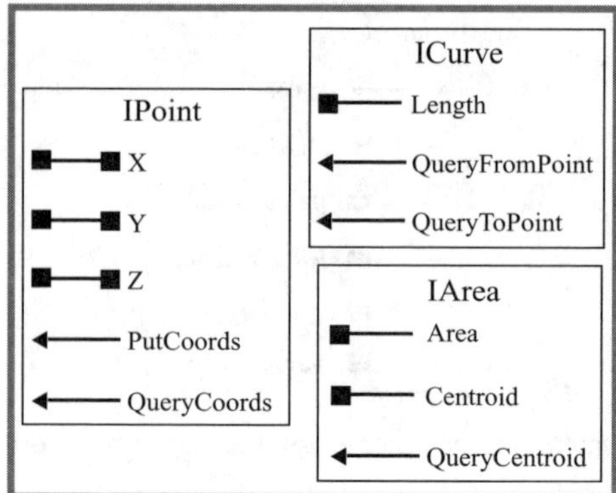

Figure 10.4   *IPoint, ICurve,* and *IArea* have members that can be important for vector data analysis.

When developing VBA (Visual Basic for Applications) macros for vector data analysis, we must be careful to separate vector data analysis from spatial data query (Chapter 9). A vector data analysis produces an output layer, whereas a spatial query produces a data subset that meets the selection criteria. Terms used in vector data analysis also appear for spatial data query, however. Among the spatial relationships that can be queried are constants such as esriSpatialRelIntersects and esriSpatialRelEnvelopeIntersects. For creating geometries for spatial query, the *ITopologicalOperator* interface actually provides methods for buffer, clip, intersect, and union. Therefore, it is important to know the context in which buffer, intersect, or union is used.

10.3 BUFFERING

There are two important considerations in writing VBA macros for buffering. First, because the distance units used in buffering may or may not be the same as the map units of the input layer, a macro must specify the spatial reference of the input layer as well as the output layer. Second, because buffering is performed for each point, each line, or each polygon, a macro must set up a feature cursor for stepping through each feature for buffering.

10.3.1 *Buffer*

Buffer creates a buffer zone around features of a line shapefile. The macro performs the same function as using the Buffer Wizard in ArcMap. *Buffer* has four parts. Part 1 creates a cursor of all features in the input dataset and gets the spatial reference of the active map. Part 2 defines the workspace and dataset name for the output. Part 3 performs buffering. Part 4 creates a feature layer from the output and adds the layer to the active map.

> **Key Interfaces:** *IFeatureClass, IFeatureCursor, ISpatialReference, IFeatureCursorBuffer2, IWorkspaceName, IDatasetName, IFeatureClassName, IName.*

Key Members: *FeatureClass, Search, SpatialReference, FeatureCursor, Dissolve, ValueDistance, BufferSpatialReference, DataFrameSpatialReference, Source-SpatialReference, TargetSpatialReference, WorkspaceFactoryProgID, PathName, Name, WorkspaceName, Buffer.*

Usage: Add *sewers.shp* to an active map. *sewers.shp* is measured in UTM (Universal Transverse Mercator) coordinates and has the spatial reference information. Import **Buffer** to Visual Basic Editor. Run the macro. The macro creates a new shapefile named *Buffer_result* and adds the shapefile as a feature layer to the active map.

```
Private Sub Buffer()
  ' Part 1: Get the map's spatial reference, and
  ' prepare a feature cursor.
  Dim pMxDoc As IMxDocument
  Dim pMap As IMap
  Dim pFeatureLayer As IFeatureLayer
  Dim pFCursor As IFeatureCursor
  Dim pSpatialReference As ISpatialReference
  Set pMxDoc = ThisDocument
  Set pMap = pMxDoc.FocusMap
  ' Set the spatial reference.
  Set pSpatialReference = pMap.SpatialReference
  ' Prepare a feature cursor of all features in
  ' the top layer.
  Set pFeatureLayer = pMap.Layer(0)
  Set pFCursor = pFeatureLayer.Search(Nothing, False)
```

Part 1 sets *pSpatialReference* to be the spatial reference of the active map. Next the code uses the *Search* method on *IFeatureLayer* to create a feature cursor referenced by *pFCursor*. Because the *Search* method does not use a query filter object, *pFCursor* contains all features of the feature layer.

```
  ' Part 2: Define the output.
  Dim pBufWSName As IWorkspaceName
  Dim pBufDatasetName As IDatasetName
  Dim pBufFCName As IFeatureClassName
  ' Define the output's workspace and name.
  Set pBufFCName = New FeatureClassName
  Set pBufDatasetName = pBufFCName
  Set pBufWSName = New WorkspaceName
  pBufWSName.WorkspaceFactoryProgID = _
  "esriCore.ShapeFileWorkspaceFactory.1"
  pBufWSName.PathName = "c:\data\chap10"
  Set pBufDatasetName.WorkspaceName = pBufWSName
  pBufDatasetName.Name = "Buffer_result"
```

Part 2 creates *pBufFCName* as an instance of the *FeatureClassName* class. Then the code performs a QueryInterface (QI) for the *IDatasetName* interface to define the workspace and name of *pBufFCName*.

```
' Part 3: Perform buffering.
Dim pFeatureCursorBuffer2 As IFeatureCursorBuffer2
' Define a feature cursor buffer object.
Set pFeatureCursorBuffer2 = New FeatureCursorBuffer
With pFeatureCursorBuffer2
  Set.FeatureCursor = pFCursor
  .Dissolve = True
  .ValueDistance = 300
  Set.BufferSpatialReference = pSpatialReference
  Set.DataFrameSpatialReference = pSpatialReference
  Set.SourceSpatialReference = pSpatialReference
  Set.TargetSpatialReference = pSpatialReference
End With
' Use the buffer method.
pFeatureCursorBuffer2.buffer pBufFCName
```

Part 3 creates *pFeatureCursorBuffer2* as an instance of the *FeatureCursorBuffer* class and defines its properties in a *With* block. The code specifies *pFCursor* for the feature cursor, True for dissolving overlapped areas, 300 (meters) for the constant buffer distance, and *pSpatialReference* for the spatial reference for buffering, the data frame, the source dataset, and the target dataset. The code then uses the *Buffer* method on *IFeatureCursorBuffer2* to buffer features in *pFCursor* and creates the output referenced by *pBufFCName*.

```
' Part 4: Create the output layer and add it to
' the active map.
Dim pName As IName
Dim pBufFC As IFeatureClass
Dim pBufFL As IFeatureLayer
Set pName = pBufFCName
Set pBufFC = pName.Open
Set pBufFL = New FeatureLayer
Set pBufFL.FeatureClass = pBufFC
pBufFL.Name = "Buffer_Result"
pMap.AddLayer pBufFL
End Sub
```

Part 4 switches to the *IName* interface and uses the *Open* method to open a feature class object referenced by *pBufFC*. The code then creates a feature layer from *pBufFC* and adds the layer to the active map.

10.3.2 Buffer Options

In the previous section, **Buffer** buffers line features with a constant distance. A variety of buffer options are available through the properties of a feature cursor buffer object. The following summarizes these options by referring to statements in **Buffer**:

- To change from a constant buffer distance to varying buffer distances, change the statement, `.ValueDistance = 300,` to `.FieldDistance = "DistanceField",` where DistanceField is the field that specifies the buffer distances.
- To create multiple buffer zones, change the statement, `.ValueDistance = 300,` to `.RingDistance(3) = 100,` where 3 means three rings and 100 means a 100-meter buffer distance for each ring.
- To create undissolved or discrete buffer zones, change the statement, `.Dissolve = True,` to `.Dissolve = False.`
- To use different buffer units (i.e., feet) than map units (i.e., meters), add the statement, `Units(esriMeters) = esriFeet,` as an additional property of *pFeatureCursorBuffer2*.

10.4 PERFORMING OVERLAY

To perform an overlay, a VBA macro must identify the input layer and the overlay layer. Among the methods offered by *IBasicGeoprocessor* are intersect, union, and clip. All three methods have exactly the same syntax. Therefore, the following macro on intersect can also be used for union and clip by simply changing the method on *IBasicGeoprocessor*.

10.4.1 *Intersect*

Intersect uses the intersect method to create an overlay output. The macro performs the same function as using the GeoProcessing/Intersect operation in Arc-Map. *Intersect* has four parts. Part 1 defines the datasets for the intersect operation. Part 2 defines the output including its workspace and name. Part 3 sets up a basic geoprocessor object and runs the intersect method. Part 4 creates a feature layer from the output and adds the layer to the active map.

> **Key Interfaces:** *IFeatureClass, IWorkspaceName, IFeatureClassName, IDataset-Name, IBasicGeoprocessor.*
> **Key Members:** *FeatureClass, WorkspaceFactoryProgID, PathName, Name, WorkspaceName, Intersect.*

Usage: Add *landsoil.shp* and *sewerbuf.shp* to an active map. *landsoil* will be used as the input layer and *sewerbuf* as the overlay layer. *landsoil* must be on top of *sewerbuf* in the table of contents. Import **Intersect** to Visual Basic Editor. Run the macro. The macro creates *Intersect_result.shp* and adds the shapefile as a feature layer to the active map.

```vba
Private Sub Intersect()
    ' Part 1: Define the datasets for intersect.
    Dim pMxDoc As IMxDocument
    Dim pMap As IMap
    Dim pInputLayer As IFeatureLayer
    Dim pOverlayLayer As IFeatureLayer
    Dim pInputFC As IFeatureClass
    Dim pOverlayFC As IFeatureClass
    Set pMxDoc = ThisDocument
    Set pMap = pMxDoc.FocusMap
    ' Define the input feature class.
    Set pInputLayer = pMap.Layer(0)
    Set pInputFC = pInputLayer.FeatureClass
    ' Define the overlay table.
    Set pOverlayLayer = pMap.Layer(1)
    Set pOverlayFC = pOverlayLayer.FeatureClass
```

Part 1 sets the input layer to be the top layer, and the overlay layer the second layer, in the active map. The code then sets *pInputFC* to be the feature class of the input layer and *pOverlayFC* to be the feature class of the overlay layer.

```vba
    ' Part 2: Define the output dataset.
    Dim pNewWSName As IWorkspaceName
    Dim pFeatClassName As IFeatureClassName
    Dim pDatasetName As IDatasetName
    Set pFeatClassName = New FeatureClassName
    Set pDatasetName = pFeatClassName
    Set pNewWSName = New WorkspaceName
    pNewWSName.WorkspaceFactoryProgID = _
    "esriCore.ShapeFileWorkspaceFactory"
    pNewWSName.PathName = "c:\data\chap10"
    Set pDatasetName.WorkspaceName = pNewWSName
    pDatasetName.name = "Intersect_result"
```

Part 2 creates *pFeatClassName* as an instance of the *FeatureClassName* class. The code then uses the *IDatasetName* interface to define the workspace and name of *pFCName*.

```
' Part 3: Perform intersect.
Dim pBGP As IBasicGeoprocessor
Dim tol As Double
Dim pOutputFC As IFeatureClass
' Define a basic geoprocessor object.
Set pBGP = New BasicGeoprocessor
' Use the default tolerance.
tol = 0#
' Run intersect.
Set pOutputFC = pBGP.Intersect(pInputFC, False, _
pOverlayFC, False, tol, pFeatClassName)
```

Part 3 creates *pBGP* as an instance of the *BasicGeoprocessor* class and uses the *Intersect* method on *IBasicGeoprocessor* to create the output referenced by *pOutputFC*. *Intersect* uses six object qualifiers and arguments. The four object qualifiers have been previously defined. The two arguments determine if only subsets of *pInputFC* and/or *pOverlayFC* are used in the overlay operation. If they are set to be false, the overlay operation ignores any selected subsets.

```
' Part 4: Create the output layer and add it to
' the active map.
Dim pOutputFeatLayer As IFeatureLayer
Set pOutputFeatLayer = New FeatureLayer
Set pOutputFeatLayer.FeatureClass = pOutputFC
pOutputFeatLayer.name = pOutputFC.AliasName
pMap.AddLayer pOutputFeatLayer
End Sub
```

Part 4 creates a feature layer from *pOutputFC*, and adds the layer to the active map.

10.4.2 Updating Area and Perimeter of a Shapefile

A basic geoprocessor object does not automatically update the area and perimeter of an output shapefile from an overlay operation. The attribute table of the output from **Intersect** in fact contains two sets of area and perimeter from the input and overlay layers. However, neither set has been updated.

If geodatabase feature classes are used as inputs to an overlay operation, the output feature class will have the fields of shape_area and shape_length with correct area and perimeter values. (Chapter 14 has sample macros that use geodatabase feature classes in overlay operations.) But if shapefiles are used as inputs, the area and perimeter of the output shapefile can be updated in two ways. The first method is to convert the shapefile to a geodatabase feature class by using a conversion macro such as ***ShapefileToAccess*** in Chapter 6. The geodatabase feature class will have the fields of shape_area and shape_length with the updated values. The second method is to add the following code fragment to ***Intersect*** as Part 5. (Intersect_Update.txt on the companion CD is a complete macro. The macro produces *Intersect_result2* with the updated area and perimeter values.)

```
' Part 5: Update area and perimeter of the output shapefile.
  Dim pField1 As IFieldEdit
  Dim pField2 As IFieldEdit
  Dim pUpdateCursor As IFeatureCursor
  Dim intArea As Integer
  Dim intLeng As Integer
  Dim pArea As IArea
  Dim pCurve As ICurve
  Dim pFeature As IFeature
  ' Define the two fields to be added.
  Set pField1 = New Field
  pField1.Name = "Shape_area"
  pField1.Type = esriFieldTypeDouble
  Set pField2 = New Field
  pField2.Name = "Shape_leng"
  pField2.Type = esriFieldTypeDouble
  ' Add the two new fields.
  pOutputFC.AddField pField1
  pOutputFC.AddField pField2
  ' Create a feature cursor to update each feature's
  ' new fields.
  Set pUpdateCursor = pOutputFC.Update(Nothing, False)
  ' Locate the new fields.
  intArea = pUpdateCursor.FindField("Shape_area")
  intLeng = pUpdateCursor.FindField("Shape_leng")
  Set pFeature = pUpdateCursor.NextFeature
  ' Loop through each feature.
  Do Until pFeature Is Nothing
     ' Update the area field.
```

```
    Set pArea = pFeature.Shape
    pFeature.Value(intArea) = pArea.Area
    ' Update the length field.
    Set pCurve = pFeature.Shape
    pFeature.Value(intLeng) = pCurve.Length
    ' Update the feature.
    pUpdateCursor.UpdateFeature pFeature
    Set pFeature = pUpdateCursor.NextFeature
  Loop
```

Part 5 first defines two new fields and adds them to *pOutputFC*. Next the code uses the *Update* method on *IFeatureClass* to create a feature cursor referenced by *pUpdateCursor*, finds the index of the new field, and sets up a *Do...Loop* to step through each feature in *pUpdateCursor*. Within the loop, the Shape_area field is given the value of the area of the feature (i.e., *pArea.Area*) and the Shape_leng is given the value of the perimeter of the feature (i.e., *pCurve.Length*). Then the *UpdateFeature* method on *IFeatureCursor* updates the feature in the database.

When ***Intersect*** is run with the additional Part 5, the attribute table of the output shapefile has the additional fields of Shape_area and Shape_leng. The values in those two fields are the updated values of area and perimeter.

10.5 JOINING DATA BY LOCATION

Join by location, or spatial join, joins attributes of two layers based on a spatial relationship between features. A common application of spatial join is to derive the distance between features using the nearest relationship.

10.5.1 *JoinByLocation*

JoinByLocation joins line features to point features using the nearest relationship. The macro performs the same function as using the option of "Join data from another layer based on spatial location" of Join Data in ArcMap. ***JoinByLocation*** has four parts. Part 1 defines the source table and the join table. Part 2 prepares the name objects of the output's workspace, feature class, and dataset. Part 3 defines a spatial join object and runs the nearest method. Part 4 creates a feature layer from the output and adds the layer to the active map.

Key Interfaces: *IFeatureClass, IWorkspaceName, IFeatureClassName, IDataset-Name, ISpatialJoin.*

Key Members: *FeatureClass, WorkspaceFactoryProgID, PathName, Name, WorkspaceName, ShowProcess, LeftOuterJoin, SourceTable, JoinTable, JoinNearest.*

Usage: Add *deer.shp* and *edge.shp* to an active map. *deer* contains deer sighting locations, and *edge* shows edges between different vegetation covers. In this spatial join operation, *deer* is the source table and *edge* is the join table. *deer* must be on top of *edge* in the table of contents. Import ***JoinByLocation*** to Visual Basic Editor. Run the macro. The macro creates a new layer named *Spatial_Join*. The attribute table of *Spatial_Join* has the same number of records as the *deer* attribute table. But each record of *Spatial_Join* has attributes from *deer* and *edge* as well as a new distance field.

```
Private Sub JoinByLocation()
  ' Part 1: Define the source and join tables.
  Dim pMxDoc As IMxDocument
  Dim pMap As IMap
  Dim pSourceLayer As IFeatureLayer
  Dim pSourceFC As IFeatureClass
  Dim pJoinLayer As IFeatureLayer
  Dim pJoinFC As IFeatureClass
  Set pMxDoc = ThisDocument
  Set pMap = pMxDoc.FocusMap
  ' Define the source feature class.
  Set pSourceLayer = pMap.Layer(0)
  Set pSourceFC = pSourceLayer.FeatureClass
  ' Define the join feature class.
  Set pJoinLayer = pMap.Layer(1)
  Set pJoinFC = pJoinLayer.FeatureClass
```

Part 1 sets *pSourceFC* to be the feature class of the top layer in the active map, and *pJoinFC* to be the feature class of the second layer.

```
  ' Part 2: Define the output dataset.
  Dim pOutWorkspaceName As IWorkspaceName
  Dim pFCName As IFeatureClassName
  Dim pDatasetName As IDatasetName
  Set pFCName = New FeatureClassName
  Set pDatasetName = pFCName
  Set pOutWorkspaceName = New WorkspaceName
  pOutWorkspaceName.WorkspaceFactoryProgID = _
    "esriCore.ShapefileWorkspaceFactory.1"
  pOutWorkspaceName.PathName = "c:\data\chap10"
  pDatasetName.Name = "Spatial_Join"
  Set pDatasetName.WorkspaceName = pOutWorkspaceName
```

Part 2 creates *pFCName* as an instance of the *FeatureClassName* class. The code then performs a QI for the *IDatasetName* interface to define the workspace and name of *pFCName*.

```
' Part 3: Perform spatial join.
Dim pSpatialJoin As ISpatialJoin
Dim pOutputFC As IFeatureClass
Dim maxMapDist As Double
' Create and define a spatial join object.
Set pSpatialJoin = New SpatialJoin
With pSpatialJoin
  .ShowProcess(True) = 0
  .LeftOuterJoin = False
  Set.SourceTable = pSourceFC
  Set.JoinTable = pJoinFC
End With
' Use infinity as the maximum map distance.
maxMapDist = -1
' Run the join nearest method.
Set pOutputFC = pSpatialJoin.JoinNearest(pFCName, _
maxMapDist)
```

Part 3 creates *pSpatialJoin* as an instance of the *SpatialJoin* class and defines its properties in a *With* block. The code specifies True for *ShowProcess* so that a message box will update the processing. The code specifies False for *LeftOuterJoin* so that a record will be added from the source table to the output only if the record has a match in the join table. Then the code uses the *JoinNearest* method on *ISpatialJoin* to create the output referenced by *pOutputFC*. The maximum map distance for the search radius for spatial join is –1, which means infinity.

```
' Part 4: Create the output layer and add it to
' the active map.
Dim pOutputFeatLayer As IFeatureLayer
Set pOutputFeatLayer = New FeatureLayer
Set pOutputFeatLayer.FeatureClass = pOutputFC
pOutputFeatLayer.Name = pOutputFC.AliasName
pMap.AddLayer pOutputFeatLayer
End Sub
```

Part 4 creates a feature layer from *pOutputFC* and adds the layer to the active map.

10.6 MANIPULATING FEATURES

The *IBasicGeoprocessor* interface offers the methods of clip, dissolve, and merge for manipulating spatial features. Clip has already been covered. This section presents dissolve and merge. Additionally, this section includes a macro that derives centroids for each polygon of a shapefile and saves the centroids into a new shapefile. The centroid is the geometric center of a polygon, a common measure that can be used as an input to vector data analysis.

10.6.1 *Dissolve*

Dissolve aggregates features of the input layer based on the values of an attribute. The macro performs the same function as using the GeoProcessing/Dissolve operation in ArcMap. *Dissolve* has four parts. Part 1 defines the input table. Part 2 defines the workspace and name of the output. Part 3 performs the dissolve operation. Part 4 creates a feature layer from the output and adds the layer to the active map.

> **Key Interfaces:** *ITable, IWorkspaceName, IFeatureClassName, IDatasetName, IBasicGeoprocessor.*
> **Key Members:** *WorkspaceFactoryProgID, PathName, Name, WorkspaceName, Dissolve.*
> **Usage:** Add *vulner.shp* to an active map. Open the attribute table of *vulner* and make sure that it contains the fields of Class and Shape_area. Class will be used as the dissolve field and Shape_area, the summary field, in *Dissolve*. Import *Dissolve* to Visual Basic Editor. Run the macro. The macro adds a new layer named *Dissolve_result* to the active map. *Dissolve_result* is a shapefile, which allows a polygon to have multiple components. Therefore, the attribute table of *Dissolve_result* has only five records, one for each class value. The field Shape_area in the attribute table summarizes the values of Shape_area in the input layer for each class value.

```
Private Sub Dissolve()
  ' Part 1: Define the input table.
  Dim pMxDoc As IMxDocument
  Dim pMap As IMap
  Dim pInputFeatLayer As IFeatureLayer
  Dim pInputTable As ITable
  Set pMxDoc = ThisDocument
  Set pMap = pMxDoc.FocusMap
  Set pInputFeatLayer = pMap.Layer(0)
  Set pInputTable = pInputFeatLayer
```

Part 1 sets *pInputFeatLayer* to be the top layer in the active map. The code then switches to the *ITable* interface and sets *pInputTable* to be the same as *pInputFeatLayer*.

```
  ' Part 2: Define the output dataset.
  Dim pNewWSName As IWorkspaceName
  Dim pFCName As IFeatureClassName
  Dim pDatasetName As IDatasetName
  Set pFCName = New FeatureClassName
```

```
Set pDatasetName = pFCName
Set pNewWSName = New WorkspaceName
pNewWSName.WorkspaceFactoryProgID = _
"esriCore.ShapefileWorkspaceFactory.1"
pNewWSName.PathName = "c:\data\chap10"
pDatasetName.Name = "Dissolve_result"
Set pDatasetName.WorkspaceName = pNewWSName
```

Part 2 creates *pFCName* as an instance of the *FeatureClassName* class and switches to the *IDatasetName* interface to define the workspace and name for *pFCName*.

```
' Part 3: Perform dissolve.
Dim pBGP As IBasicGeoprocessor
Dim pOutputFC As IFeatureClass
Set pBGP = New BasicGeoprocessor
' Run dissolve.
Set pOutputFC = pBGP.Dissolve _
(pInputTable, False, "Class", _
"Dissolve.Shape, Sum.Shape_area", pFCName)
```

Part 3 creates *pBGP* as an instance of the *BasicGeoprocessor* class and uses the *Dissolve* method on *IBasicGeoprocessor* to create the output referenced by *pOutputFC. Dissolve* requires three arguments in addition to two object qualifiers already defined. The code specifies False for *useSelected* so that all records in the input table will be dissolved. The code specifies Class for *dissolveField*, or the field to dissolve. The argument *summaryFields* has a different syntax from the others. The first entry in the comma delimited string, Dissolve.Shape, is required if the output is a shapefile (instead of a summary table). The second entry, Sum.Shape_area, specifies that the Shape_area values be summed for each value of the dissolve field. Besides sum, other available statistics are count, minimum, maximum, average, variance, and standard deviation.

```
' Part 4: Create the output feature layer and add
' the layer to the active map.
Dim pOutputFeatLayer As IFeatureLayer
Set pOutputFeatLayer = New FeatureLayer
Set pOutputFeatLayer.FeatureClass = pOutputFC
pOutputFeatLayer.Name = pOutputFC.AliasName
pMap.AddLayer pOutputFeatLayer
End Sub
```

Part 4 creates a feature layer from *pOutputFC* and adds the layer to the active map.

10.6.2 *Merge*

Merge combines two line shapefiles into a single shapefile. The macro performs the same function as using the GeoProcessing/Merge operation in ArcMap. *Merge* has four parts. Part 1 defines the two tables to be merged. Part 2 defines the output's workspace and name. Part 3 sets up a basic geoprocessor object and runs the merge method. Part 4 creates a feature layer from the output and adds the layer to the active map.

> **Key Interfaces:** *ITable, IWorkspaceName, IFeatureClassName, IDatasetName, IBasicGeoprocessor, IArray, IFeatureClass.*
>
> **Key Members:** *WorkspaceFactoryProgID, PathName, Name, WorkspaceName, Array, Add, Merge, FeatureClass.*
>
> **Usage:** Add *mtroads_idtm.shp* and *idroads.shp* to an active map. *mtroads_idtm* is a major road shapefile of Montana, and *idroads* is a major road shapefile of Idaho. Both shapefiles are projected onto the Idaho transverse Mercator (IDTM) coordinate system. Because only one of the shapefiles can define the attributes for the output, *Merge* will use *mtroads_idtm*'s attributes for the output. Make sure that *mtroads_idtm* is on top of *idroads* in the table of contents. Import *Merge* to Visual Basic Editor. Run the macro. The macro creates *Merge_result.shp* and adds the shapefile to the active map.

```
Private Sub Merge()
  ' Part 1: Define the two input tables.
  Dim pMxDoc As IMxDocument
  Dim pMap As IMap
  Dim pFirstLayer As IFeatureLayer
  Dim pFirstTable As ITable
  Dim pSecondLayer As IFeatureLayer
  Dim pSecondTable As ITable
  Set pMxDoc = ThisDocument
  Set pMap = pMxDoc.FocusMap
  ' Define the first table.
  Set pFirstLayer = pMap.Layer(0)
  Set pFirstTable = pFirstLayer
  ' Define the second table.
  Set pSecondLayer = pMap.Layer(1)
  Set pSecondTable = pSecondLayer
```

Part 1 sets *pFirstLayer* to be the top layer in the active map and sets *pFirstTable* to be the same as *pFirstLayer*. Next the code sets *pSecondLayer* to be the second layer and sets *pSecondTable* to be the same as *pSecondLayer*.

```
  ' Part 2: Define the output dataset.
    Dim pNewWSName As IWorkspaceName
```

```
Dim pFCName As IFeatureClassName
Dim pDatasetName As IDatasetName
Set pFCName = New FeatureClassName
Set pDatasetName = pFCName
Set pNewWSName = New WorkspaceName
pNewWSName.WorkspaceFactoryProgID = _
"esriCore.ShapefileWorkspaceFactory.1"
pNewWSName.PathName = "c:\data\chap10"
Set pDatasetName.WorkspaceName = pNewWSName
pDatasetName.Name = "Merge_result"
```

Part 2 creates *pFCName* as an instance of the *FeatureClassName* class. The code then switches to the *IDatasetName* interface to define the workspace and name for *pFCName*.

```
' Part 3: Perform merge.
Dim pBGP As IBasicGeoprocessor
Dim inputArray As IArray
Dim pOutputFC As IFeatureClass
' Add the two input tables to an array.
Set inputArray = New esriCore.Array
inputArray.Add pFirstTable
inputArray.Add pSecondTable
Set pBGP = New BasicGeoprocessor
Set pOutputFC = pBGP.Merge(inputArray, pFirstTable, _
pFCName)
```

Part 3 first creates *inputArray* as an instance of the *esriCore.Array* class and uses the *Add* method on *IArray* to add the two input tables to the array variable. Next the code creates *pBGP* as an instance of the *BasicGeoprocessor* class and uses the *Merge* method to create the merged output referenced by *pOutputFC*. The object qualifier *pFirstTable* determines that the fields in the output correspond to those in the first input table.

```
' Part 4: Create the output layer and add it to
' the active map.
Dim pOutputFeatLayer As IFeatureLayer
Set pOutputFeatLayer = New FeatureLayer
Set pOutputFeatLayer.FeatureClass = pOutputFC
pOutputFeatLayer.Name = pOutputFC.AliasName
pMxDoc.FocusMap.AddLayer pOutputFeatLayer
End Sub
```

Part 4 creates a feature layer from *pOutputFC* and adds the layer to the active map.

10.6.3 *Centroid*

Centroid derives a centroid for each polygon of a shapefile and creates a new point shapefile that contains the centroids. *Centroid* has three parts. Part 1 defines the input dataset and calls the *CreateNewShapefile* function to create an empty output feature class. Part 2 derives the centroid of each polygon and stores the centroids in the feature class passed from *CreateNewShapefile*. Part 3 creates a feature layer from the output and adds the layer to the active map.

CreateNewShapefile requires the spatial reference of the input dataset as an argument and returns a feature class object to *Centroid*. *CreateNewShapefile* has three parts. Part 1 defines the output's workspace. Part 2 edits and defines a shape field and adds the field to a field collection. Part 3 creates a new feature class and returns the feature class to *Centroid*. Two constants used in *CreateNewShapefile*, one for the output's workspace and the other for the output's name, are declared at the start of the module.

> **Key Interfaces:** *ISpatialReference, IFeatureClass, IFeatureCursor, IArea, IFeature, IPoint, IWorkspaceFactory, IFeatureWorkspace, IFieldsEdit, IFieldEdit, IGeometryDef, IGeometryDefEdit.*
>
> **Key Members:** *SpatialReference, Search, NextFeature, Shape, QueryCentroid, CreateFeature, Store, OpenFromFile, Name, Type, GeometryType, GeometryDef, AddField, CreateFeatureClass.*
>
> **Usage:** Add *idcounty.shp* to an active map. *idcounty* shows 44 counties in Idaho. Import **Centroid** to Visual Basic Editor. Run the module. The module adds a new shapefile named *Centroid* to the active map. The attribute table of *Centroid* has three fields: FID, Shape, and Id. ArcGIS automatically adds Id because the software requires at least one field in addition to the object ID (FID) and the geometry field (Shape). Id has zeroes for all 44 records but, if necessary, the values can be calculated to be FID + 1.

```
Const strFolder As String = "c:\data\chap10"
Const strName As String = "Centroid"
Private Sub Centroid()
    ' Part 1: Define the input dataset, and call a
    ' function to create the output.
  Dim pMxDoc As IMxDocument
  Dim pMap As IMap
  Dim pInputFeatLayer As IFeatureLayer
  Dim pGeoDataset As IGeoDataset
  Dim pSpatialReference As ISpatialReference
  Dim pOutputFeatClass As IFeatureClass
  Set pMxDoc = ThisDocument
  Set pMap = pMxDoc.FocusMap
```

```
' Define the input dataset.
Set pInputFeatLayer = pMap.Layer(0)
Set pGeoDataset = pInputFeatLayer
Set pSpatialReference = pGeoDataset.SpatialReference
' Call the CreateNewShapefile function to create
' the output feature class.
Set pOutputFeatClass = CreateNewShapefile _
(pSpatialReference)
```

Part 1 sets *pInputFeatLayer* to be the top layer in the active map and uses the *IGeoDataset* interface to derive its spatial reference, which is referenced by *pSpatialReference*. The code then calls the **CreateNewShapefile** function, which uses *pSpatialReference* as an argument and returns a feature class object referenced by *pOutputFeatClass*.

```
' Part 2: Derive centroids for each polygon and
' store them.
Dim pFCursor As IFeatureCursor
Dim pInputFeature As IFeature
Dim pCentroidTemp As IPoint
Dim pArea As IArea
Dim pOutputFeature As IFeature
' Set up a cursor for all features.
Set pCentroidTemp = New Point
Set pFCursor = pInputFeatLayer.Search(Nothing, True)
Set pInputFeature = pFCursor.NextFeature
' Loop over each polygon feature.
Do Until pInputFeature Is Nothing
  Set pArea = pInputFeature.Shape
  ' Get the centroid.
  pArea.QueryCentroid pCentroidTemp
  Set pOutputFeature = pOutputFeatClass.CreateFeature
  ' Store the centroid.
  Set pOutputFeature.Shape = pCentroidTemp
  pOutputFeature.Store
  Set pInputFeature = pFCursor.NextFeature
Loop
```

Part 2 derives centroids for each polygon and stores them as features in *pOutputFeatClass*. The code first creates *pCentroidTemp* as an instance of the *Point* class and creates a feature cursor referenced by *pFCursor*. Next the code uses a *Do...Loop* to step through each feature in the cursor, derive its centroid, and store the centroid as a new feature in *pOutputFeatClass*. The method for deriving the centroid is *QueryCentroid* on *IArea*, which is implemented by a polygon object. The method to store the centroid is *Store* on *IFeature*, which is implemented by a feature object.

```vba
' Part 3: Create the output feature layer, and add
' the layer to the active map.
Dim pFeatureLayer As IFeatureLayer
Set pFeatureLayer = New FeatureLayer
Set pFeatureLayer.FeatureClass = pOutputFeatClass
pFeatureLayer.Name = "Centroid"
pMap.AddLayer pFeatureLayer
End Sub
```

Part 3 creates a new feature layer from *pOutputFeatClass* and adds the layer to the active map.

```vba
Public Function CreateNewShapefile(pSpatialReference _
As ISpatialReference) As IFeatureClass
  ' Part 1: Define the output's workspace.
  Dim pFWS As IFeatureWorkspace
  Dim pWorkspaceFactory As IWorkspaceFactory
  Set pWorkspaceFactory = New ShapefileWorkspaceFactory
  Set pFWS = pWorkspaceFactory.OpenFromFile(strFolder, 0)
```

Part 1 uses the *OpenFromFile* method and the constant *strFolder* to open a feature workspace for the new shapefile.

```vba
' Part 2: Edit and define a Shape field.
Dim pFields As IFieldsEdit
Dim pField As IFieldEdit
Dim pGeomDef As IGeometryDef
Dim pGeomDefEdit As IGeometryDefEdit
' Make the shape field.
Set pField = New Field
pField.Name = "Shape"
pField.Type = esriFieldTypeGeometry
' Define the geometry of the shape field.
Set pGeomDef = New GeometryDef
Set pGeomDefEdit = pGeomDef
With pGeomDefEdit
  .GeometryType = esriGeometryPoint
```

```
    Set.SpatialReference = pSpatialReference
End With
Set pField.GeometryDef = pGeomDef
' Add the shape field to the field collection.
Set pFields = New Fields
pFields.AddField pField
```

Part 2 defines the shape field and adds it to a field collection. The code first creates *pField* as an instance of the *Field* class and defines its name and type properties. Next the code creates *pGeomDef* as an instance of the *GeometryDef* class and uses the *GeometryDefEdit* interface to define the geometry type and spatial reference of the new geometry definition. After assigning *pGeomDef* to be the geometry definition of *pField*, the code creates *pFields* as an instance of the *Fields* class and adds *pField* to the new field collection.

```
' Part 3: Create the new feature class and return it.
Dim pFeatClass As IFeatureClass
Set pFeatClass = pFWS.CreateFeatureClass _
(strName, pFields, Nothing, Nothing, esriFTSimple, _
"Shape", "")
' Return the feature class.
Set CreateNewShapefile = pFeatClass
End Function
```

Part 3 uses the *CreateFeatureClass* method on *IFeatureWorkspace* to create a feature class object referenced by *pFeatClass*. **CreateNewShapefile** then returns *pFeatClass* to **Centroid**.

Raster Data Operations

The simple data structure of raster data is computationally efficient and well suited for a large variety of raster data operations. One typically starts a raster data operation by setting up an analysis environment. The analysis environment includes the area for analysis and the output cell size. Both parameters are important because the inputs to an operation may have different area extents and different cell sizes.

Raster data query selects cells that meet a query expression. The query expression may involve one factor from a single raster or multiple factors from multiple rasters. The output from a query is a new raster, which separates cells that meet the query expression from those that do not.

Raster data analysis is traditionally grouped into local, neighborhood, zonal, and distance measure operations. A local operation computes the cell values of a new raster by using a function that relates the input to the output. The operation is performed on a cell-by-cell basis. A neighborhood operation uses a focal cell and a neighborhood. The operation computes the focal cell value from the cell values within the neighborhood. A zonal operation works with zones or groups of cells of same values. Given a single input raster, a zonal operation calculates the geometry of zones in the raster. Given a zonal raster and a value raster, a zonal operation summarizes the cell values in the value raster for each zone in the zonal raster. A distance measure operation involves an entire raster. The operation calculates for each cell the distance away from the closest source cell. If the distance is measured in cell units, the operation produces a series of wave-like distance zones over the raster. If the distance is measured in cost units, the operation produces for each cell the least accumulative cost to a source cell.

This chapter covers raster data operations. Section 11.1 reviews raster data analysis in ArcGIS. Section 11.2 discusses objects that are related to raster data operations. Section 11.3 includes macros for saving, extracting, and querying raster data. Section 11.4 offers macros for cell-by-cell operations. Section 11.5 has a macro for a neighborhood operation. Section 11.6 has a macro for a zonal operation. Section

11.7 offers macros for physical distance and cost distance measure operations. All macros start with the listing of key interfaces and key members (i.e., properties and methods) and the usage.

11.1 ANALYZING RASTER DATA IN ARCGIS

The Spatial Analyst extension to ArcGIS is designed for raster data operations. Spatial Analyst has an Options command that lets the user specify an analysis mask, an area for analysis, and an output cell size. An analysis mask limits analysis to cells that do not carry the cell value of no data. The area for analysis may correspond to a specific raster, an extent defined by a set of minimum and maximum x-, y-coordinates, a combination of rasters, or a mask. The output cell size can be at any scale deemed suitable by the user although, conceptually, it should be equal to the largest cell size among the input rasters for analysis.

The Raster Calculator incorporates arithmetic operators, logical operators, Boolean connectors, and mathematical functions in its dialog box. The Raster Calculator is therefore useful for raster data queries as well as for cell-by-cell operations. A raster data query uses an SQL (Structured Query Language) statement on a single raster or multiple rasters. The query result is saved into a raster, in which cells that meet the condition are coded one and other cells are coded zero.

Besides the Raster Calculator, the Cell Statistics and Reclassify commands also perform local operations. The Cell Statistics command computes summary statistics such as maximum, minimum, and mean from multiple rasters. The Reclassify command reclassifies the values of the input cells on a cell-by-cell basis.

The Neighborhood Statistics command performs neighborhood operations. The command uses a dialog to gather the input data, field, statistic type, and neighborhood for computation. A neighborhood may be a rectangle, circle, annulus, or wedge. The statistic type includes maximum, minimum, range, sum, mean, standard deviation, variety, majority, and minority.

The Zonal Statistics command performs zonal operations on a zonal raster and a value raster. The command uses a dialog to gather the zone dataset, zone field, and value raster. Given a single raster with zones, one can use the Raster Calculator with such functions as area, perimeter, centroid, and thickness to compute the zonal geometry.

The Distance command has the following selections: Straight Line Distance, Allocation, Cost Weighted, and Shortest Path. The first two use physical distance measures, and the last two use cost distance measures. Straight Line Distance creates an output raster containing continuous distance measures away from the source cells in a raster. Allocation creates a raster in which each cell is assigned the value of its closest source cell. Cost Weighted calculates for each cell the least accumulative cost, over a cost raster, to its closest source cell. Additionally, the Cost Weighted command can also create a direction raster, which shows the direction of the least cost path from each cell to a source, and an allocation raster, which shows the assignment of each cell to a source cell. Shortest Path uses the results from the Cost Weighted command to generate the least cost path from any cell or zone.

11.2 ARCOBJECTS FOR RASTER DATA ANALYSIS

ArcObjects includes objects for raster data operations in two subsystems: Raster and Spatial Analyst Extension. Raster objects provide access to raster data and interfaces to manage and visualize raster data. Spatial Analysis objects control the analysis environment and perform raster data operations. To have access to both Raster and Spatial Analyst objects in programming, we must check the Spatial Analyst extension in ArcMap's Tools/Extensions dialog as well as the ESRI Spatial Analyst Extension Object Library and ESRI Spatial Analyst Shared Object Library in Visual Basic's Tools/References dialog. Unless both steps are taken, errors can occur in running a VBA (Visual Basic for Applications) macro for raster data analysis.

11.2.1 Raster Objects

The basic raster objects are *RasterDataset*, *RasterBand*, *Raster*, and *RasterLayer* (Figure 11.1). A raster dataset object represents an existing dataset stored on disk in a particular raster format (e.g., ESRI grid, TIFF). A raster band object represents an individual band of a raster dataset. The number of bands in a raster may vary; an ESRI grid typically contains a single band, whereas a satellite image has multiple bands. A raster object is a virtual representation of a raster dataset, useful for raster data operations. Created from an existing raster or a raster dataset, a raster layer object is a visual display of raster data.

One other raster object needs to be mentioned is *RasterDescriptor*. When used for data conversion (Chapter 6), a raster descriptor object represents a raster that uses a field other than the value field. When used for raster data query, a raster descriptor object is associated with a query filter and can be created from a raster or a raster's selection set (Figure 11.2).

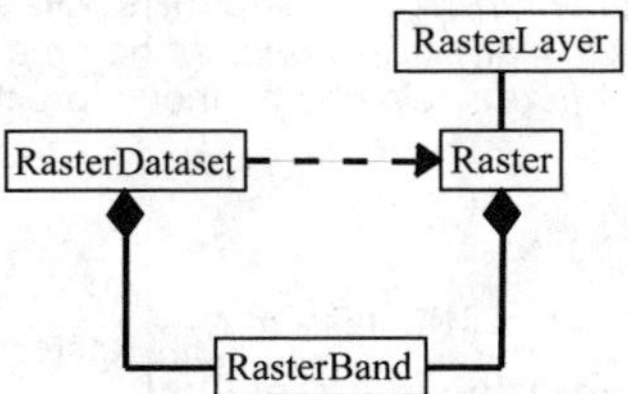

Figure 11.1 The diagram shows the relationship between *RasterDataset*, *RasterBand*, *Raster*, and *RasterLayer*. See text for explanation.

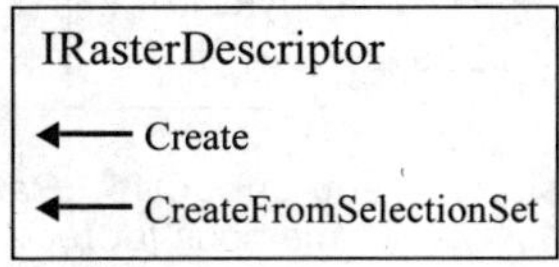

Figure 11.2 *IRasterDescriptor* has methods to create a raster descriptor from a raster or a selection set.

11.2.2 Operator Objects

The majority of Spatial Analyst objects are operator objects that provide methods for raster data analysis. A good reference for Spatial Analyst objects is the Spatial Analyst Functional Reference in the ArcGIS Desktop Help. The reference groups Spatial Analyst objects by type of analysis and offers the ArcObjects syntax and examples for each object.

This section focuses on Spatial Analyst objects that are used in this chapter's sample macros. A *RasterExtractionOp* object supports *IExtractOp*, which has methods for data extraction based on points, a polygon, or a raster (Figure 11.3). A *RasterReclassOp* object implements *IReclassOp*, which has methods for reclassifying raster data (Figure 11.4). A *RasterMathOps* object supports *ILogicalOp* and *IMathOp*. *ILogicalOp* has methods for logical (Boolean) operations, and *IMathOp* has methods for mathematical operations (Figure 11.5). Notice each of the above operator objects also supports *IRasterAnalysisEnvironment*, which has members for defining the analysis environment.

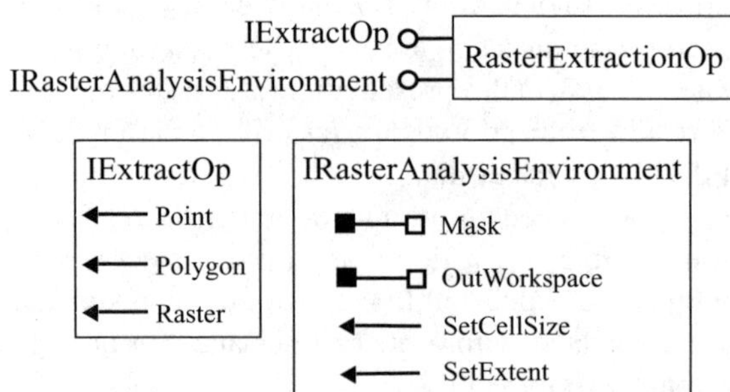

Figure 11.3 A *RasterExtractionOp* object supports *IRasterAnalysisEnvironment* and *IExtractOp*. *IRasterAnalysisEnvironment* has members for defining the analysis environment, and *IExtractOp* has methods to extract raster data.

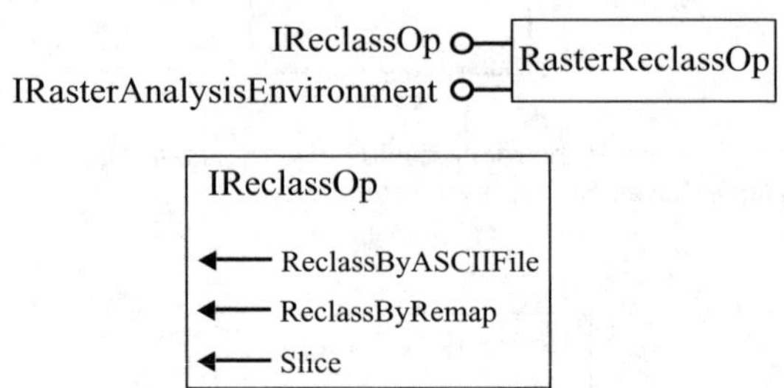

Figure 11.4 A *RasterReclassOp* object supports *IRasterAnalysisEnvironment* and *IReclassOp*. *IReclassOp* has methods for reclassifying and slicing raster data.

Local, neighborhood, zonal, and distance measure operations are each covered by an operator object. A *RasterLocalOp* object supports *ILocalOp* (Figure 11.6), a *RasterNeighborhoodOp* object *INeighborhoodOp* (Figure 11.7), a *RasterZonalOp* object *IZonalOp* (Figure 11.8), and a *RasterDistanceOp* object *IDistanceOp* (Figure 11.9). Each of these operator objects also implements *IRasterAnalysisEnvironment* for access to the analysis environment.

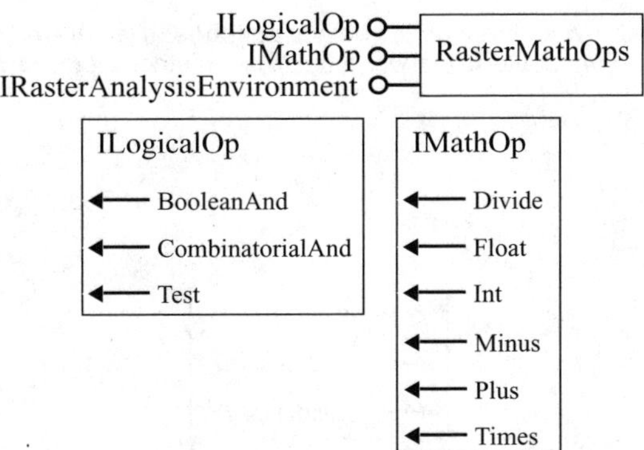

Figure 11.5 A *RasterMathOp* object supports *IRasterAnalysisEnvironment*, *ILogicalOp*, and *IMathOp*. *ILogicalOp* has methods for logical operations, and *IMathOp* has methods for mathematical operations.

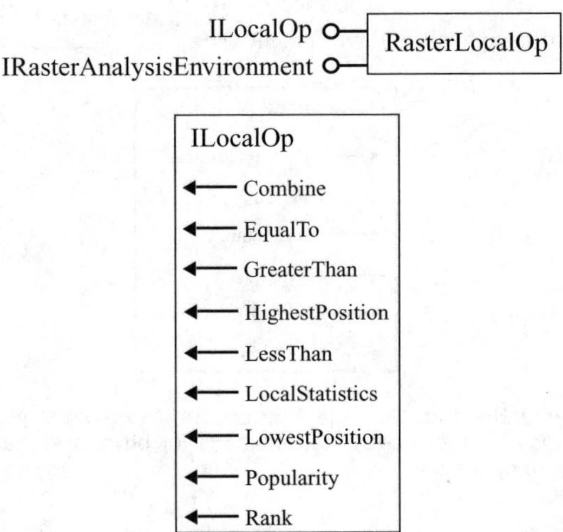

Figure 11.6 A *RasterLocalOp* object supports *IRasterAnalysisEnvironment* and *ILocalOp*. *ILocalOp* has methods for cell-by-cell operations.

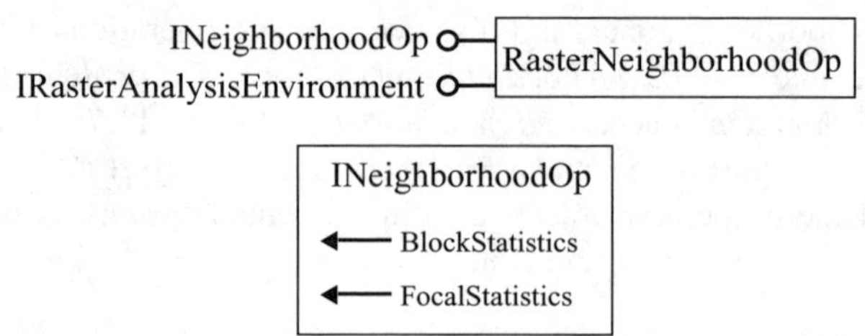

Figure 11.7   A *RasterNeighborhoodOp* object supports *IRasterAnalysisEnvironment* and *INeighborhoodOp*. *INeighborhoodOp* has methods for focal and block operations.

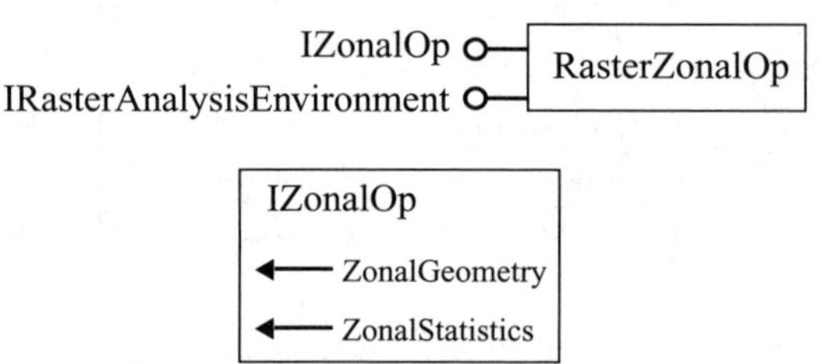

Figure 11.8   A *RasterZonalOp* object supports *IRasterAnalysisEnvironment* and *IZonalOp*. *IZonalOp* has methods for zonal operations using a single raster or a zonal raster and a value raster.

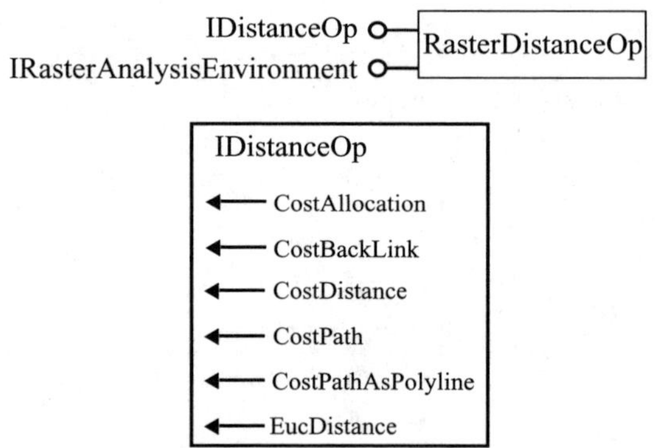

Figure 11.9   A *RasterDistanceOp* object supports *IRasterAnalysisEnvironment* and *IDistanceOp*. *IDistanceOp* has methods for physical distance or cost distance measure operations.

11.3 MANAGING RASTER DATA

This section covers the tasks of saving raster data, extracting raster data, and querying raster data.

11.3.1 *MakePermanent*

An output from a raster data operation is a temporary dataset. The temporary dataset is lost unless a location on disk and a file name are set up to make it permanent. *MakePermanent* takes a raster layer, which represents a temporary dataset, and saves it as a permanent raster. The macro performs the same function as using the Make Permanent command from the layer's context menu in ArcMap. *MakePermanent* has two parts. Part 1 defines the temporary raster dataset. Part 2 specifies a workspace and makes permanent the temporary raster dataset.

> **Key Interfaces:** *IRasterLayer, IRaster, IRasterBandCollection, IRasterDataset, IWorkspaceFactory, IWorkspace, ITemporaryDataset, IDataset.*
> **Key Members:** *Raster, Item(), RasterDataset, OpenFromFile, MakePermanentAS, Name.*
> **Usage:** Add *emidalat*, an elevation raster, to an active map. Click Spatial Analyst, point to Surface Analysis, and select Slope. *Slope of emidalat* is added to the active map. Right-click *Slope of emidalat* and select Properties. The Source tab of the Layer Properties dialog shows the raster has the temporary status and a name of SLOPEx. Import *MakePermanent* to Visual Basic Editor. Run the macro. The macro creates a permanent raster from SLOPEx.

```
Private Sub MakePermanent()
  ' Part 1: Define the temporary raster data.
  Dim pMxDoc As IMxDocument
  Dim pMap As IMap
  Dim pRasterLy As IRasterLayer
  Dim pRaster As IRaster
  Dim pRasBandC As IRasterBandCollection
  Dim pRasterDS As IRasterDataset
  Set pMxDoc = ThisDocument
  Set pMap = pMxDoc.FocusMap
  ' Get the raster from the raster layer.
  Set pRasterLy = pMap.Layer(0)
  Set pRaster = pRasterLy.Raster
  ' Set the raster dataset to be the first band of
  ' the raster.
  Set pRasBandC = pRaster
  Set pRasterDS = pRasBandC.Item(0).RasterDataset
```

The code first sets *pRaster* to be the raster of the top layer in the active map. Next the code perform a QueryInterface (QI) for the *IRasterBandCollection* interface and assigns the raster dataset associated with the first band of *pRaster* to *pRasterDB* (Figure 11.10).

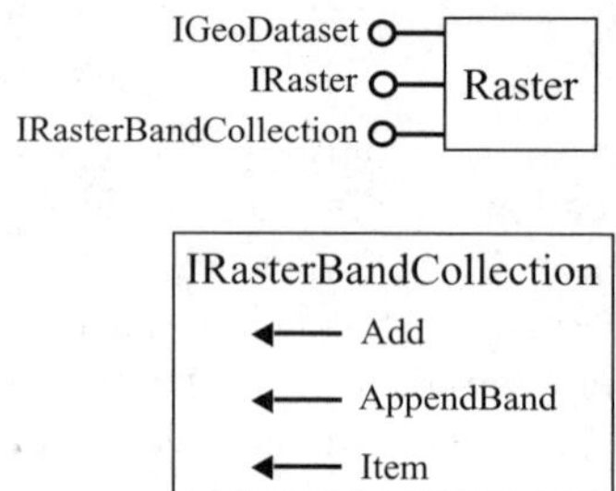

Figure 11.10 The diagram shows how to use QI on *IRaster* for *IRasterBandCollection* so that a raster band can be added, appended, or extracted from a raster band collection object.

```
' Part 2: Make permanent the raster dataset.
Dim pWSF As IWorkspaceFactory
Dim pWS As IRasterWorkspace
Dim pTempDS As ITemporaryDataset
Dim pDataset As IDataset
Dim Name As String
' Define the workspace.
Set pWSF = New RasterWorkspaceFactory
Set pWS = pWSF.OpenFromFile("c:\data\chap11\", 0)
' Define the temporary dataset.
Set pDataset = pRasterDS
Name = pDataset.Name
Set pTempDS = pRasterDS
Set pRasterDS = pTempDS.MakePermanentAs(Name, _
pWS, "GRID")
End Sub
```

Part 2 makes *pRasterDS* a permanent raster. To do that, the code must first define a workspace and a name for the permanent raster. The workspace is specified using the *OpenFromFile* method on *IWorkspaceFactory*. The name is specified via the *Name* property on *IDataset*. Then the code performs a QI for the *ITemporaryDataset* interface and uses the *MakePermanentAs* method to make *pRasterDS* a permanent ESRI grid (Figure 11.11).

11.3.2 *ExtractByMask*

ExtractByMask uses a mask raster to extract a data subset from an input raster. The macro performs the same function as using an analysis mask in Spatial Analyst to extract a new raster from an input raster. ***ExtractByMask*** has three parts. Part 1 defines the input raster. Part 2 prepares a mask raster and uses it to extract data. Part 3 creates a raster layer from the extracted data and adds the layer to the active map.

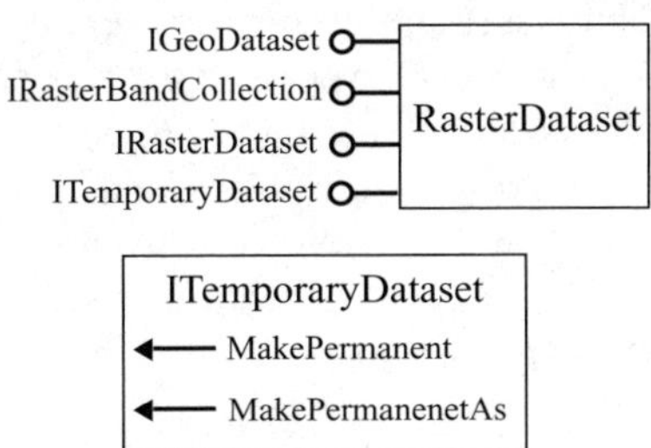

Figure 11.11 The diagram shows how to QI on *IRasterDataset* for *ITemporaryDataset* so that a temporary raster can be made permanent.

Key Interfaces: *IRasterLayer, IRaster, IGeoDataset, IExtractionOp.*
Key Members: *Raster.*
Usage: Add *splinegd* and *idoutlgd* to an active map. *splinegd* is an interpolated precipitation raster, whose extent is determined by the extent of *x*- and *y*-coordinates of the weather stations used in interpolation. *idoutlgd* is a raster showing the outline of Idaho. *splinegd* must be on top of *idoutlgd*. Import **ExtractByMask** to Visual Basic Editor. **ExtractByMask** requires the use of Spatial Analyst Extension objects. Select References from Visual Basic's Tools menu, and check the box for ESRI Spatial Analyst Extension Object Library. Run the macro. The macro uses *idoutlgd* as an analysis mask to extract a temporary raster from *splinegd*.

```
Private Sub ExtractByMask()
  ' Part 1: Define the input raster.
  Dim pMxDoc As IMxDocument
  Dim pInRasterLayer As IRasterLayer
  Dim pInRaster As IRaster
  Set pMxDoc = ThisDocument
  Set pInRasterLayer = pMxDoc.FocusMap.Layer(0)
  Set pInRaster = pInRasterLayer.Raster
```

Part 1 sets *pInRaster* to be the raster of the top layer in the active map.

```
  ' Part 2: Use a mask to extract the input raster.
  Dim pRasterLayer As IRasterLayer
  Dim pMaskDataset As IGeoDataset
  Dim pOutRaster As IRaster
  Dim pExtrOp As IExtractionOp
  ' Define the mask dataset.
  Set pRasterLayer = pMxDoc.FocusMap.Layer(1)
  Set pMaskDataset = pRasterLayer.Raster
  Set pExtrOp = New RasterExtractionOp
  ' Perform extraction.
  Set pOutRaster = pExtrOp.Raster(pInRaster, pMaskDataset)
```

Part 2 defines the mask dataset and performs an extraction operation. To define the mask raster, the code sets *pRasterLayer* to be the second layer of the active map and *pMaskDataset* to be the raster of *pRasterLayer*. Next the code creates *pExtrOp* as an instance of the *RasterExtractionOp* class and uses the *Raster* method on *IExtractOp* to create a raster referenced by *pOutRaster*.

```
' Part 3: Create the output layer and add it to
' the active map.
Dim pRL As IRasterLayer
Set pRL = New RasterLayer
pRL.CreateFromRaster pOutRaster
pMxDoc.FocusMap.AddLayer pRL
End Sub
```

Part 3 uses the *CreateFromRaster* method on *IRasterLayer* to create a raster layer from *pOutRaster*. The code then adds the raster layer to the active map.

11.3.3 *RasterQuery*

RasterQuery queries an input raster and produces an output raster that shows the query result. If the input raster has an attribute table, the output raster contains the value of one for cells that meet the expression and zero for cells that do not. If the input raster does not have an attribute table, the output raster retains the original cell values for those cells that meet the expression and no data for cells that do not. In the case of an integer raster (i.e., a raster with an attribute table), the macro performs a similar function as using the Raster Calculator in Spatial Analyst to query a raster.

RasterQuery has three parts. Part 1 defines the input raster. Part 2 prepares a query filter and performs the data query. Part 3 creates a raster layer from the output and adds the layer to the active map.

Key Interfaces: *IRaster, IQueryFilter, IRasterDescriptor, ILogicalOp, IGeoDataset.*
Key Members: *Raster, WhereClause, Create, Test, CreateFromRaster.*
Usage: Add *slopegrd*, a raster with four slope classes, to an active map. Import *RasterQuery* to Visual Basic Editor. Make sure that the ESRI Spatial Analyst Extension Object Library is available. Run the macro. The macro creates a temporary raster that shows cells in the slope class of two and adds it as a raster layer to the active map.

```
Private Sub RasterQuery()
    ' Part 1: Define the raster for query.
    Dim pMxDoc As IMxDocument
    Dim pMap As IMap
    Dim pLayer As IRasterLayer
    Dim pRaster As IRaster
```

```
Set pMxDoc = ThisDocument
Set pMap = pMxDoc.FocusMap
Set pLayer = pMap.Layer(0)
Set pRaster = pLayer.Raster
```

Part 1 sets *pRaster* to be the raster of the top layer in the active map.

```
' Part 2: Perform raster query.
Dim pQFilter As IQueryFilter
Dim pRasDes As IRasterDescriptor
Dim pLogicalOp As ILogicalOp
Dim pOutputRaster As IGeoDataset
' Prepare a query filter.
Set pQFilter = New QueryFilter
pQFilter.WhereClause = "Value = 2"
' Prepare a raster descriptor.
Set pRasDes = New RasterDescriptor
pRasDes.Create pRaster, pQFilter, "value"
' Run a logical operation.
Set pLogicalOp = New RasterMathOps
Set pOutputRaster = pLogicalOp.Test(pRasDes)
```

Part 2 performs a raster data query and transforms the result into a raster with one and zero cell values. The code first creates *pQFilter* as an instance of the *QueryFilter* class and defines its *WhereClause* condition. Next the code sets *pRasDes* to be an instance of the *RasterDescriptor* class and uses the *Create* method to create the new raster descriptor from *pRaster*. Then the code creates *pLogicalOp* as an instance of the *RasterMathOps* class and uses the *Test* method on *ILogicalOp* to create a raster referenced by *pOutputRaster*. The output raster contains only ones and zeros.

```
' Part 3: Create the output raster layer, and add
' the layer to the active map.
Dim pOutputLayer As IRasterLayer
Set pOutputLayer = New RasterLayer
pOutputLayer.CreateFromRaster pOutputRaster
pOutputLayer.Name = "QueryOutput"
pMap.AddLayer pOutputLayer
End Sub
```

Part 3 creates a new raster layer from *pOutputRaster* and adds the layer to the active map. The temporary layer has the value of one for cells that have the slope value of two and zero for other cells.

11.3.4 *Query2Rasters*

Query2Rasters queries two rasters and produces an output raster with the cell values of one and zero. The macro performs the same function as using the Raster Calculator to query two rasters in Spatial Analyst. *Query2Rasters* has four parts. Part 1 queries the first input raster and saves the result into a dataset. Part 2 queries the second raster and saves the result into a dataset. Part 3 uses the two datasets from the previous queries in a logical operation to create an output raster. Part 4 creates a raster layer from the output and adds the layer to the active map. Parts 1 and 2 contain basically the same code as *RasterQuery*.

> **Key Interfaces:** *IRaster, IQueryFilter, IRasterDescriptor, ILogicalOp, IGeoDataset.*
> **Key Members:** *Raster, WhereClause, Create, Test, BooleanAnd, CreateFromRaster, Name.*
> **Usage:** Add *slopegrd* and *aspectgrd* to an active map, with *slopegrd* on top in the table of contents. *slopegrd* contains four slope classes, and *aspectgrd* contains five aspect classes. Import *Query2Rasters* to Visual Basic Editor. Make sure that the ESRI Spatial Analyst Extension Object Library is available. Run the macro. The macro adds to the active map a temporary raster, which has the cell values of one and zero. Cells with the value of one have the slope class of two and the aspect class of two.

```vba
Private Sub Query2Rasters()
    ' Part 1: Query the first raster.
    Dim pMxDoc As IMxDocument
    Dim pMap As IMap
    Dim pRLayer1 As IRasterLayer
    Dim pRaster1 As IRaster
    Dim pFilt1 As IQueryFilter
    Dim pDesc1 As IRasterDescriptor
    Dim pLogicalOp As ILogicalOp
    Dim pOutputRaster1 As IGeoDataset
    Set pMxDoc = ThisDocument
    Set pMap = pMxDoc.FocusMap
    ' Define the first raster.
    Set pRLayer1 = pMap.Layer(0)
    Set pRaster1 = pRLayer1.Raster
    ' Create the first query filter.
    Set pFilt1 = New QueryFilter
    pFilt1.WhereClause = "value = 2"
    ' Create the first raster descriptor.
    Set pDesc1 = New RasterDescriptor
    pDesc1.Create pRaster1, pFilt1, "value"
```

```
' Create the first output with 1's and 0's.
Set pLogicalOp = New RasterMathOps
Set pOutputRaster1 = pLogicalOp.Test(pDesc1)
```

Part 1 uses the same code as in **RasterQuery** to create *pOutputRaster1* from the first input raster and a query filter.

```
' Part 2: Query the second raster.
Dim pRLayer2 As IRasterLayer
Dim pRaster2 As IRaster
Dim pFilt2 As IQueryFilter
Dim pDesc2 As IRasterDescriptor
Dim pOutputRaster2 As IGeoDataset
' Define the second raster.
Set pRLayer2 = pMap.Layer(1)
Set pRaster2 = pRLayer2.Raster
' Create the second query filter.
Set pFilt2 = New QueryFilter
pFilt2.WhereClause = "value = 2"
' Create the second raster descriptor.
Set pDesc2 = New RasterDescriptor
pDesc2.Create pRaster2, pFilt2, "value"
' Create the second output with 1's and 0's.
Set pLogicalOp = New RasterMathOps
Set pOutputRaster2 = pLogicalOp.Test(pDesc2)
```

Part 2 creates *pOutputRaster2* from the second input raster and a query filter.

```
' Part 3: Run a logical operation on the two query results.
Dim pOutputRaster3 As IGeoDataset
Set pLogicalOp = New RasterMathOps
Set pOutputRaster3 = pLogicalOp.BooleanAnd _
(pOutputRaster1, pOutputRaster2)
```

Like Parts 1 and 2, Part 3 also performs a logical operation. But instead of using the *Test* method, which is limited to one input raster, the code uses the *BooleanAnd* method, which accepts two input rasters. The output referenced by *pOutputRaster3* has the cell value of one, where both *pOutputRaster1* and *pOutputRaster2* have the cell value of one, and zero elsewhere. The *BooleanAnd* method can use *pDesc1* and *pDesc2*, instead of *pOutputRaster1* and *pOutputRaster2*, as the object qualifiers. But the output will have the cell values of one and no data instead of one and zero.

```
' Part 4: Create the output raster layer, and add
' the layer to the active map.
Dim pRLayer As IRasterLayer
Set pRLayer = New RasterLayer
pRLayer.CreateFromRaster pOutputRaster3
pRLayer.Name = "QueryOutput2"
pMap.AddLayer pRLayer
End Sub
```

Part 4 creates a raster layer from *pOutputRaster3* and adds the layer to the active map.

11.4 PERFORMING LOCAL OPERATIONS

Local operations constitute the core of raster data analysis. A large variety of local operations are available in Spatial Analyst. Reclassify and combine are examples of local operations in this section. Reclassify operates on a single raster, whereas combine operates on two or more rasters.

11.4.1 *ReclassNumberField*

ReclassNumberField uses a number remap to reclassify an input raster. A remap has two columns. The first column lists a cell value or a range of cell values to be reclassified, and the second column lists the output value, including no data. A remap object in ArcObjects can be either a number remap or a string remap.

ReclassNumberField performs the same function as using Reclassify in Spatial Analyst to create a classified integer raster. The macro has three parts. Part 1 defines the input raster. Part 2 performs the reclassification. Part 3 creates a raster layer from the reclassify output and adds the layer to the active map.

Key Interfaces: *IGeoDataset, IReclassOp, INumberRemap, IRaster.*
Key Members: *Raster, MapRange, ReclassByRemap, CreateFromRaster.*
Usage: Add *slope*, a continuous slope raster, to an active map. Import *Reclass-NumberField* to Visual Basic Editor. Make sure that the ESRI Spatial Analyst Shared Object Library is available. Run the macro. The macro produces a temporary raster with the reclassification result and adds the raster to the active map.

```
Private Sub ReclassNumberField()
  ' Part 1: Define the raster for reclassify.
  Dim pMxDoc As IMxDocument
  Dim pMap As IMap
  Dim pRasterLy As IRasterLayer
  Dim pGeoDs As IGeoDataset
  Set pMxDoc = ThisDocument
```

```
Set pMap = pMxDoc.FocusMap
Set pRasterLy = pMap.Layer(0)
Set pGeoDs = pRasterLy.Raster
```

Part 1 sets *pGeoDs* to be the raster of the top layer in the active map.

```
' Part 2: Reclassify the input raster.
Dim pReclassOp As IReclassOp
Dim pNRemap As INumberRemap
Dim pOutRaster As IRaster
' Prepare a number remap.
Set pNRemap = New NumberRemap
pNRemap.MapRange 0, 10#, 1
pNRemap.MapRange 10.1, 20#, 2
pNRemap.MapRange 20.1, 30#, 3
pNRemap.MapRange 30.1, 90#, 4
' Run the reclass operation.
Set pReclassOp = New RasterReclassOp
Set pOutRaster = pReclassOp.ReclassByRemap _
(pGeoDs, pNRemap, False)
```

Part 2 reclassifies *pGeoDS* by using a remap that is built in code. The code first creates *pNRemap* as an instance of the *NumberRemap* class and uses the *MapRange* method on *INumberRemap* to set the output value based on a numeric range of the input values (Figure 11.12). A cell within the numeric range of 0 to 10.0 is assigned an output value of 1, 10.1 to 20.0 an output value of 2, and so on. Then, after having created *pReclassOp* as an instance of the *RasterReclassOp* class, the code uses the *ReclassByRemap* method to create a reclassified raster referenced by *pOutRaster*.

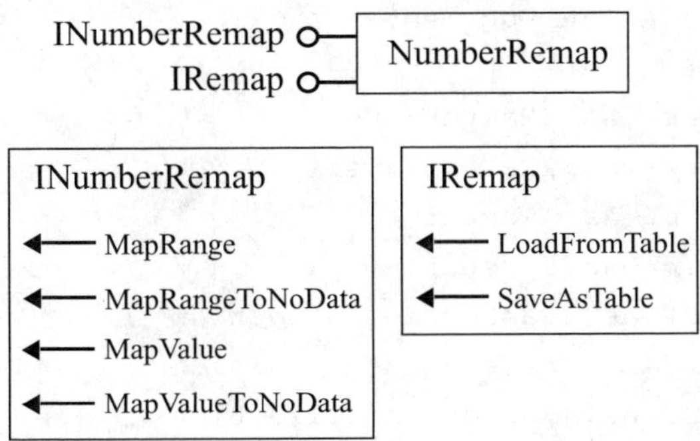

Figure 11.12 A *NumberRemap* object supports *INumberRemap* and *IRemap*. These two interfaces have methods to define a remap object for reclassifying a raster.

```
' Part 3: Create the output layer, and add it to
' the active map.
Dim pReclassLy As IRasterLayer
Set pReclassLy = New RasterLayer
pReclassLy.CreateFromRaster pOutRaster
pMap.AddLayer pReclassLy
End Sub
```

Part 3 creates a new raster layer from *pOutRaster* and adds the layer to the active map.

11.4.2 *Combine2Rasters*

Combine2Rasters combines two input rasters and produces an output raster with its cell values representing each unique combination of the input cell values. The macro performs the same function as using the combine function in Spatial Analyst's Raster Calculator. *Combine2Rasters* has three parts. Part 1 defines the two input rasters. Part 2 creates a new raster by adding to it two bands from the input rasters and then performs a combine operation on the new raster. Part 3 creates a new raster layer from the output raster and adds the layer to the active map.

> **Key Interfaces:** *IRaster, IRasterBandCollection, IRasterBand, ILocalOp.*
> **Key Members:** *Raster, Item(), Add, Combine, CreateFromRaster.*
> **Usage:** Add *slopegrd* and *aspectgrd* to an active map. *slopegrd* shows four slope classes, and *aspectgrd* shows four principal directions and a fifth class for flat areas. Import *Combine2Rasters* to Visual Basic Editor. Make sure that the ESRI Spatial Analyst Extension Object Library is available. Run the macro. The macro adds to the active map a new temporary raster showing 19 unique combinations of slope and aspect values.

```
Private Sub Combine2Rasters()
  ' Part 1: Define the two rasters for combine.
  Dim pMxDoc As IMxDocument
  Dim pMap As IMap
  Dim pRLayer1 As IRasterLayer
  Dim pRLayer2 As IRasterLayer
  Dim pRaster1 As IRaster
  Dim pRaster2 As IRaster
  Dim pRasBC1 As IRasterBandCollection
  Dim pBand1 As IRasterBand
  Dim pRasBC2 As IRasterBandCollection
  Dim pBand2 As IRasterBand
  Set pMxDoc = ThisDocument
  Set pMap = pMxDoc.FocusMap
```

```
' Derive the first band from the first raster.
Set pRLayer1 = pMap.Layer(0)
Set pRaster1 = pRLayer1.Raster
Set pRasBC1 = pRaster1
Set pBand1 = pRasBC1.Item(0)
' Derive the first band from the second raster.
Set pRLayer2 = pMap.Layer(1)
Set pRaster2 = pRLayer2.Raster
Set pRasBC2 = pRaster2
Set pBand2 = pRasBC2.Item(0)
```

Part 1 defines the raster bands for the combine operation. To define the first raster band, the code executes the following steps: sets *pRaster1* to be the raster of the top layer; switches to the *IRasterBandCollection* interface; and sets *pBand1* to be the first band of *pRaster1*. The same steps are taken to derive *pBand2*.

```
' Part 2: Create a new raster and perform combine.
Dim pRasBC As IRasterBandCollection
Dim pLocalOp As ILocalOp
Dim pOutRaster As IRaster
' Define a new raster and add to it two input bands.
Set pRasBC = New Raster
pRasBC.Add pBand1, 0
pRasBC.Add pBand2, 1
' Run the combine local operation.
Set pLocalOp = New RasterLocalOp
Set pOutRaster = pLocalOp.Combine(pRasBC)
```

Part 2 uses a multi-band raster to perform the combine operation. The code first creates *pRasBC* as an instance of the *Raster* class. Next the code uses the *Add* method on *IRasterBandCollection* to add *pBand1* and *pBand2* to *pRasBC*. The code then creates *pLocalOp* as an instance of the *RasterLocalOp* class and uses the *Combine* method on *ILocalOp* to create a combine raster referenced by *pOutRaster*.

```
' Part 3: Create the output layer, and add it to
' the active map.
Dim pRLayer As IRasterLayer
Set pRLayer = New RasterLayer
pRLayer.CreateFromRaster pOutRaster
pRLayer.Name = "slp_asp"
pMap.AddLayer pRLayer
End Sub
```

Part 3 creates *pRLayer* as a new raster layer from *pOutRaster* and adds the layer to the active map.

11.4.3 Other Local Operations

Macros for other local operations involving multiple rasters have the same code structure as **Combine2Rasters**. A local mean operation (i.e., deriving the mean on a cell-by-cell basis from two or more rasters) would change the last line of Part 2 in **Combine2Rasters** to read:

```
Set pOutRaster = pLocalOp.LocalStatistics _
(pRaster, esriGeoAnalysisStatsMean)
```

The line uses the *LocalStatistics* method on *ILocalOp* to compute the local mean from a raster with multiple raster bands and to save the results into an output raster.

Likewise, a local maximum operation would change the last line of Part 2 in **Combine2Rasters** to read:

```
Set pOutRaster = pLocalOp.HighestPosition(pRaster)
```

The line uses the *HighestPosition* method on *ILocalOp* to derive the highest value among the input rasters, which are stored in a multi-band raster, and to save the results into an output raster.

11.5 PERFORMING NEIGHBORHOOD OPERATIONS

A VBA macro for a neighborhood operation must define the neighborhood to be used and the statistic to be derived. The statistics are computed by using either an overlapping neighborhood or a nonoverlapping neighborhood. An overlapping neighborhood occurs when the operation moves from cell to cell. A nonoverlapping neighborhood occurs when the operation moves from block to block.

11.5.1 *FocalMean*

FocalMean computes the mean using the cell values within a 3 ∞ 3 neighborhood and assigns the mean to the focal cell. The macro performs the same function as using Neighborhood Statistics in Spatial Analyst. *FocalMean* has three parts. Part 1 defines the input raster. Part 2 prepares a neighborhood operator and a 3 ∞ 3 neighborhood and runs the focal mean operation. Part 3 creates a raster layer from the output raster and adds it to the active map.

Key Interfaces: *IRaster, INeighborhoodOp, IRasterNeighborhood.*
Key Members: *Raster, FocalStatistics, CreateFromRaster.*
Usage: Add *emidalat* to an active map. Import **FocalMean** to Visual Basic Editor. Make sure that the ESRI Spatial Analyst Extension Object Library is available. Run the macro. The macro creates a temporary raster showing the neighborhood mean and adds the raster to the active map.

```
Private Sub FocalMean()
  ' Part 1: Define the input raster.
  Dim pMxDoc As IMxDocument
  Dim pMap As IMap
  Dim pRLayer As IRasterLayer
  Dim pRaster As IRaster
  Set pMxDoc = ThisDocument
  Set pMap = pMxDoc.FocusMap
  Set pRLayer = pMap.Layer(0)
  Set pRaster = pRLayer.Raster
```

Part 1 sets *pRaster* to be the raster of the top layer in the active map.

```
  ' Part 2: Perform focal mean.
  Dim pNbrOp As INeighborhoodOp
  Dim pNbr As IRasterNeighborhood
  Dim pOutputRaster As IRaster
  ' Define the neighborhood.
  Set pNbr = New RasterNeighborhood
  pNbr.SetRectangle 3, 3, esriUnitsCells
  ' Run the focal mean neighborhood operation.
  Set pNbrOp = New RasterNeighborhoodOp
  Set pOutputRaster = pNbrOp.FocalStatistics(pRaster, _
  esriGeoAnalysisStatsMean, pNbr, False)
```

Part 2 creates *pNbr* as an instance of the *RasterNeighborhood* class and uses the *SetRectangle* method on *IRasterNeighborhood* to define a rectangular neighborhood with three cells for both its height and width. The code then creates *pNbrOp* as an instance of the *RasterNeighborhoodOp* class and uses the *FocalStatistics* method to create a focal mean raster referenced by *pOutputRaster*. (The *FocalStatistics* method uses an overlapping neighborhood, whereas the *BlockStatistics* method, also on *INeighborhoodOp*, uses a nonoverlapping neighborhood.)

```
  ' Part 3: Create the output layer and add it to the
  ' active map.
  Dim pOutputLayer As IRasterLayer
  Set pOutputLayer = New RasterLayer
  pOutputLayer.CreateFromRaster pOutputRaster
  pOutputLayer.Name = "FocalMean"
  pMap.AddLayer pOutputLayer
End Sub
```

Part 3 creates a raster layer from *pOutputRaster* and adds the layer to the active map.

11.6 PERFORMING ZONAL OPERATIONS

A VBA macro for a zonal operation involving two rasters must define a value raster and a zonal raster so that the cell values of the value raster can be summarized by zone. The zonal raster must be an integer raster. Like a neighborhood operation, a zonal operation offers various statistics.

11.6.1 *ZonalMean*

ZonalMean computes the mean precipitation for each watershed in Idaho. The macro performs the same function as using Zonal Statistics in Spatial Analyst. *ZonalMean* has three parts. Part 1 defines the rasters to be used in the zonal operation and adds the zonal and value rasters to the active map. Part 2 defines and runs the zonal mean operation. Part 3 creates the raster layer from the output and adds the layer to the active map.

Key Interfaces: *IRaster, IGeoDataset, IZonalOp.*
Key Members: *Raster, CreateFromFilePath, ZonalStatistics, CreateFromRaster.*
Usage: Import *ZonalMean* to Visual Basic Editor in ArcMap. Make sure that the ESRI Spatial Analyst Extension Object Library is available. Run the macro. The macro adds *precipgd* and *hucgd*, the two input rasters, as well as *ZonalMean*, the output raster, to the active map. In this zonal operation, *precipgd* is the value raster, which shows annual precipitation in hundredths of an inch, and *hucgd* is the zonal raster, which shows 13 six-digit watersheds in Idaho. *ZonalMean* contains the mean precipitation in each of the 13 watersheds.

```
Private Sub ZonalMean()
  ' Part 1: Define the zonal and value raster datasets.
  Dim pMxDocument As IMxDocument
  Dim pMap As IMap
  Dim pZoneRL As IRasterLayer
  Dim pZoneRaster As IRaster
  Dim pValueRL As IRasterLayer
  Dim pValueRaster As IRaster
  Set pMxDocument = Application.Document
  Set pMap = pMxDocument.FocusMap
  ' Define the zonal raster dataset.
  Set pZoneRL = New RasterLayer
  pZoneRL.CreateFromFilePath "c:\data\chap11\hucgd"
  Set pZoneRaster = pZoneRL.Raster
  ' Define the value raster dataset.
  Set pValueRL = New RasterLayer
  pValueRL.CreateFromFilePath "c:\data\chap11\precipgd"
```

```
Set pValueRaster = pValueRL.Raster
' Add the zonal and value raster layers to the active map.
pMap.AddLayer pZoneRL
pMap.AddLayer pValueRL
```

Part 1 defines the zonal and value raster datasets to be used in the zonal operation. Instead of referring to raster layers in the active map, the code uses the *CreateFromFilePath* method on *IRasterLayer* to open a zonal layer referenced by *pZoneRL* and sets *pZoneRaster* to be its raster. The same procedure is followed to open *pValueRL* and to set *pValueRaster*. The code then adds *pZoneRL* and *pValueRL* to the active map.

```
' Part 2: Perform zonal mean.
Dim pZonalOp As IZonalOp
Dim pOutputRaster As IGeoDataset
' Run the zonal mean operation.
Set pZonalOp = New RasterZonalOp
Set pOutputRaster = pZonalOp.ZonalStatistics _
(pZoneRaster, pValueRaster, _
esriGeoAnalysisStatsMean, True)
```

Part 2 creates *pZonalOp* as an instance of the *RasterZonalOp* class and uses the *ZonalStatistics* method on *IZonalOp* to create a zonal mean raster referenced by *pOutputRaster*.

```
' Part 3: Create the output layer and add it to
' the active map.
Dim pOutputLayer As IRasterLayer
Set pOutputLayer = New RasterLayer
pOutputLayer.CreateFromRaster pOutputRaster
pOutputLayer.Name = "ZonalMean"
pMap.AddLayer pOutputLayer
End Sub
```

Part 3 creates a raster layer from *pOutputRaster* and adds the layer to the active map.

11.7 PERFORMING DISTANCE MEASURE OPERATIONS

Distance measure operations can be based on physical distance or cost distance. Both types of operations calculate the distance from each cell of a raster to a source cell. The main difference is that cost distance measures are based on a cost raster. This section covers both types of distance measures.

11.7.1 *EucDist*

EucDist calculates continuous Euclidean distance measures away from a stream (source) raster. The macro performs the same function as the Distance/Straight Line option in Spatial Analyst. *EucDist* has three parts. Part 1 defines the source raster. Part 2 defines and runs a distance measure operation. Part 3 creates a raster layer from the output and adds the layer to the active map.

> **Key Interfaces:** *IRaster, IGeoDataset, IDistanceOp.*
> **Key Members:** *Raster, EucDistance, CreateFromRaster.*
> **Usage:** Add *emidastrmgd* to an active map. Import *EucDist* to Visual Basic Editor. Make sure that the ESRI Spatial Analyst Extension Object Library is available. Run the macro. The macro creates and adds a new temporary raster named *EuclideanDistance* to the active map.

```
Private Sub EucDist()
  ' Part 1: Define the input raster dataset.
  Dim pMxDoc As IMxDocument
  Dim pMap As IMap
  Dim pSourceRL As IRasterLayer
  Dim pSourceRaster As IRaster
  Set pMxDoc = ThisDocument
  Set pMap = pMxDoc.FocusMap
  Set pSourceRL = pMap.Layer(0)
  Set pSourceRaster = pSourceRL.Raster
```

Part 1 sets the raster of the top layer to be the source raster, and references the raster by *pSourceRaster*.

```
  ' Part 2: Perform distance measures.
  Dim pDistanceOp As IDistanceOp
  Dim pOutputRaster As IGeoDataset
  ' Run the Euclidean distance operation.
  Set pDistanceOp = New RasterDistanceOp
  Set pOutputRaster = pDistanceOp.EucDistance _
  (pSourceRaster)
```

Part 2 creates *pDistanceOp* as an instance of the *RasterDistanceOp* class and uses the *EucDistance* method on *IDistanceOp* to create a continuous distance measure raster referenced by *pOutputRaster*.

```
  ' Part 3: Create the output layer and add it to
  ' the active map.
  Dim pOutputLayer As IRasterLayer
  Set pOutputLayer = New RasterLayer
```

```
      pOutputLayer.CreateFromRaster pOutputRaster
      pOutputLayer.Name = "EuclideanDistance"
      pMap.AddLayer pOutputLayer
   End Sub
```

Part 3 creates a new raster layer from *pOutputRaster* and adds the layer to the active map.

11.7.2 Use of a Feature Layer as the Source in *EucDist*

EucDist uses a raster as the source. With some modification in Part 1, *EucDist* can use a shapefile (i.e., *emidastrm.shp*) as the source. The shapefile must be converted into a raster before the distance measure operation starts. The following shows the change in Part 1 of *EucDist* to accommodate the use of a shapefile source. (EucDistToFeatLayer.txt on the companion CD is a complete macro to run the distance measure operation from *emidastrm.shp*.)

```
' Part 1: Convert the input feature layer to a
' raster dataset.
 Dim pMxDoc As IMxDocument
 Dim pMap As IMap
 Dim pSourceLayer As IFeatureLayer
 Dim pSourceFC As IFeatureClass
 Dim pConOp As IConversionOp
 Dim pEnv As IRasterAnalysisEnvironment
 Dim pWSF As IWorkspaceFactory
 Dim pWS As IWorkspace
 Dim pSourceDataset As IGeoDataset
 Set pMxDoc = ThisDocument
 Set pMap = pMxDoc.FocusMap
 ' Define the source feature class.
 Set pSourceLayer = pMap.Layer(0)
 Set pSourceFC = pSourceLayer.FeatureClass
 ' Define the workspace for the raster.
 Set pWSF = New RasterWorkspaceFactory
 Set pWS = pWSF.OpenFromFile("c:\data\chap11\", 0)
 ' Prepare a new conversion operator.
 Set pConOp = New RasterConversionOp
 Set pEnv = pConOp
 pEnv.SetCellSize esriRasterEnvValue, 30
 ' Run the conversion operation.
 Set pSourceDataset = pConOp.ToRasterDataset _
 (pSourceFC, "GRID", pWS, "SourceGrid")
```

This revised Part 1 of ***EucDist*** first sets *pSourceFC* to be the feature class of the source layer. The conversion of *pSourceFC* to a raster involves three steps. First, the code defines the workspace for the raster. Second, the code creates *pConOp* as an instance of the *RasterConversionOp* class and uses the *IRasterAnalysis-Environment* interface to define the output cell size. Third, the code uses the *ToRasterDataset* method on *IConversionOp* to create the source raster referenced by *pSourceDataset*.

11.7.3 *Slice*

Slice calculates continuous Euclidean distance measures away from a stream (source) raster and reclassifies the distance measure raster into five equal intervals. ***Slice*** has three parts. Part 1 defines the source raster. Part 2 runs a distance measure operation and then a reclassify operation. Part 3 creates new raster layers from the output rasters and adds them to the active map.

> **Key Interfaces:** *IRaster, IGeoDataset, IDistanceOp, IReclassOp.*
> **Key Members:** *Raster, EucDistance, Slice, CreateFromRaster.*
> **Usage:** Add *emidastrmgd* to an active map. Import ***Slice*** to Visual Basic Editor. Make sure that both the ESRI Spatial Analyst Extension Object Library and the ESRI Spatial Analyst Shared Object Library are available. Run the macro. The macro creates and adds two new temporary rasters named *EuclideanDistance* and *EqualInterval* respectively to the active map.

```
Private Sub Slice()
  ' Part 1: Define the source raster.
  Dim pMxDoc As IMxDocument
  Dim pMap As IMap
  Dim pSourceRL As IRasterLayer
  Dim pSourceRaster As IRaster
  Set pMxDoc = ThisDocument
  Set pMap = pMxDoc.FocusMap
  Set pSourceRL = pMap.Layer(0)
  Set pSourceRaster = pSourceRL.Raster
```

Part 1 sets *pSourceRaster* to be the raster of the first layer in the active map.

```
  ' Part 2: Define and run the distance measure operation.
  Dim pDistanceOp As IDistanceOp
  Dim pOutputRaster As IGeoDataset
  Dim pReclassOp As IReclassOp
  Dim pSliceRaster As IGeoDataset
  ' Run an EucDistance operation.
  Set pDistanceOp = New RasterDistanceOp
```

```
Set pOutputRaster = pDistanceOp.EucDistance _
(pSourceRaster)
' Run a slice operation.
Set pReclassOp = New RasterReclassOp
Set pSliceRaster = pReclassOp.Slice(pOutputRaster, _
esriGeoAnalysisSliceEqualInterval, 5)
```

Part 2 first creates *pDistanceOp* as an instance of the *RasterDistanceOp* class and uses the *EucDistance* method to create a continuous distance measure raster referenced by *pOutputRaster*. Then the code creates *pReclassOp* as an instance of the *RasterReclassOp* class and uses the *Slice* method on *IReclassOp* to create an equal-interval raster with five classes. The equal-interval or sliced raster is referenced by *pSliceRaster*.

```
' Part 3: Create the new raster layers and add them
' to the active map.
Dim pOutputLayer As IRasterLayer
Dim pSliceLayer As IRasterLayer
' Create and add the distance measure layer.
Set pOutputLayer = New RasterLayer
pOutputLayer.CreateFromRaster pOutputRaster
pOutputLayer.Name = "EuclideanDistance"
pMap.AddLayer pOutputLayer
' Create and add the slice layer.
Set pSliceLayer = New RasterLayer
pSliceLayer.CreateFromRaster pSliceRaster
pSliceLayer.Name = "EqualInterval"
pMap.AddLayer pSliceLayer
End Sub
```

Part 3 creates new raster layers from *pOutputRaster* and *pSliceRaster*, respectively, and adds these two layers to the active map.

11.7.4 *CostDist*

CostDist uses a source raster and a cost raster to calculate the least accumulative cost distance. The macro performs the same function as using the Distance/Cost Weighted option in Spatial Analyst. *CostDist* has three parts. Part 1 defines the source and cost rasters. Part 2 runs a cost distance measure operation. Part 3 creates a new raster layer from the output and adds it to the active map.

Key Interfaces: *IRaster, IDistanceOp.*
Key Members: *Raster, CostDistance, CreateFromRaster.*

Usage: Add *emidastrmgd*, a source raster, and *emidacostgd*, a cost raster, to an active map. The source raster must be on top of the cost raster in the active map. Import ***CostDist*** to Visual Basic Editor. Make sure that the ESRI Spatial Analyst Extension Object Library is available. Run the macro. The macro creates and adds a new temporary raster named *CostDistance* to the active map.

```vba
Private Sub CostDist()
  ' Part 1: Define the source and cost rasters.
  Dim pMxDoc As IMxDocument
  Dim pMap As IMap
  Dim pSourceRL As IRasterLayer
  Dim pSourceRaster As IRaster
  Dim pCostRL As IRasterLayer
  Dim pCostRaster As IRaster
  Set pMxDoc = ThisDocument
  Set pMap = pMxDoc.FocusMap
  ' Define the source dataset.
  Set pSourceRL = pMap.Layer(0)
  Set pSourceRaster = pSourceRL.Raster
  ' Define the cost dataset.
  Set pCostRL = pMap.Layer(1)
  Set pCostRaster = pCostRL.Raster
```

Part 1 sets *pSourceRaster* to be the raster of the top layer in the active map and *pCostRaster* to be the raster of the second layer.

```vba
  ' Part 2: Perform cost distance measures.
  Dim pDistanceOp As IDistanceOp
  Dim pOutputRaster As IRaster
  ' Run the cost distance operation.
  Set pDistanceOp = New RasterDistanceOp
  Set pOutputRaster = pDistanceOp.CostDistance _
  (pSourceRaster, pCostRaster)
```

Part 2 creates *pDistanceOp* as an instance of the *RasterDistanceOp* class and uses the *CostDistance* method on *IDistanceOp* to create a least accumulative cost distance raster referenced by *pOutputRaster*.

```vba
  ' Part 3: Create the output layer and add it to
  ' the active map.
  Dim pRLayer As IRasterLayer
  Set pRLayer = New RasterLayer
  pRLayer.CreateFromRaster pOutputRaster
```

```
    pRLayer.Name = "CostDistance"
  pMap.AddLayer pRLayer
End Sub
```

Part 3 creates a new raster layer from *pOutputRaster* and adds the layer to the active map.

11.7.5 *CostDistFull*

CostDistFull uses a source raster and a cost raster to calculate a back link raster and an allocation raster in addition to a least accumulative cost distance raster. The macro performs the same function as the Distance/Cost Weighted command, with both Create direction and Create allocation checked, in Spatial Analyst. *CostDistFull* has four parts. Part 1 defines the source and cost rasters. Part 2 defines and runs a cost distance measure operation. From the output of Part 2, Part 3 derives the least accumulative cost, back link, and allocation rasters. Part 4 creates new raster layers of the least accumulative cost, back link, and allocation and adds these layers to the active map.

> **Key Interfaces:** *IRaster, IGeoDataset, IDistanceOp, IRasterBandCollection, IRasterband.*
> **Key Members:** *Raster, CostDistanceFull, Item(), AppendBand, CreateFromRaster, Name.*
> **Usage:** Add *emidastrmgd*, a source raster, and *emidacostgd*, a cost raster, to an active map. The source raster must be an integer raster for calculating the allocation output. In the active map, *emidastrmgd* must be on top of *emidacostgd*. Import *CostDistFull* to Visual Basic Editor in ArcMap. Make sure that the ESRI Spatial Analyst Extension Object Library is available. Run the macro. The macro creates and adds three temporary rasters: *CostDistance* for the least accumulative cost distance, *BackLink* for the back link, and *Allocation* for the allocation. *BackLink* contains cell values from zero through eight, which define the next neighboring cell along the least accumulative cost path from a cell to reach its closest source cell. *Allocation* assigns each cell to its closest source cell.

```
Private Sub CostDistFull()
  ' Part 1: Define the source and cost rasters.
  Dim pMxDoc As IMxDocument
  Dim pMap As IMap
  Dim pSourceRL As IRasterLayer
  Dim pSourceRaster As IRaster
  Dim pCostRL As IRasterLayer
  Dim pCostRaster As IRaster
  Set pMxDoc = ThisDocument
  Set pMap = pMxDoc.FocusMap
  ' Define the source dataset.
```

```
Set pSourceRL = pMap.Layer(0)
Set pSourceRaster = pSourceRL.Raster
' Define the cost dataset.
Set pCostRL = pMap.Layer(1)
Set pCostRaster = pCostRL.Raster
```

Part 1 sets *pSourceRaster* to be the raster of the source layer and *pCostRaster* to be the raster of the cost layer.

```
' Part 2: Perform cost distance measures.
Dim pDistanceOp As IDistanceOp
Dim pOutputRaster As IGeoDataset
' Run the cost distance operation.
Set pDistanceOp = New RasterDistanceOp
Set pOutputRaster =
pDistanceOp.CostDistanceFull(pSourceRaster, _
pCostRaster, True, True, True)
```

Part 2 creates *pDistanceOp* as an instance of the *RasterDistanceOp* class and uses the *CostDistanceFull* method on *IDistanceOp* to create the output referenced by *pOutputRaster*. The *CostDistanceFull* method creates the least accumulative cost distance, back link, and allocation rasters all at once.

```
' Part 3: Derive the least accumulative cost
' distance, backlink, and allocation rasters.
Dim pRasterBandCollection As IRasterBandCollection
Dim pCostDistRB As IRasterband
Dim pCostDistRBCollection As IRasterBandCollection
Dim pBackLinkRB As IRasterband
Dim pBackLinkRBCollection As IRasterBandCollection
Dim pAllocationRB As IRasterband
Dim pAllocationRBCollection As IRasterBandCollection
' Extract and create the least accumulative cost
' distance raster.
Set pRasterBandCollection = pOutputRaster
Set pCostDistRB = pRasterBandCollection.Item(0)
Set pCostDistRBCollection = New Raster
pCostDistRBCollection.AppendBand pCostDistRB
' Extract and create the backlink raster.
Set pBackLinkRB = pRasterBandCollection.Item(1)
Set pBackLinkRBCollection = New Raster
pBackLinkRBCollection.AppendBand pBackLinkRB
```

```
' Extract and create the allocation raster.
Set pAllocationRB = pRasterBandCollection.Item(2)
Set pAllocationRBCollection = New Raster
pAllocationRBCollection.AppendBand pAllocationRB
```

To extract the three rasters created in Part 2, Part 3 first performs a QI for the *IRasterBandCollection* interface and assigns the first band, which contains the least accumulative cost distance output to *pCostDistRB*. Next the code creates *pCostDistRBCollection* as an instance of the *Raster* class and uses the *AppendBand* method on *IRasterBandCollection* to append *pCostDistRB* to *pCostDistRBCollection*, thus completing the creation of the least accumulative cost distance raster. Part 3 repeats the same procedure to extract and create the back link raster referenced by *pBackLinkRBCollection* and the allocation raster referenced by *pAllocationRB-Collection*.

```
' Part 4: Create the output layers and add them to
' the active map.
Dim pCostDistLayer As IRasterLayer
Dim pBackLinkLayer As IRasterLayer
Dim pAllocationLayer As IRasterLayer
' Add the least accumulative cost distance layer.
Set pCostDistLayer = New RasterLayer
pCostDistLayer.CreateFromRaster pCostDistRBCollection
pCostDistLayer.Name = "CostDistance"
pMap.AddLayer pCostDistLayer
' Add the back link layer.
Set pBackLinkLayer = New RasterLayer
pBackLinkLayer.CreateFromRaster pBackLinkRBCollection
pBackLinkLayer.Name = "BackLink"
pMap.AddLayer pBackLinkLayer
' Add the allocation layer.
Set pAllocationLayer = New RasterLayer
pAllocationLayer.CreateFromRaster _
pAllocationRBCollection
pAllocationLayer.Name = "Allocation"
pMap.AddLayer pAllocationLayer
End Sub
```

Part 4 creates new raster layers from *pCostDistRBCollection, pBackLinkRBCollection*, and *pAllocationRBCollection*, and adds these layers to the active map.

Terrain Mapping and Analysis

Terrain mapping refers to use of techniques such as contours, hill shading, and perspective views to depict the land surface. Terrain analysis provides measures of the land surface such as slope and aspect. Terrain analysis also includes viewshed analysis and watershed analysis. A viewshed analysis predicts areas of the land surface that are visible from one or more observation points. A watershed analysis can derive watersheds from an elevation raster. Terrain mapping and analysis is useful for a wide variety of applications.

A common data source for terrain mapping and analysis is the digital elevation model (DEM). A DEM consists of a regular array of elevation points compiled from air photos, satellite images, radar data, and other data sources. For terrain mapping and analysis, a DEM is first converted to an elevation raster. The simple data structure of an elevation raster makes it relatively easy to perform computations that are necessary for deriving slope, aspect, and other topographic parameters.

An alternative to the DEM is the triangulated irregular network (TIN). A TIN approximates the land surface with a series of non-overlapping triangles. Elevation values and x-, y-coordinates are stored at nodes that make up the triangles. Many users of geographic information systems (GIS) compile an initial TIN from a DEM or LIDAR (light detection and ranging) data and then use other data sources such as a stream network to modify and improve the TIN. Besides being flexible in terms of data sources, TIN is also an excellent data model for terrain mapping and three-dimensional display. The triangular facets of a TIN create a sharper image of the terrain than a DEM does.

This chapter covers terrain mapping and analysis. Section 12.1 reviews terrain mapping and analysis in ArcGIS. Section 12.2 discusses objects that are related to terrain mapping and analysis. Section 12.3 includes macros for deriving contour, slope, aspect, and hillshade from an elevation raster. Section 12.4 has a macro for viewshed analysis. Section 12.5 has a macro for watershed analysis. Section 12.6 offers macros for compiling and modifying a TIN and for deriving features of a TIN All macros start with the listing of key interfaces and key members (i.e., properties and methods) and the usage.

12.1 PERFORMING TERRAIN MAPPING AND ANALYSIS IN ARCGIS

ArcGIS Desktop incorporates commands for terrain mapping and analysis in the Spatial Analyst and 3D Analyst extensions. Both extensions have a surface analysis menu, which includes contour, slope, aspect, hillshade, and viewshed. The input to these analysis functions can be either an elevation raster or a TIN, and the output is in raster format except for the contour, which is in shapefile format.

The Raster Calculator in Spatial Analyst is an important tool for terrain mapping and analysis because it can evaluate many surface analysis functions. For example, it can evaluate the slope function and create slope rasters directly without going through the surface analysis menu. The Raster Calculator can also evaluate watershed analysis functions that are not incorporated into Spatial Analyst's menu selections.

3D Analyst has menu selections for creating or modifying TINs. We can create an initial TIN from a DEM or feature data such as LIDAR data and contour lines, and then add point, line, and area features to modify the TIN. Additional point data may include surveyed elevation points and GPS (global positioning system) data. Line data may include breaklines such as streams, shorelines, ridges, and roads that represent changes of the land surface. And area data may include lakes and reservoirs. 3D Analyst also has a three-dimensional viewing application called ArcScene that lets the user prepare and manipulate perspective views, three-dimensional draping, and three-dimensional animation.

12.2 ARCOBJECTS FOR TERRAIN MAPPING AND ANALYSIS

ArcObjects organizes objects for terrain mapping and analysis within the Spatial Analyst and 3D Analyst subsystems. Two primary components in the Spatial Analyst subsystem for terrain mapping and analysis are *RasterSurfaceOp* and *RasterHydrologyOp*.

A *RasterSurfaceOp* object supports *IRasterAnalysisEnvironment* and *ISurfaceOp*. *IRasterAnalysisEnvironment* controls the analysis environment such as the output workspace and the output cell size. *ISurfaceOp* has methods for creating contour; calculating hillshade, slope, aspect, and curvature; and performing viewshed analysis (Figure 12.1).

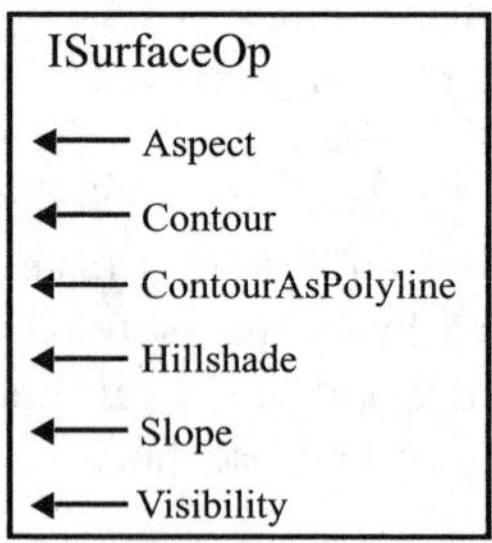

Figure 12.1 *ISurfaceOp* has methods for deriving contour, slope, aspect, hillshade, and view-shed from an elevation raster.

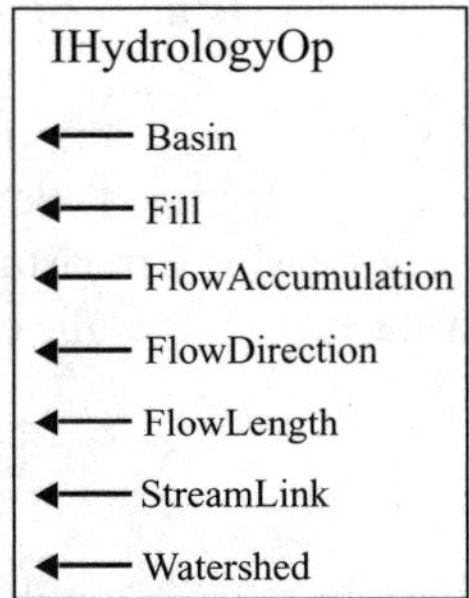

Figure 12.2   *IHydrologyOp* has methods for deriving hydrologic parameters for watershed analysis.

A *RasterHydrologyOp* object supports *IRasterAnalysisEnvironment* and *IHydrologyOp*. *IHydrologyOp* has methods for filling sinks in a surface, creating flow direction, creating flow accumulation, assigning stream links, and delineating watersheds (Figure 12.2).

The primary component in the 3D Analyst subsystem is *TIN*. A TIN object implements *ITinEdit*, *ITinAdvanced*, and *ITinSurface* (Figure 12.3). *ITinEdit* has methods for constructing and editing TINs. *ITinAdvanced* has access to the underlying data structure of a TIN such as numbers of nodes, edges, and triangles. And *ITinSurface* provides surface analysis functions such as contouring and deriving slope and aspect from a TIN.

Raster objects covered in Chapter 11 may also be involved in terrain mapping and analysis. For example, to convert an elevation raster to a TIN requires use of basic raster data objects.

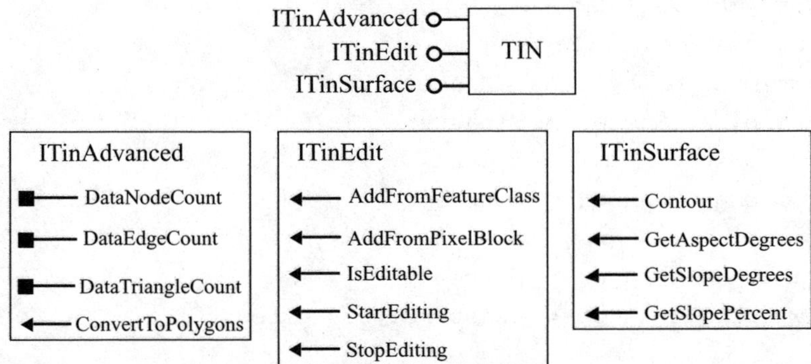

Figure 12.3   A *TIN* object supports *ITinAdvanced*, *ITinEdit*, and *ITinSurface*. These interfaces have members to create and edit a TIN, to derive topographic measures from a TIN, and to derive numbers of nodes, edges, and triangles from a TIN.

12.3 DERIVING CONTOUR, SLOPE, ASPECT, AND HILLSHADE

This section covers contour, slope, aspect, and hillshade. Besides showing how to derive these topographic measures from an elevation raster, this section also includes code fragments for selecting slope measurement units, classifying slope measures, and displaying aspect measures in eight principal directions.

12.3.1 *Contour*

Contour creates contour lines from an elevation raster by connecting points of equal elevation. The macro performs the same function as using the Surface Analysis/Contour command in Spatial Analyst to create a contour line shapefile. *Contour* has three parts. Part 1 defines the input elevation raster. Part 2 performs contouring and saves the output in a specified workspace. Part 3 creates a feature layer from the output, prepares the contour line labels, and adds the layer to the active map.

> **Key Interfaces:** *ISurfaceOp, IRasterAnalysisEnvironment, IWorkspaceFactory, IGeoDataset, IGeoFeatureLayer.*
> **Key Members:** *Raster, OpenFromFile, OutWorkspace, Contour, FeatureClass, DisplayField, DisplayAnnotation.*
> **Usage:** Add *emidalat*, an elevation raster, to an active map. Import *Contour* to Visual Basic Editor. Make sure that the ESRI Spatial Analyst Shared Object Library is checked in the References dialog. Run the macro. The macro adds *Contour* with labels to the active map.

```
Private Sub Contour()
  ' Part 1: Define the input raster.
  Dim pMxDoc As IMxDocument
  Dim pMap As IMap
  Dim pLayer As ILayer
  Dim pRasterLayer As IRasterLayer
  Dim pInputRaster As IRaster
  Set pMxDoc = ThisDocument
  Set pMap = pMxDoc.FocusMap
  Set pLayer = pMap.Layer(0)
  Set pRasterLayer = pLayer
  Set pInputRaster = pRasterLayer.Raster
```

Part 1 sets *pInputRaster* to be the raster of the top layer in the active map.

```
  ' Part 2: Perform contour.
  Dim pSurfaceOp As ISurfaceOp
  Dim pEnv As IRasterAnalysisEnvironment
  Dim pWS As IWorkspace
```

```
Dim pWSF As IWorkspaceFactory
Dim pOutput As IDataset
' Define a surface operation.
Set pSurfaceOp = New RasterSurfaceOp
Set pEnv = pSurfaceOp
Set pWSF = New RasterWorkspaceFactory
Set pWS = pWSF.OpenFromFile("c:\data\chap12", 0)
Set pEnv.OutWorkspace = pWS
' Run the contour surface operation.
Set pOutput = pSurfaceOp.Contour(pInputRaster, 50, 850)
POutput.Rename ("Contour")
```

Part 2 first creates *pSurfaceOp* as an instance of the *RasterSurfaceOp* class. Next the code performs a QueryInterface (QI) for *IRasterAnalysisEnvironment* and uses the *OutWorkspace* property to assign *pWS* as the output workspace. The code then runs the *Contour* method on *ISurfaceOp* to create a geographic dataset referenced by *pOutput*. The *Contour* method uses the arguments of 50 for the contour interval and 850 for the base contour. The contour interval represents the vertical distance between contour lines, and the base contour is the contour line of the lowest elevation. The *Rename* method on *IDataset* changes the name of *pOutput* to *Contour*. (Without renaming, the contour dataset will have a default name of Shape*.shp.)

```
' Part 3: Create the output layer, and add it to
' the active map.
Dim pOutLayer As IFeatureLayer
Dim pGeoFeatureLayer As IGeoFeatureLayer
' Create the output feature layer.
Set pOutLayer = New FeatureLayer
Set pOutLayer.FeatureClass = pOutput
pOutLayer.Name = "Contour"
' Label the contour lines.
Set pGeoFeatureLayer = pOutLayer
pGeoFeatureLayer.DisplayField = "CONTOUR"
pGeoFeatureLayer.DisplayAnnotation = True
pMap.AddLayer pOutLayer
End Sub
```

Part 3 first creates a new feature layer from *pOutput* and names the layer Contour. The code then switches to the *IGeoFeatureLayer* interface to set up the contour labels before adding the layer to the active map.

12.3.2 *Slope*

Slope derives a temporary slope raster from an elevation raster. The macro performs the same function as using the Surface Analysis/Slope command in Spatial Analyst. *Slope* has three parts. Part 1 defines the input raster. Part 2 runs the slope operation and saves the output in a specified workspace. Part 3 creates a raster layer from the output and adds the layer to the active map.

> **Key Interfaces:** *ISurfaceOp, IRasterAnalysisEnvironment, IWorkspace, IWorkspace-Factory.*
> **Key Members:** *Raster, OpenFromFile, OutWorkspace, Slope, CreateFromRaster.*
> **Usage:** Add *emidalat*, an elevation raster, to an active map. Import *Slope* to Visual Basic Editor. Make sure that the ESRI Spatial Analyst Shared Object Library is checked in the References dialog. Run the macro. The macro adds a slope layer to the active map.

```
Private Sub Slope()
  ' Part 1: Define the input raster.
  Dim pMxDoc As IMxDocument
  Dim pMap As IMap
  Dim pLayer As ILayer
  Dim pRasterLayer As IRasterLayer
  Dim pInputRaster As IRaster
  Set pMxDoc = ThisDocument
  Set pMap = pMxDoc.FocusMap
  Set pLayer = pMap.Layer(0)
  Set pRasterLayer = pLayer
  Set pInputRaster = pRasterLayer.Raster
```

Part 1 sets *pInputRaster* to be the raster of the top layer in the active map.

```
  ' Part 2: Perform slope.
  Dim pSurfaceOp As ISurfaceOp
  Dim pEnv As IRasterAnalysisEnvironment
  Dim pWS As IWorkspace
  Dim pWSF As IWorkspaceFactory
  Dim pOutRaster As IRaster
  ' Prepare a raster surface operation.
  Set pSurfaceOp = New RasterSurfaceOp
  Set pEnv = pSurfaceOp
  Set pWSF = New RasterWorkspaceFactory
  Set pWS = pWSF.OpenFromFile("c:\data\chap12", 0)
  Set pEnv.OutWorkspace = pWS
```

```
' Run the slope surface operation.
Set pOutRaster = pSurfaceOp.Slope(pInputRaster, _
esriGeoAnalysisSlopePercentrise)
```

Part 2 first creates *pSurfaceOp* as an instance of the *RasterSurfaceOp* class. Next the code accesses the *IRasterAnalysisEnvironment* interface and sets the workspace for the output. Then the code uses the *Slope* method on *ISurfaceOp* to create a slope raster referenced by *pOutRaster*. The argument of esriGeoAnalysisSlopePercentrise specifies that the slope raster is measured in percentage of rise.

```
' Part 3: Create the output raster layer, and add
' it to the active map.
Set pSlopeLayer = New RasterLayer
pSlopeLayer.CreateFromRaster pOutRaster
pSlopeLayer.Name = "Slope"
pMap.AddLayer pSlopeLayer
End Sub
```

Part 3 creates a new raster layer from *pOutRaster* and adds the layer to the active map.

12.3.3 Choice of Slope Measure

A slope raster can be measured in either degree slope or percent slope. Percent slope is 100 times the ratio of rise (vertical distance) over run (horizontal distance), whereas degree slope is the arc tangent of the ratio of rise over run. Because different projects may require different slope measures, a macro can use an input box and let the user choose one of these two options before running the slope operation. To offer these two options, one can replace Part 2 of *Slope* by the following code fragment (ChooseSlopeMeasure.txt on the companion CD incorporates the revision):

```
' Part 2: Perform slope.
Dim pSurfaceOp As ISurfaceOp
Dim pEnv As IRasterAnalysisEnvironment
Dim pWS As IWorkspace
Dim pWSF As IWorkspaceFactory
Dim Message As String
Dim Default As String
Dim Choice As String
Dim pOutRaster As IRaster
' Prepare a raster surface operation.
Set pSurfaceOp = New RasterSurfaceOp
Set pEnv = pSurfaceOp
Set pWSF = New RasterWorkspaceFactory
```

```
Set pWS = pWSF.OpenFromFile("c:\data\chap12", 0)
Set pEnv.OutWorkspace = pWS
' Choose slope measures in degrees or in percentage
' of rise.
Message = _
"Enter D for degrees or P for percentage of rise"
Default = "D"
Choice = InputBox(Message,, Default)
' Run the slope surface operation according to the choice.
If Choice = "D" Or Choice = "d" Then
  Set pOutRaster = pSurfaceOp.Slope(pInputRaster, _
  esriGeoAnalysisSlopeDegrees)
    ElseIf Choice = "P" Or Choice = "p" Then
      Set pOutRaster = pSurfaceOp.Slope(pInputRaster, _
      esriGeoAnalysisSlopePercentrise)
    Else: Exit Sub
  End If
```

The code fragment runs the *Slope* method with the user's choice of the slope type (D for degrees and P for percentage of rise). D for degrees is the default.

12.3.4 *ReclassifySlope*

A slope raster is often classified into slope classes before it is used in a GIS project. ***ReclassifySlope*** is a sub that can be called at the end of ***Slope*** to group slope measures into five classes (ReclassifySlope.txt on the companion CD is a module that has both ***Slope*** and ***ReclassifySlope***):

```
' Call the sub to display the slope layer in a
' defined color scheme.
Call ReclassifySlope(pSlopeLayer)
```

ReclassifySlope has three parts. Part 1 gets *pSlopeLayer* from ***Slope*** for the input raster. Part 2 performs reclassification. Part 3 creates a raster layer from the output and adds the layer to the active map.

```
Private Sub ReclassifySlope(pSlopeLayer As IRasterLayer)
  ' Part 1: Define the raster for reclassify.
  Dim pMxDoc As IMxDocument
  Dim pMap As IMap
  Dim pRasterLy As IRasterLayer
  Dim pGeoDs As IGeoDataset
  Set pMxDoc = ThisDocument
```

```
Set pMap = pMxDoc.FocusMap
Set pRasterLy = pSlopeLayer
Set pGeoDs = pRasterLy.Raster
```

Part 1 gets *pSlopeLayer* from **Slope** and assigns the layer to *pRasterLy*. The code then sets *pGeoDs* to be the raster of *pRasterLy* via *IGeoDataset*.

```
' Part 2: Reclassify the input raster.
Dim pReclassOp As IReclassOp
Dim pNRemap As INumberRemap
Dim pOutRaster As IRaster
' Prepare a number remap.
Set pNRemap = New NumberRemap
pNRemap.MapRange 0, 10#, 1
pNRemap.MapRange 10.1, 20#, 2
pNRemap.MapRange 20.1, 30#, 3
pNRemap.MapRange 30.1, 40#, 4
pNRemap.MapRange 40.1, 90#, 5
' Run the reclass operation.
Set pReclassOp = New RasterReclassOp
Set pOutRaster = pReclassOp.ReclassByRemap _
(pGeoDs, pNRemap, False)
```

Part 2 first creates *pNRemap* as an instance of the *NumberRemap* class and uses the *MapRange* method on *INumberRemap* to set the output value based on a numeric range of the input values. A cell within the numeric range of 0 to 10.0 is assigned an output value of 1, 10.1 to 20.0 an output value of 2, and so on. Next the code creates *pReclassOp* as an instance of the *RasterReclassOp* class and uses the *ReclassByRemap* method on *IReclassOp* to create a reclassified raster referenced by *pOutRaster*. The reclassified raster has five slope classes.

```
' Part 3: Create the output layer, and add it to
' the active map.
Dim pReclassLy As IRasterLayer
Set pReclassLy = New RasterLayer
pReclassLy.CreateFromRaster pOutRaster
pReclassLy.Name = "Classified Slope"
pMap.AddLayer pReclassLy
End Sub
```

Part 3 creates a new raster layer from *pOutRaster* and adds the classified slope layer to the active map.

12.3.5 Aspect

To derive a temporary aspect raster from an elevation raster, one can use the same macro as **Slope** but change the last line statement in Part 2 to:

```
' Run the aspect surface operation.
Set pOutRaster = pSurfaceOp.Aspect(pInputRaster)
```

The *Aspect* method on *ISurfaceOp* uses *pInputRaster* as the only object qualifier.

12.3.6 *Aspect_Symbol*

Aspect is a circular measure, which starts with $0°$ at the north, moves clockwise, and ends with $360°$ also at the north. An aspect raster is typically classified into four or eight principal directions and an additional class for flat areas. **Aspect_Symbol** uses a random color ramp to display an aspect raster with eight principal directions plus flat areas. One way to use **Aspect_Symbol** is to call the sub at the end of a macro that has derived an aspect layer. The macro can then pass the aspect layer as an argument to **Aspect_Symbol.** (Aspect_Symbol.txt on the companion CD includes **Aspect** for deriving aspect measures from *emidalat* and **Aspect_Symbol** for displaying aspect measures.)

Aspect_Symbol has three parts. Part 1 creates a raster renderer. Part 2 specifies the color, break, and label for each aspect class. Part 3 assigns the renderer to the raster layer and refreshes the active view.

```
Private Sub Aspect_Symbol(pAspectLayer As IRasterLayer)
  ' Part 1: Create a raster renderer.
  Dim pMxDoc As IMxDocument
  Dim pClassRen As IRasterClassifyColorRampRenderer
  Dim pRasRen As IRasterRenderer
  Dim pRaster As IRaster
  Set pRaster = pAspectLayer.Raster
  ' Prepare a raster classify renderer.
  Set pClassRen = New RasterClassifyColorRampRenderer
  pClassRen.ClassCount = 10
  ' Define the raster and update the renderer.
  Set pRasRen = pClassRen
  Set pRasRen.Raster = pRaster
  pRasRen.Update
```

Part 1 first sets *pRaster* to be the raster of the aspect layer passed to the sub as an argument. Next the code creates *pClassRen* as an instance of the *RasterClassifyColorRampRenderer* class and specifies 10 for the number of classes. The code then switches to the *IRasterRenderer* interface, assigns *pRaster* to be the raster for *pRasRen*, and updates *pRasRen*.

```
' Part 2: Specify the color, break, and label for
' each aspect class.
Dim pRamp As IRandomColorRamp
Dim pColors As IEnumColors
Dim pFSymbol As ISimpleFillSymbol
' Prepare a random color ramp.
Set pRamp = New RandomColorRamp
pRamp.Size = 10
pRamp.Seed = 100
pRamp.CreateRamp (True)
Set pColors = pRamp.Colors
' Define the symbol, break, and label for 10
' aspect classes.
Set pFSymbol = New SimpleFillSymbol
pFSymbol.Color = pColors.Next
pClassRen.Symbol(0) = pFSymbol
pClassRen.Break(0) = -1
pClassRen.Label(0) = "Flat(-1)"
pFSymbol.Color = pColors.Next
pClassRen.Symbol(1) = pFSymbol
pClassRen.Break(1) = -0.01
pClassRen.Label(1) = "North(0-22.5)"
pFSymbol.Color = pColors.Next
pClassRen.Symbol(2) = pFSymbol
pClassRen.Break(2) = 22.5
pClassRen.Label(2) = "Northeast(22.5-67.5)"
pFSymbol.Color = pColors.Next
pClassRen.Symbol(3) = pFSymbol
pClassRen.Break(3) = 67.5
pClassRen.Label(3) = "East(67.5-112.5)"
pFSymbol.Color = pColors.Next
pClassRen.Symbol(4) = pFSymbol
pClassRen.Break(4) = 112.5
pClassRen.Label(4) = "Southeast(112.5-157.5)"
pFSymbol.Color = pColors.Next
pClassRen.Symbol(5) = pFSymbol
pClassRen.Break(5) = 157.5
pClassRen.Label(5) = "South(157.5-202.5)"
pFSymbol.Color = pColors.Next
pClassRen.Symbol(6) = pFSymbol
```

```
pClassRen.Break(6)  =  202.5
pClassRen.Label(6)  =  "Southwest(202.5-247.5)"
pFSymbol.Color = pColors.Next
pClassRen.Symbol(7)  =  pFSymbol
pClassRen.Break(7)  =  247.5
pClassRen.Label(7)  =  "West(247.5-292.5)"
pFSymbol.Color = pColors.Next
pClassRen.Symbol(8)  =  pFSymbol
pClassRen.Break(8)  =  292.5
pClassRen.Label(8)  =  "Northwest(292.5-337.5)"
pFSymbol.Color = pColors.Next
pClassRen.Symbol(9)  =  pFSymbol
pClassRen.Break(9)  =  337.5
pClassRen.Label(9)  =  "North(337.5-360)"
' Set Symbol 9 for north to be the same as
' Symbol 1 for north.
pClassRen.Symbol(9)  =  pClassRen.Symbol(1)
```

Part 2 creates *pRamp* as an instance of the *RandomColorRamp* class and specifies 10 for the number of colors to be generated, 100 for the seed of the generator, and True to generate the color ramp. The rest of Part 2 assigns a random color, a class break, and a label for each of the 10 classes. Because the north aspect includes classes (1) and (9) (0–22.5° and 337.5–360°), the symbol for class (9) is reset to be the same as the symbol for class (1).

```
' Part 3: Assign the renderer to the aspect layer,
' and refresh the active view.
pRasRen.Update
Set pAspectLayer.Renderer = pRasRen
Set pMxDoc = ThisDocument
pMxDoc.ActiveView.Refresh
pMxDoc.UpdateContents
End Sub
```

Part 3 assigns the updated *pRasRen* to *pAspectLayer*, refreshes the active view, and updates the contents of the map document.

12.3.7 Hillshade

To derive a temporary hillshade raster from an elevation raster, one can use the same macro as **Slope** but change the last line statement in Part 2 to:

```
' Run the hillshade surface operation.
```

```
Set pOutRaster = pSurfaceOp.Hillshade(pInputRaster, _
315, 30, True)
```

Besides the object qualifier *pInputRaster*, the *Hillshade* method on *ISurfaceOp* uses three arguments: 315 for the Sun's azimuth, 30 for the Sun's altitude, and True for the shaded relief type to include shadows. The Sun's azimuth is the direction of the incoming light, ranging from 0° (due north) to 360° in a clockwise direction. The Sun's altitude is the angle of the incoming light measured above the horizon between 0 and 90°.

12.4 PERFORMING VIEWSHED ANALYSIS

This section covers viewshed analysis. Inputs to a viewshed analysis include an elevation raster and a feature class containing one or more observation points. The analysis derives areas of the land surface that are visible from the observation point(s).

12.4.1 *Visibility*

Visibility creates a viewshed using an elevation raster and two observation points. The macro performs the same function as using the Surface Analysis/Viewshed command in Spatial Analyst. *Visibility* has three parts. Part 1 defines the elevation and lookout datasets. Part 2 runs the visibility analysis and saves the output in a specified workspace. Part 3 creates a raster layer from the output and adds the layer to the active map.

Key Interfaces: *IGeoDataset, ISurfaceOp, IRasterAnalysisEnvironment, IWorkspace, IWorkspaceFactory.*

Key Members: *FeatureClass, Raster, OpenFromFile, OutWorkspace, Visibility, CreateFromRaster.*

Usage: Add *plne*, an elevation raster, and *lookouts.shp*, a lookout point shapefile, to an active map. Make sure that *lookouts* is on top of *plne* in the table of contents. Import *Visibility* to Visual Basic Editor. Make sure that the ESRI Spatial Analyst Shared Object Library is checked in the References dialog. Run the macro. The macro adds *Viewshed* to the active map.

```
Private Sub Visibility()
   ' Part 1: Define the elevation and lookout datasets.
   Dim pMxDoc As IMxDocument
   Dim pMap As IMap
   Dim pRasterLayer As IRasterLayer
   Dim pRaster As IRaster
   Dim pFeatureLayer As IFeatureLayer
   Dim pLookoutDataset As IGeoDataset
```

```
Set pMxDoc = ThisDocument
Set pMap = pMxDoc.FocusMap
' Define the elevation raster.
Set pRasterLayer = pMap.Layer(1)
Set pRaster = pRasterLayer.Raster
' Define the lookout dataset.
Set pFeatureLayer = pMap.Layer(0)
Set pLookoutDataset = pFeatureLayer.FeatureClass
```

Part 1 defines the elevation and lookout point datasets. The elevation dataset referenced by *pRaster* is the raster of the second layer in the active map. The lookout dataset referenced by *pLookoutDataset* is the feature class of the top layer.

```
' Part 2: Perform visibility analysis.
Dim pSurfaceOp As ISurfaceOp
Dim pEnv As IRasterAnalysisEnvironment
Dim pWS As IWorkspace
Dim pWSF As IWorkspaceFactory
Dim pOutRaster As IGeoDataset
' Define a raster surface operation.
Set pSurfaceOp = New RasterSurfaceOp
Set pEnv = pSurfaceOp
Set pWSF = New RasterWorkspaceFactory
Set pWS = pWSF.OpenFromFile("c:\data\chap12", 0)
Set pEnv.OutWorkspace = pWS
' Run the visibility surface operation.
Set pOutRaster = pSurfaceOp.Visibility(pRaster, _
pLookoutDataset, esriGeoAnalysisVisibilityFrequency)
```

Part 2 creates *pSurfaceOp* as an instance of the *RasterSurfaceOp* class. Next the code uses the *IRasterAnalysisEnvironment* interface to set the workspace for the output. The *Visibility* method on *ISurfaceOp* uses *pRaster* and *pLookoutDataset* as the object qualifiers. In this case, the lookout dataset is a point shapefile. But it can also be a shapefile or coverage containing point or line features. The only other argument used by *Visibility* is the visibility type, which can be one of the following four choices:

- esriGeoAnalysisVisibilityFrequency to record number of times each cell can be seen
- esriGeoAnalysisVisibilityObservers to record which lookout points can be seen
- esriGeoAnalysisVisibilityFrequencyUseCurvature to specify whether Earth curvature corrections will be used with frequency
- esriGeoAnalysisVisibilityObserversUseCurvature to specify whether Earth curvature corrections will be used with observers

Visibility uses the first visibility type. Therefore, the output shows number of times each cell can be seen from the two lookout points.

```
' Part 3: Create the output layer and add it to
' the active map.
Dim pRLayer As IRasterLayer
Set pRLayer = New RasterLayer
pRLayer.CreateFromRaster pOutRaster
pRLayer.Name = "Viewshed"
pMap.AddLayer pRLayer
End Sub
```

Part 3 creates a new raster layer from *pOutRaster* and adds the layer to the active map.

12.5 PERFORMING WATERSHED ANALYSIS

To derive watersheds from an elevation raster requires creation of the following intermediate rasters: a filled elevation raster, a flow direction raster, a flow accumulation raster, and a stream links raster. A filled elevation raster is void of depressions. A flow direction raster shows the direction water will flow out of each cell of a filled elevation raster. A flow accumulation raster tabulates for each cell the number of cells that will flow to it. In other words, a flow accumulation raster shows how many upstream cells that will contribute drainage to each cell. Cells having high accumulation values generally correspond to stream channels. Therefore, a stream links raster can be derived from a flow accumulation raster by using some threshold accumulation value. Stream links and the flow direction raster are the inputs for deriving area-wide watersheds.

12.5.1 *Watershed*

Watershed delineates area-wide watersheds from an elevation raster. In the process, the macro also creates a filled elevation raster, a flow direction raster, a flow accumulation raster, a source raster, and a stream links raster. *Watershed* has eight parts. Part 1 defines the input elevation raster. Part 2 creates a hydrologic operation object and specifies the output workspace. Parts 3 though 8 perform the various hydrologic operations for the purpose of delineating watersheds. As each operation is completed, a raster layer is created and added to the active map. It will take a while to execute the macro, especially if the elevation dataset is large.

Key Interfaces: *IHydrologyOp, IRasterAnalysisEnvironment, IWorkspace, IWorkspaceFactory, IQueryFilter, IRasterDescriptor, ILogicalOp, IRaster.*

Key Members: *Raster, OpenFromFile, OutWorkspace, Fill, CreateFromRaster, Flowdirection, FlowAccumulation, WhereClause, Create, Test, StreamLink, Watershed.*

Usage: Add *emidalat*, an elevation raster, to an active map. Import **Watershed** to Visual Basic Editor. Make sure that both the ESRI Spatial Analyst Shared Object Library and the ESRI Spatial Analysis Extension Object Library are checked in the References dialog. Run the macro. The macro adds *Filled DEM, Flowdirection, Flowaccumulation, Source, Stream link*, and *Watershed* to the active map.

```
Private Sub Watershed()
  ' Part 1: Define the input raster.
  Dim pMxDoc As IMxDocument
  Dim pMap As IMap
  Dim pLayer As ILayer
  Dim pRasterLayer As IRasterLayer
  Dim pInputRaster As IRaster
  Set pMxDoc = ThisDocument
  Set pMap = pMxDoc.FocusMap
  Set pLayer = pMap.Layer(0)
  Set pRasterLayer = pLayer
  Set pInputRaster = pRasterLayer.Raster
```

Part 1 sets *pInputRaster* to be the raster of the top layer in the active map.

```
  ' Part 2: Create a new hydrology operation.
  Dim pHydrologyOp As IHydrologyOp
  Dim pEnv As IRasterAnalysisEnvironment
  Dim pWS As IWorkspace
  Dim pWSF As IWorkspaceFactory
  Set pHydrologyOp = New RasterHydrologyOp
  Set pEnv = pHydrologyOp
  Set pWSF = New RasterWorkspaceFactory
  Set pWS = pWSF.OpenFromFile("c:\data\chap12", 0)
  Set pEnv.OutWorkspace = pWS
```

Part 2 creates *pHydrologyOp* as an instance of the *RasterHydrologyOp* class and uses the *IRasterAnalysisEnvironment* interface to set the workspace for the output.

```
  ' Part 3: Fill sinks.
  Dim pFillDS As IRaster
  Set pFillDS = pHydrologyOp.Fill(pInputRaster)
  ' Add the filled DEM to the active map.
  Set pRasterLayer = New RasterLayer
  pRasterLayer.CreateFromRaster pFillDS
  pRasterLayer.Name = "Filled DEM"
  pMap.AddLayer pRasterLayer
```

Part 3 uses the *Fill* method on *IHydrologyOp* to create a filled elevation raster referenced by *pFillDS*. The code then creates a new raster layer from *pFillDS* and adds the layer to the active map.

```
' Part 4: Derive flow direction.
Dim pFlowdirectionDS As IRaster
Set pFlowdirectionDS = pHydrologyOp.Flowdirection _
(pFillDS, True, True)
' Add the flow direction layer to the active map.
Set pRasterLayer = New RasterLayer
pRasterLayer.CreateFromRaster pFlowdirectionDS
pRasterLayer.Name = "Flowdirection"
pMap.AddLayer pRasterLayer
```

Part 4 uses the *Flowdirection* method on *IHydrologyOp* to create a flow direction raster referenced by *pFlowdirectionDS*. Besides the object qualifier *pFillDS*, *Flowdirection* uses two other arguments: the first specifies whether or not an output raster will be created, and the second determines the flow direction at the edges of the elevation dataset. The code then adds a new flow direction raster layer to the active map.

```
' Part 5: Derive flow accumulation.
Dim pFlowAccumulationDS As IRaster
Set pFlowAccumulationDS = pHydrologyOp.FlowAccumulation _
(pFlowdirectionDS)
' Add the flow accumulation layer to the active map.
Set pRasterLayer = New RasterLayer
pRasterLayer.CreateFromRaster pFlowAccumulationDS
pRasterLayer.Name = "Flowaccumulation"
pMap.AddLayer pRasterLayer
```

Part 5 uses the *FlowAccumulation* method on *IHydrologyOp* to create a flow accumulation raster referenced by *pFlowAccumulationDS*. The code then adds a new flow accumulation raster layer to the active map.

```
' Part 6: Derive the source raster.
Dim pQFilter As IQueryFilter
Dim pRasDes As IRasterDescriptor
Dim pLogicalOp As ILogicalOp
Dim pSourceDS As IRaster
' Use a minimum of 500 cells.
Set pQFilter = New QueryFilter
```

```vba
pQFilter.WhereClause = "Value > 500"
Set pRasDes = New RasterDescriptor
pRasDes.Create pFlowAccumulationDS, pQFilter, "value"
' Run a logical operation.
Set pLogicalOp = New RasterMathOps
Set pSourceDS = pLogicalOp.Test(pRasDes)
' Add the source layer to the active map.
Set pRasterLayer = New RasterLayer
pRasterLayer.CreateFromRaster pSourceDS
pRasterLayer.Name = "Source"
pMap.AddLayer pRasterLayer
```

Part 6 determines which cells in the flow accumulation raster are to be included in the source dataset. The code first creates *pQFilter* as an instance of the *QueryFilter* class and defines its *WhereClause* condition as "Value > 500." Next the code uses the flow accumulation dataset, *pQFilter*, and the field name of value to create an instance of the *RasterDescriptor* class referenced by *pRasDes*. Then the code creates *pLogicalOp* as an instance of the *RasterMathOps* class and uses the *Test* method to create a source dataset referenced by *pSourceDS*. The source dataset has one for those cells that have value > 500 and no data for those cells that have value ≤ 500. A new source raster layer is then added to the active map.

```vba
' Part 7: Derive stream links.
Dim pStreamLinkDS As IRaster
Set pStreamLinkDS = pHydrologyOp.StreamLink _
(pSourceDS,pFlowdirectionDS)
' Add the stream link layer to the active map.
Set pRasterLayer = New RasterLayer
pRasterLayer.CreateFromRaster pStreamLinkDS
pRasterLayer.Name = "Stream link"
pMap.AddLayer pRasterLayer
```

Part 7 uses the *StreamLink* method on *IHydrologyOp* to create a stream links raster referenced by *pStreamLinkDS*. The code then adds a new stream links raster layer to the active map.

```vba
' Part 8: Derive watersheds.
Dim pWatershed As IRaster
Set pWatershed = pHydrologyOp.Watershed _
(pFlowdirectionDS, pStreamLinkDS)
' Add the watershed layer to the active map.
Set pRasterLayer = New RasterLayer
```

```
    pRasterLayer.CreateFromRaster pWatershed
    pRasterLayer.Name = "Watershed"
    pMap.AddLayer pRasterLayer
End Sub
```

Part 8 applies the *Watershed* method on *IHydrologyOp* to create a watershed raster referenced by *pWatershed*. The code then adds a new watershed raster layer to the active map.

12.6 CREATING AND EDITING TIN

This section includes three sample macros. The first macro converts an elevation raster (i.e., a DEM) to a TIN. The second macro modifies the initial TIN by using streams as breaklines. And the third macro reports the numbers of nodes and triangles that make up the modified TIN.

12.6.1 *RasterToTin*

RasterToTin converts an elevation raster to a TIN. The macro performs the same task as using the Convert/Raster to TIN command in 3D Analyst. *RasterToTin* has three parts. Part 1 defines the input raster band and initiates a new TIN. Part 2 creates a pixel block and reads into it raw pixels from the raster band. Part 3 uses the pixel block to populate the TIN and adds the TIN layer to the active map.

Key Interfaces: *IRasterBandCollection, IRasterBand, ITinEdit, IRawPixels, IRasterProps, IPixelBlock, IEnvelope, ITinLayer.*
Key Members: *Item(), InitNew, SetCoords, CreatePixelBlock, Read, SafeArray(), Extent, AddFromPixelBlock, Dataset.*
Usage: Add *emidalat*, an elevation raster, to an active map. Import *RasterToTin* to Visual Basic Editor. Make sure that the ESRI Tin Object Library is checked in the References dialog. Run the macro. The macro adds a new TIN to the active map.

```
Private Sub RasterToTin()
    ' Part 1: Define the input raster band and initiate
    ' a new TIN.
    Dim pMxDoc As IMxDocument
    Dim pMap As IMap
    Dim pInputRL As IRasterLayer
    Dim pInRaster As IRaster
    Dim pRasGDS As IGeoDataset
    Dim pRasterBandColl As IRasterBandCollection
    Dim pRasterBand As IRasterBand
    Dim pTin As ITinEdit
```

```
Set pMxDoc = ThisDocument
Set pMap = pMxDoc.FocusMap
Set pInputRL = pMap.Layer(0)
Set pInRaster = pInputRL.Raster
' Extract the first raster band.
Set pRasterBandColl = pInRaster
Set pRasterBand = pRasterBandColl.Item(0)
' Initiate a new TIN.
Set pTin = New Tin
Set pRasGDS = pInRaster
pTin.InitNew pRasGDS.Extent
```

Part 1 sets *pInRaster* to be the raster of the first layer in the active map. Next the code performs a QI for the *IRasterBandCollection* interface and uses the *Item* method to assign the first raster band of *pInRaster* to *pRasterBand*. The code then uses the *InitNew* method on *ITinEdit* to create *pTin* as an instance of the *TIN* class. The initialization of a TIN object requires an object qualifier that defines the data area, which in this case corresponds to the extent of *pInRaster*. The *InitNew* method also places *pTin* in edit mode.

```
' Part 2: Create a pixel block.
Dim pRawPixels As IRawPixels
Dim pProps As IRasterProps
Dim pBlockSize As IPnt
Dim pPixelBlock As IPixelBlock
Dim pBlockOrigin As IPnt
' Define a block size.
Set pProps = pRasterBand
Set pBlockSize = New DblPnt
pBlockSize.SetCoords pProps.Width, pProps.Height
' Allocate a pixel block.
Set pRawPixels = pRasterBand
Set pPixelBlock = pRawPixels.CreatePixelBlock _
(pBlockSize)
' Define the block origin.
Set pBlockOrigin = New DblPnt
pBlockOrigin.SetCoords 0, 0
' Read into the pixel block from the input raster band.
pRawPixels.Read pBlockOrigin, pPixelBlock
```

Part 2 reads *pRasterBand* from Part 1 into a block of pixels. The process involves defining a block size, allocating a pixel block, and reading the cell values into the

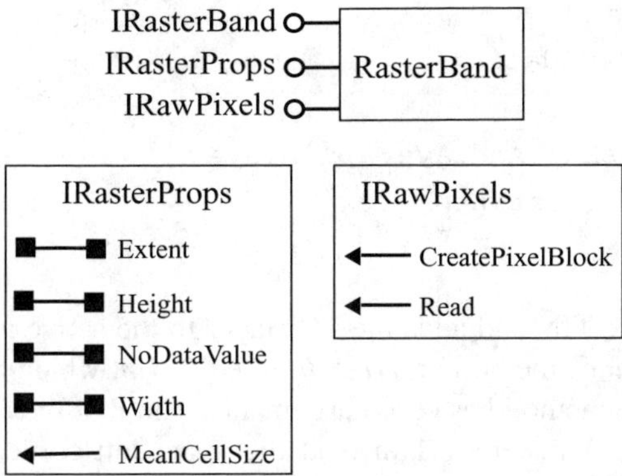

Figure 12.4 A *RasterBand* object supports *IRasterBand*, *IRasterProps*, and *IRawPixels*. These interfaces can be used to convert an elevation raster into a TIN.

pixel block. The process also involves *IRawPixels* and *IRasterProps* that a *RasterBand* object supports (Figure 12.4). The code first performs a QI for the *IRasterProps* interface and uses the width and height of *pRasterBand* to define a block size referenced by *pBlockSize*. Next the code switches to the *IRawPixels* interface and uses the *CreatePixelBlock* method to allocate a pixel block referenced by *pPixelBlock*. The allocation is based on the width and height of *pBlockSize*. After defining the block origin at (0, 0), the code uses the *Read* method on *IRawPixels* to read *pRasterBand* into *pPixelBlock*.

```
' Part 3: Create the TIN and add the TIN layer to
' the active map.
Dim dataArray As Variant
Dim pEnv As IEnvelope
Dim zTol As Double
Dim pTinLayer As ITinLayer
' Use the pixel block and a z tolerance to create the TIN.
dataArray = pPixelBlock.SafeArray(0)
Set pEnv = pProps.Extent
zTol = 5
pTin.AddFromPixelBlock pEnv.XMin, pEnv.YMax, _
pProps.MeanCellSize.X, pProps.MeanCellSize.Y, _
pProps.NoDataValue, dataArray, zTol
pTin.SaveAs ("c:\data\chap12\emidatin")
' Create the TIN layer and add it to the active map.
Set pTinLayer = New TinLayer
With pTinLayer
  Set.Dataset = pTin
```

```
    .Visible = True
    .Name = "emidatin"
  End With
  pMxDoc.FocusMap.AddLayer pTinLayer
  pMxDoc.UpdateContents
End Sub
```

Part 3 creates the TIN and adds the TIN layer to the active map. The code uses the *AddFromPixelBlock* method on *ITinEdit* to edit *pTin*, which has been set in edit mode in Part 1. The method has seven arguments. The first two are the coordinates of the TIN's origin. A raster's origin is at the upper left corner, whereas a TIN's origin is at the lower left corner. Therefore, the values entered are the xmin and ymax of the extent of *pRawPixels*. The third and fourth arguments are the pixel sizes in the *X* and *Y* dimensions, which are set to be the same as the mean cell sizes in *pRawPixels*. The fifth argument is the value of no data, which is set to be the same as the no data value in *pRawPixels*. The sixth argument is an array of pixel values to be read, which, in this case, is *dataArray* or the *SafeArray* property of *pPixelBlock*. The last argument is the z tolerance used for converting a raster to a TIN. The z tolerance determines the accuracy of a TIN in depicting the original elevation raster. The larger the z tolerance is, the less accurate the TIN becomes. In this case, the code uses 5 meters as the z tolerance. After *pTin* is created, it is saved as *emidatin* on disk. Finally, Part 3 creates a new TIN layer from *pTin* and adds the layer to the active map.

12.6.2 *EditTin*

EditTin modifies an existing TIN by adding streams as hard breaklines. The macro performs the same task as using the Create/Modify TIN/Add Features to TIN command in 3D Analyst. *EditTin* has two parts. Part 1 defines the TIN and the breaklines dataset. Part 2 adds streams as breaklines to the TIN and refreshes the map with the modified TIN.

> **Key Interfaces:** *ITin, ITinEdit*.
> **Key Members:** *Dataset, StartEditing, AddFromFeatureClass, StopEditing*.
> **Usage:** Add *newtin*, a new TIN, and *emidastrm.shp*, a stream shapefile for modifying the TIN, to an active map. The stream shapefile must be on top of *newtin* in the table of contents. Import *EditTin* to Visual Basic Editor. Make sure that the ESRI Tin Object Library is available. Run the macro. The macro modifies *newtin* by adding *emidatstrm* as hardlines. The code then adds the modified TIN as a new layer to the active map.

```
Private Sub EditTin()
  ' Part 1: Define the TIN and the layer for modifying
  ' the TIN.
  Dim pMxDoc As IMxDocument
```

```
Dim pMap As IMap
Dim pTinLayer As ITinLayer
Dim pStreamFL As IFeatureLayer
Dim pStreamFC As IFeatureClass
Dim pTin As ITin
Set pMxDoc = ThisDocument
Set pMap = pMxDoc.FocusMap
' Define the TIN.
Set pTinLayer = pMap.Layer(1)
Set pTin = pTinLayer.Dataset
' Define the stream layer for modifying the TIN.
Set pStreamFL = pMap.Layer(0)
Set pStreamFC = pStreamFL.FeatureClass
```

Part 1 sets *pTin* to be the dataset of the TIN layer in the active map, and *pStreamFC* to be the feature class of the stream feature layer.

```
' Part 2: Modify the TIN and add the modified TIN to
' the active map.
Dim pTinEdit As ITinEdit
Set pTinEdit = pTin
' Edit the TIN.
pTinEdit.StartEditing
pTinEdit.AddFromFeatureClass pStreamFC, Nothing, _
Nothing, Nothing, esriTinHardLine
pTinEdit.StopEditing (True)
' Add the modified TIN to the active map.
Set pTinLayer = New TinLayer
With pTinLayer
  Set.Dataset = pTin
  .Visible = True
  .Name = "Modified Tin"
End With
pMxDoc.FocusMap.AddLayer pTinLayer
pMxDoc.UpdateContents
End Sub
```

Part 2 performs a QI for the *ITinEdit* interface and uses the *StartEditing* method to start the editing process, the *AddFromFeatureClass* method to add features from a feature class to the TIN, and the *StopEditing* method to stop editing. Besides the object qualifier *pStreamFC*, the *AddFromFeatureClass* method uses four arguments.

The first argument is a query filter object, which, if specified, uses selected features to modify the TIN. The second and third arguments are the fields for height and tag value, if they exist. The last is the surface type option. **EditTin** opts for hard breaklines as the surface type. There are 18 other surface types to choose from in ArcObjects. Finally, the code adds to the active map a new TIN layer showing the modified TIN.

12.6.3 *TinNodes*

TinNodes reports the numbers of nodes and triangles of an existing TIN. The macro performs the same function as using the Source tab on the Layer Properties dialog. *TinNodes* has two parts. Part 1 defines the TIN dataset. Part 2 derives and reports the numbers of nodes and triangles.

> **Key Interfaces:** *ITinAdvanced.*
> **Key Members:** *Dataset, DataNodeCount, DataTriangleCount.*
> **Usage:** Add *plnetin* to an active map. Import **TinNodes** to Visual Basic Editor. Make sure that the ESRI Tin Object Library is available. Run the macro. The macro reports the numbers of nodes and triangles in a message box.

```
Private Sub TinNodes()
  ' Part 1: Define the TIN.
  Dim pMxDoc As IMxDocument
  Dim pMap As IMap
  Dim pTinLayer As ITinLayer
  Dim pTin As ITin
  Set pMxDoc = ThisDocument
  Set pMap = pMxDoc.FocusMap
  Set pTinLayer = pMap.Layer(0)
  Set pTin = pTinLayer.Dataset
```

Part 1 defines *pTin* as the dataset of the TIN layer in the active map.

```
  ' Part 2: Derive and report numbers of nodes and
  ' triangles.
  Dim pTinAdvanced As ITinAdvanced
  Dim pNodeCount As Double
  Dim pTriangleCount As Double
  Set pTinAdvanced = pTin
  pNodeCount = pTinAdvanced.DataNodeCount
  pTriangleCount = pTinAdvanced.DataTriangleCount
```

```
    MsgBox "The number of nodes is: " & pNodeCount & _
    "The number of triangles is: " & pTriangleCount
End Sub
```

Part 2 first performs a QI for the *ITinAdvanced* interface. Next the code derives the number of nodes from the *DataNodeCount* property on *ITinAdvanced* and assigns the number to *pNodeCount*. The code also derives the number of triangles from the *DataTriangleCount* property and assigns the number to *pTriangleCount*. A message box then reports these two numbers.

Spatial Interpolation

Spatial interpolation is the process of using points with known values to estimate values at other points. A geographic information system (GIS) typically applies spatial interpolation to a raster with estimates made for all cells. Spatial interpolation is therefore a means for converting point data to surface data so that the surface data can be used with other surfaces for analysis and modeling.

A variety of methods have been proposed for spatial interpolation. These methods can be categorized in several ways:

First, they can be grouped into global and local methods. A global interpolation method uses every known point available to estimate an unknown value, whereas a local interpolation method uses a sample of known points to estimate an unknown value.

Second, spatial interpolation methods can be grouped into exact and inexact interpolation. Exact interpolation predicts a value at the point location that is the same as its known value, whereas inexact interpolation does not.

Third, spatial interpolation methods may be deterministic or stochastic. A deterministic interpolation method provides no assessment of errors with predicted values, whereas a stochastic interpolation method does.

This chapter covers spatial interpolation. Section 13.1 reviews spatial interpolation using ArcGIS. Section 13.2 discusses objects that are related to spatial interpolation. Section 13.3 includes sample macros for using the methods of inverse distance weighted (IDW), spline, trend surface, and kriging for spatial interpolation. Both the IDW and spline methods are local, exact, and deterministic. Trend surface is a global, inexact, and deterministic method. Kriging is a local, exact, and stochastic method. Section 13.4 offers a macro for comparing different interpolation methods. All macros start with the listing of key interfaces and key members (i.e., properties and methods) and the usage.

13.1 RUNNING SPATIAL INTERPOLATION IN ARCGIS

ArcGIS Desktop offers spatial interpolation through its extensions of Spatial Analyst and Geostatistical Analyst. Spatial Analyst has menu access to IDW, Spline (spline with tension and regularized spline), and Kriging (ordinary and universal). Geostatistical Analyst's main menu has the selections of Explore Data, Geostatistical Wizard, and Create Subsets. The Explore Data command offers histogram, semivariogram, QQ (quantile-quantile) plot, and others for exploratory data analysis. The Geostatistical Wizard offers a large variety of global and local interpolation methods including IDW, trend surface, local polynomial, radial basis function, kriging, and cokriging. The Create Subsets command handles model validation.

Most ArcGIS users will probably use Geostatistical Analyst for spatial interpolation and take advantage of its data exploration and model validation capabilities. But, as of ArcGIS 8.3, objects used by the Geostatistical Analyst extension are not yet exposed to software developers. Therefore, macros on spatial interpolation are currently limited to objects in the Spatial Analyst subsystem.

13.2 ARCOBJECTS FOR SPATIAL INTERPOLATION

As of ArcGIS 8.3, ArcObjects treats spatial interpolation as one type of raster data analysis and offers the *RasterInterpolationOp* coclass to manage spatial interpolation. A *RasterInterpolationOp* object supports *IRasterAnalysisEnvironment* and *IInterpolationOp* (Figure 13.1). *IRasterAnalysisEnvironment* controls the analysis environment such as use of a mask and the setup of the output cell size. *IInterpolationOp* has the following methods for spatial interpolation: *IDW* for inverse distance weighted, *Krige* for kriging, *Spline* for spline, *Trend* for trend surface, and *Variogram* for kriging. The difference between *Krige* and *Variogram* is that the former uses a predefined type of semivariogram, and the latter uses a user-defined semivariogram.

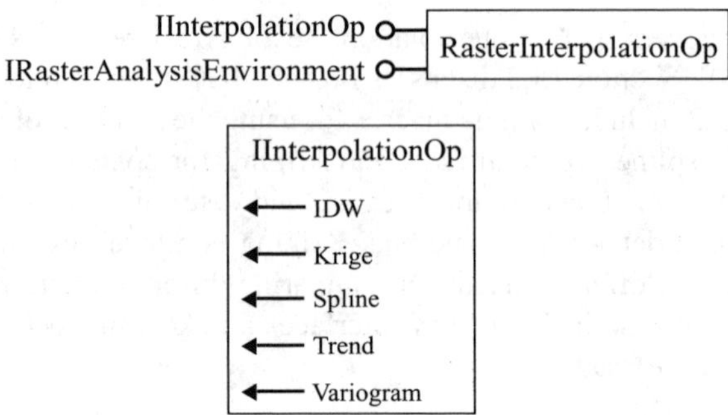

Figure 13.1 A *RasterInterpolationOp* object supports *IInterpolationOp* and *IRasterAnalysisEnvironment*. *IInterpolationOp* has various methods for spatial interpolation.

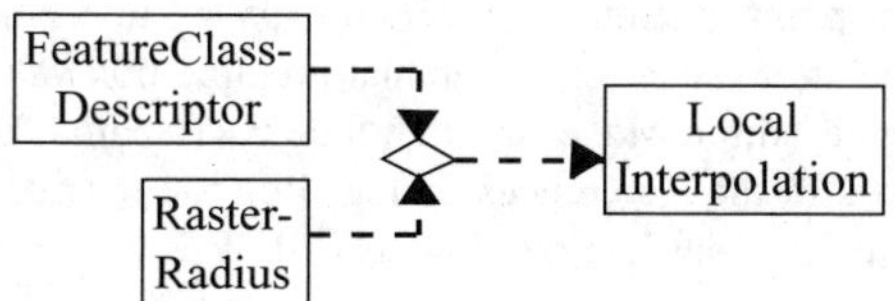

Figure 13.2 A *FeatureClassDescriptor* object and a *RasterRadius* object together create a *LocalInterpolation* object.

A local interpolation operation using ArcObjects typically requires two object qualifiers to create the output (Figure 13.2). The first is a feature class descriptor, which provides the input data for interpolation. A feature class descriptor is a feature class based on a specific field. For example, the feature class descriptor to be used in this chapter's sample macros is based on a numeric field that records the annual precipitation at weather stations. The second is a search radius, expressed as either a distance or a number, for selecting the sample points to be used in local interpolation.

Because spatial interpolation operations produce rasters, these rasters can be further analyzed by using objects covered in Chapter 11. For example, mathematical operations on rasters can compare and evaluate the results from two different interpolation methods.

13.3 PERFORMING SPATIAL INTERPOLATIONS

This section covers the spatial interpolation methods of IDW, spline, trend surface, and kriging. The first three methods have the same code structure. As the only stochastic method in the group, kriging produces an interpolated surface and an error measure surface.

13.3.1 *Idw*

Idw uses the IDW method to interpolate a precipitation surface from a point shapefile. The macro performs the same function as using the Interpolate to Raster/Inverse Distance Weighted command in Spatial Analyst. *Idw* has three parts. Part 1 defines the point shapefile and a mask dataset. Part 2 creates a raster interpolation object, sets the analysis environment, and runs the interpolation method. Part 3 creates a raster layer from the output and adds the layer to the active map.

Key Interfaces: *IFeatureClassDescriptor, IGeoDataset, IInterpolationOp, IRasterAnalysisEnvironment, IRasterRadius, IRaster.*

Key Members: *Create, Raster, SetCellSize, Mask, SetVariable, IDW, CreateFromRaster.*

Usage: Add *idoutlgd* and *stations.shp* to an active map. *stations* must be on top of *idoutlgd* in the table of contents. *idoutlgd* is a raster showing the outline of Idaho, and *stations* is a point shapefile containing 105 weather stations in Idaho. The Ann_prec field in *stations* stores the annual precipitation values in inches. Import **Idw** to Visual Basic Editor. Make sure that the ESRI Spatial Analyst Shared Object Library is checked in the References dialog. Run the macro. The macro creates a temporary precipitation surface raster named IDW.

```
Private Sub Idw()
  ' Part 1: Define the input and mask datasets.
  Dim pMxDoc As IMxDocument
  Dim pMap As IMap
  Dim pFeatureLayer As IFeatureLayer
  Dim pFeatureClass As IFeatureClass
  Dim sFieldName As String
  Dim pFCDescr As IFeatureClassDescriptor
  Dim pRasterLayer As IRasterLayer
  Dim pMaskDataset As IGeoDataset
  Set pMxDoc = ThisDocument
  Set pMap = pMxDoc.FocusMap
  Set pFeatureLayer = pMap.Layer(0)
  Set pFeatureClass = pFeatureLayer.FeatureClass
  ' Use a value field to create a feature class descriptor.
  sFieldName = "Ann_prec"
  Set pFCDescr = New FeatureClassDescriptor
  pFCDescr.Create pFeatureClass, Nothing, sFieldName
  ' Define a mask dataset.
  Set pRasterLayer = pMap.Layer(1)
  Set pMaskDataset = pRasterLayer.Raster
```

Part 1 defines the input and mask datasets. The code first sets *pFeatureClass* to be the feature class of the top layer in the active map. Next the code creates *pFCDescr* as an instance of the *FeatureClassDescriptor* class and uses the *Create* method to create the feature class descriptor based on the Ann_prec field. The code also sets *pMaskDataset* to be the raster of the second layer in the active map.

```
  ' Part 2: Perform interpolation using IDW.
  Dim pIntOp As IInterpolationOp
  Dim pEnv As IRasterAnalysisEnvironment
  Dim pRadius As IRasterRadius
  Dim pOutRaster As IRaster
  ' Create a raster interpolation operation.
```

```
Set pIntOp = New RasterInterpolationOp
Set pEnv = pIntOp
pEnv.SetCellSize esriRasterEnvValue, pMaskDataset
Set pEnv.Mask = pMaskDataset
' Set the search radius for the input points.
Set pRadius = New RasterRadius
pRadius.SetVariable 12
' Run the IDW method.
Set pOutRaster = pIntOp.IDW(pFCDescr, 2, pRadius)
```

Part 2 creates *pIntOp* as an instance of the *RasterInterpolationOp* class and performs a QueryInterface (QI) for the *IRasterAnalysisEnvironment* interface to set the analysis mask and cell size. Next the code sets the search radius, referenced by *pRadius*, to include 12 sample points. The code then uses the *IDW* method on *IInterpolationOp* to create the interpolated raster referenced by *pOutRaster*. The *IDW* method uses *pFCDescr* and *pRadius* for the object qualifiers and specifies two for the weight (i.e., exponent) of distance.

```
' Part 3: Create the output layer and add it to
' the active map.
Dim pRLayer As IRasterLayer
Set pRLayer = New RasterLayer
pRLayer.CreateFromRaster pOutRaster
pRLayer.Name = "IDW"
pMap.AddLayer pRLayer
End Sub
```

Part 3 creates a new raster layer from *pOutRaster* and adds the layer to the active map.

13.3.2 Spline

To use spline as an interpolation method, the only change from *Idw* is the following line statement:

```
Set pOutRaster = pIntOp.Spline _
(pFCDescr, esriGeoAnalysisRegularizedSpline)
```

The *Spline* method on *IInterpolationOp* requires a feature class descriptor as an object qualifier and a selection on the type of spline. The two types of splines are esriGeoAnalysisRegularizedSpline for regularized spline and esriGeoAnalysis-TensionSpline for spline with tension. The *Spline* method has two optional arguments for the weight to be used for interpolation and the number of sample points (with 12 as the default).

13.3.3 Trend Surface

To use trend surface as an interpolation method, the only change from *Idw* is the following line statement:

```
Set pOutRaster = pIntOp.Trend _
(pFCDescr, esriGeoAnalysisLinearTrend, 2)
```

The *Trend* method on *IInterpolationOp* requires a feature class descriptor as an object qualifier, a selection on the type of trend surface, and the order of the polynomial. For the type of trend surface, ArcObjects offers esriGeoAnalysis-LinearTrend for least squares surface and esriGeoAnalysisLogisticTrend for logistic surface. The order of the polynomial can range from 1 through 12.

13.3.4 *Kriging*

Kriging uses the kriging method to interpolate a precipitation surface from a point shapefile. The macro performs the same function as using the Interpolate to Raster/Kriging command in Spatial Analyst. *Kriging* has four parts. Part 1 defines the input and mask datasets. Part 2 creates a raster interpolation object, sets the analysis environment, and runs kriging. Part 3 extracts the kriged and variance datasets from the interpolation output. Part 4 creates raster layers from the output datasets and adds the layers to the active map.

Unlike the previous interpolation methods, the kriging method in ArcObjects apparently ignores the mask dataset. But a mask dataset is still included in *Kriging* to specify the output cell size. A macro such as *ExtractByMask* in Chapter 11 can clip the output datasets from kriging to fit the study area.

> **Key Interfaces:** *IFeatureClassDescriptor, IGeoDataset, IInterpolationOp, IRaster-AnalysisEnvironment, IRasterRadius, IRasterBandCollection, IRasterBand.*
>
> **Key Members:** *Create, Raster, SetCellSize, SetVariable, Krige, Item( ), AppendBand, CreateFromRaster.*
>
> **Usage:** Add *idoutlgd* and *stations.shp* to an active map. *stations* must be on top of *idoutlgd* in the table of contents. Import *Kriging* to Visual Basic Editor. Make sure that the ESRI Spatial Analyst Shared Object Library is available. Run the macro. The macro creates two temporary rasters: *Krige* represents the kriged surface, and *Variance* represents the variance surface.

```
Private Sub Kriging()
  ' Part 1: Define the input dataset.
  Dim pMxDoc As IMxDocument
  Dim pMap As IMap
  Dim pFeatureLayer As IFeatureLayer
  Dim pFeatureClass As IFeatureClass
  Dim sFieldName As String
  Dim pFCDescr As IFeatureClassDescriptor
```

```
Dim pRasterLayer As IRasterLayer
Dim pMaskDataset As IGeoDataset
Set pMxDoc = ThisDocument
Set pMap = pMxDoc.FocusMap
Set pFeatureLayer = pMap.Layer(0)
Set pFeatureClass = pFeatureLayer.FeatureClass
' Use a value field to create a feature class descriptor.
sFieldName = "Ann_prec"
Set pFCDescr = New FeatureClassDescriptor
pFCDescr.Create pFeatureClass, Nothing, sFieldName
' Define a mask dataset.
Set pRasterLayer = pMap.Layer(1)
Set pMaskDataset = pRasterLayer.Raster
```

Part 1 sets *pFeatureClass* to be the feature class of the top layer, and creates a
feature class descriptor referenced by *pFCDescr*. Next the code sets *pMaskDataset*
to be the raster of the second layer.

```
' Part 2: Perform interpolation using kriging.
Dim pIntOp As IInterpolationOp
Dim pEnv As IRasterAnalysisEnvironment
Dim pRadius As IRasterRadius
Dim pOutRaster As IGeoDataset
' Create a raster interpolation operation.
Set pIntOp = New RasterInterpolationOp
Set pEnv = pIntOp
pEnv.SetCellSize esriRasterEnvValue, pMaskDataset
' Set the search radius for the input points.
Set pRadius = New RasterRadius
pRadius.SetVariable 12
' Run the Krige method.
Set pOutRaster = pIntOp.Krige(pFCDescr, _
esriGeoAnalysisCircularSemiVariogram, pRadius, True)
```

Part 2 creates *pIntOp* as an instance of the *RasterInterpolationOp* class, and switches
to the *IRasterAnalysisEnvironment* interface to set the analysis cell size by using *pMask-
Dataset* as the cell size provider. Next the code sets the search radius to include 12
sample points. The code then uses the *Krige* method on *IInterpolationOp* to create the
output raster referenced by *pOutRaster*. The *Krige* method requires a feature class
descriptor, a selection on the type of semivariogram, a search radius, and the option for
creation of a variance dataset. ArcObjects offers the following types of semivariograms:
circular, exponential, Gaussian, linear, spherical, and universal.

```
' Part 3: Extract the kriged surface and variance rasters.
Dim pRasterBandCollection As IRasterBandCollection
Dim pKrigeRB As IRasterBand
Dim pKrigeRBCollection As IRasterBandCollection
Dim pVarianceRB As IRasterBand
Dim pVarianceRBCollection As IRasterBandCollection
Set pRasterBandCollection = pOutRaster
' Extract the kriged surface raster.
Set pKrigeRB = pRasterBandCollection.Item(0)
Set pKrigeRBCollection = New Raster
pKrigeRBCollection.AppendBand pKrigeRB
' Extract the variance raster.
Set pVarianceRB = pRasterBandCollection.Item(1)
Set pVarianceRBCollection = New Raster
pVarianceRBCollection.AppendBand pVarianceRB
```

Because Part 2 asks to create a variance dataset, the output raster referenced by *pOutRaster* actually contains two datasets. Part 3 is therefore designed to extract the two datasets (Figure 13.3). The code performs a QI for the *IRasterBandCollection* interface, extracts the first raster band of *pOutRaster*, and assigns the band to *pKrigeRB*. The code then creates *pKrigeRBCollection* as an instance of the *Raster* class and uses the *AppendBand* method on *IRasterBandCollection* to append *pKrigeRB* to *pKrigeRBCollection*. The raster band collection thus created has one raster band and represents the kriged surface. The code uses the same procedure to create *pVarianceRBCollection* for the variance surface.

```
' Part 4: Create the output layers and add them to
' the active map.
Dim pKrigeLayer As IRasterLayer
Dim pVarianceLayer As IRasterLayer
Set pKrigeLayer = New RasterLayer
pKrigeLayer.CreateFromRaster pKrigeRBCollection
pKrigeLayer.Name = "Krige"
pMap.AddLayer pKrigeLayer
Set pVarianceLayer = New RasterLayer
```

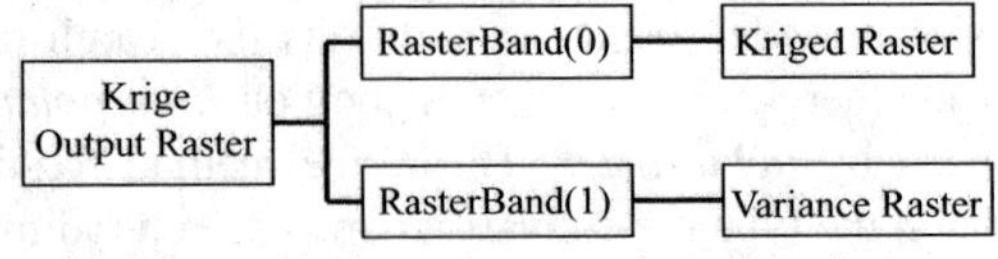

Figure 13.3 The output raster from *Krige* contains two raster bands, one for the kriged surface and the other for the variance surface.

```
    pVarianceLayer.CreateFromRaster pVarianceRBCollection
    pVarianceLayer.Name = "Variance"
    pMap.AddLayer pVarianceLayer
End Sub
```

Part 4 creates new raster layers from *pKrigeRBCollection* and *pVarianceRB-Collection*, and adds the layers to the active map.

13.4 COMPARING INTERPOLATION METHODS

A variety of factors can influence the result of spatial interpolation. They include the interpolation method, number of sample points, distribution of sample points, and the quality of input data. This section shows how to create an output so that the results from two different interpolation methods can be compared visually. A more vigorous comparison of interpolation methods would involve cross validation and model validation techniques.

13.4.1 *Compare*

Compare compares the interpolated surfaces from the IDW method and the spline method and highlights areas with large differences between the two (Figure 13.4). *Compare* has four parts. Part 1 defines the input and mask datasets. Part 2 performs the IDW and spline interpolation methods. Part 3 subtracts one interpolated surface from the other and reclassifies the surface representing the difference. Part 4 creates new raster layers from the outputs and adds them to the active map.

Key Interfaces: *IFeatureClassDescriptor, IGeoDataset, IInterpolationOp, IRaster-AnalysisEnvironment, IRasterRadius, IRaster, IMathOp, IRasterBandCollection, IRasterBand, IReclassOp, IRemap, INumberRemap.*

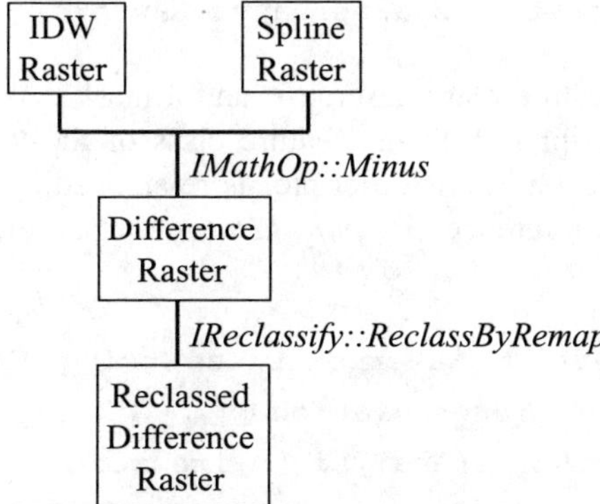

Figure 13.4 The flow chart shows how the *Compare* sub works.

Key Members: *Create, Raster, SetCellSize, Mask, SetVariable, IDW, Spline, Minus, Statistics, Minimum, Maximum, MapRange, ReclassByRemap, CreateFromRaster.*

Usage: Add *idoutlgd* and *stations.shp* to an active map. *stations* must be on top of *idoutlgd* in the table of contents. Import **Compare** to Visual Basic Editor. Make sure that both the ESRI Spatial Analyst Shared Object Library and the ESRI Spatial Analyst Extension Object Library are available. Run the macro. The macro creates four temporary rasters named *IDW, Spline, Difference*, and *Reclassed Difference* and adds them to the active map.

```vba
Private Sub Compare()
  ' Part 1: Define the input and mask datasets.
  Dim pMxDoc As IMxDocument
  Dim pMap As IMap
  Dim pFeatureLayer As IFeatureLayer
  Dim pFeatureClass As IFeatureClass
  Dim sFieldName As String
  Dim pFCDescr As IFeatureClassDescriptor
  Dim pRasterLayer As IRasterLayer
  Dim pMaskDataset As IGeoDataset
  Set pMxDoc = ThisDocument
  Set pMap = pMxDoc.FocusMap
  Set pFeatureLayer = pMap.Layer(0)
  Set pFeatureClass = pFeatureLayer.FeatureClass
  ' Use a value field to create a feature class descriptor.
  sFieldName = "Ann_prec"
  Set pFCDescr = New FeatureClassDescriptor
  pFCDescr.Create pFeatureClass, Nothing, sFieldName
  ' Define a mask dataset.
  Set pRasterLayer = pMap.Layer(1)
  Set pMaskDataset = pRasterLayer.Raster
```

Part 1 defines a feature class descriptor and a mask dataset. The code creates the feature class descriptor from the feature class of the top layer based on the Ann_prec field. The feature class descriptor is referenced by *pFCDescr*. The code sets the mask dataset, referenced by *pMaskDataset*, to be the raster of the second layer.

```vba
  ' Part 2: Perform interpolations using IDW and Spline.
  Dim pIntOp As IInterpolationOp
  Dim pEnv As IRasterAnalysisEnvironment
  Dim pRadius As IRasterRadius
  Dim pIdwRaster As IRaster
```

```
Dim pSplineRaster As IRaster
' Create a raster interpolation operation.
Set pIntOp = New RasterInterpolationOp
Set pEnv = pIntOp
pEnv.SetCellSize esriRasterEnvValue, pMaskDataset
Set pEnv.Mask = pMaskDataset
' Set the search radius for the input points.
Set pRadius = New RasterRadius
pRadius.SetVariable 12
' Run the IDW method.
Set pIdwRaster = pIntOp.IDW(pFCDescr, 2, pRadius)
' Run the Spline method.
Set pSplineRaster = pIntOp.Spline _
(pFCDescr, esriGeoAnalysisRegularizedSpline)
```

Before running the interpolation methods, Part 2 creates *pIntOp* as an instance of the *RasterInterpolationOp* class, sets the analysis cell size and mask, and defines the search radius. The code then runs the *IDW* method to create *pIdwRaster* and the *Spline* method to create *pSplineRaster*.

```
' Part 3: Compare the two interpolated rasters.
Dim pMathOp As IMathOp
Dim pDiffRaster As IRaster
Dim pRasBC As IRasterBandCollection
Dim pBand1 As IRasterBand
Dim pMinimum As Double
Dim pMaximum As Double
Dim pReclassOp As IReclassOp
Dim pRemap As IRemap
Dim pNRemap As INumberRemap
Dim pOutRaster As IRaster
' Subtract one raster from the other to create
' the difference raster.
Set pMathOp = New RasterMathOps
Set pDiffRaster = pMathOp.Minus(pIdwRaster, _
pSplineRaster)
' Derive the minimum and maximum values from the
' difference raster.
Set pRasBC = pDiffRaster
Set pBand1 = pRasBC.Item(0)
pMinimum = pBand1.Statistics.Minimum
```

```
pMaximum = pBand1.Statistics.Maximum
' Prepare a number remap.
Set pNRemap = New NumberRemap
pNRemap.MapRange pMinimum, -3.1, 1
pNRemap.MapRange -3, 3, 2
pNRemap.MapRange 3.1, pMaximum, 3
Set pRemap = pNRemap
' Reclassify the difference raster.
Set pReclassOp = New RasterReclassOp
Set pOutRaster = pReclassOp.ReclassByRemap _
(pDiffRaster, pRemap, False)
```

Part 3 derives a difference raster from *pIdwRaster* and *pSplineRaster* and reclassifies the difference raster into three classes showing high negative difference, low difference, and high positive difference. To create the difference raster referenced by *pDiffRaster*, the code uses the *Minus* method on *IMathOp* and *pIdwRaster* and *pSplineRaster* as the object qualifiers. To reclassify *pDiffRaster*, the code first derives the minimum and maximum values in *pDiffRaster* by using the *Statistics* property on *IRasterBand*. The minimum and maximum values are then entered in a number remap referenced by *pNRemap* to set up the following three classes: minimum to −3.1 (high negative difference), −3 to 3 (low difference), and 3.1 to maximum (high positive difference). The result of applying the *ReclassByRemap* method on *IReclassOp* is a reclassified difference raster referenced by *pOutRaster*.

```
' Part 4: Create the output raster layers, and add
' them to the active map.
Dim pIdwLayer As IRasterLayer
Dim pSplineLayer As IRasterLayer
Dim pDiffLayer As IRasterLayer
Dim pOutLayer As IRasterLayer
' Add the IDW layer.
Set pIdwLayer = New RasterLayer
pIdwLayer.CreateFromRaster pIdwRaster
pIdwLayer.Name = "IDW"
pMap.AddLayer pIdwLayer
' Add the Spline layer.
Set pSplineLayer = New RasterLayer
pSplineLayer.CreateFromRaster pSplineRaster
pSplineLayer.Name = "Spline"
pMap.AddLayer pSplineLayer
' Add the Difference layer.
Set pDiffLayer = New RasterLayer
```

```
pDiffLayer.CreateFromRaster pDiffRaster
pDiffLayer.Name = "Difference"
pMap.AddLayer pDiffLayer
' Add the reclassed layer.
Set pOutLayer = New RasterLayer
pOutLayer.CreateFromRaster pOutRaster
pOutLayer.Name = "Reclassed Difference"
pMap.AddLayer pOutLayer
End Sub
```

Part 4 creates new raster layers from *pIdwRaster*, *pSplineRaster*, *pDiffRaster*, and *pOutRaster*, respectively, and adds these layers to the active map.

Binary and Index Models

A model is a simplified representation of a phenomenon or a system. Many types of models are used in different disciplines. Two types of models that a geographic information system (GIS) can build are binary and index models, either vector-based or raster-based.

A binary model selects spatial features that meet a set of criteria. Those features that meet the selection criteria are coded 1 (True) and those that do not are coded 0 (False). A common application of binary models is site analysis. A vector-based binary model requires that overlay operations be performed to combine attributes (criteria) to be queried. A raster-based binary model, on the other hand, can be derived directly from querying multiple rasters, with each raster representing a criterion.

An index model calculates the index value for each unit area and produces a ranked map based on the index values. Index models are commonly used for suitability analysis and vulnerability analysis. Like a binary model, an index model also involves multi-criteria evaluation. The weighted linear combination method is a popular method for developing an index model. The method calculates the index value by summing the weighted criterion values. The weight represents the relative importance of a criterion against other criteria, and the criterion values represent the standardized values, such as 0 to 1, 1 to 5, or 0 to 100, for each criterion.

This chapter covers binary models and index models. Section 14.1 reviews important commands in ArcGIS for building models. Section 14.2 discusses objects for building models. Almost all of these objects have been covered in previous chapters. Section 14.3 includes sample modules for building binary and index models, both vector- and raster-based. Each module has a description of its usage.

14.1 BUILDING MODELS IN ARCGIS

ArcGIS does not offer menu choices for building a binary or index model. The ModelBuilder extension to ArcView 3.x can build raster-based index models. The extension is not available in ArcGIS 8.3, however. Until ModelBuilder is reintroduced into ArcGIS, modelers must construct binary and index models by using various commands in ArcGIS.

Commands important to building a vector-based binary or index model are overlay (e.g., union, intersect, and spatial join), attribute data manipulation (e.g., adding fields and calculating field values), and attribute data query. Raster Calculator and Reclassify in Spatial Analyst are important commands for building raster-based binary or index models. If selection criteria involve buffer zones, then buffering using vector data or distance measuring using raster data becomes part of the model-building process.

14.2 ARCOBJECTS FOR GIS MODELS

ArcObjects does not have model objects. Instead, objects in the ArcMap, Raster, and Spatial Analyst subsystems can be used to build models. Many of these objects have already been covered in Chapters 9, 10, and 11.

14.3 BUILDING BINARY AND INDEX MODELS

This section covers sample VBA (Visual Basic for Applications) modules for building binary and index models, both vector- and raster-based. Because a model typically involves several separate tasks, each of the following modules consists of several subs and functions. And because many objects used have already been covered in previous chapters, they do not need detailed explanation.

14.3.1 *VectorBinaryModel*

Using an elevation dataset and a stream dataset as inputs, *VectorBinaryModel* produces a vector-based binary model that selects areas that are in elevation zone 2 and within 200 meters of streams. Both input datasets are stored as feature classes (*elevzone* and *stream*) in a personal geodatabase (*Binary.mdb*). By using geodatabase feature classes as inputs to an overlay operation, we do not have to update the area and perimeter values on the output.

VectorBinaryModel has two subs and two functions (Figure 14.1):

Start: a sub for managing the input and output layers and for attribute data query
Intersect: a sub for the overlay/intersect operation
Buffer: a function for creating a buffer zone
SelectDataset: a function for selecting the input to the buffering and intersect operations

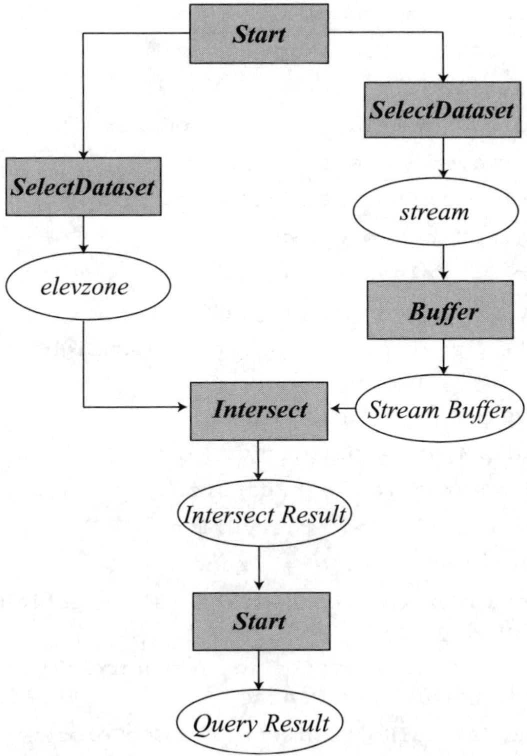

Figure 14.1 The flow chart shows the modular structure of *VectorBinaryModel*.

Usage: Import *VectorBinaryModel* to Visual Basic Editor. Run the module. Select *stream* from the dialog box for the layer to be buffered. Then select *elevzone* for the overlay layer. The macro adds *stream*, *elevzone*, and the output feature classes (*Buffer_Result* and *Intersect_Result*) to the active map and highlights those areas that meet the selection criteria.

```
Private Sub Start()
  Dim pMxDoc As IMxDocument
  Dim pMap As IMap
  Dim Id As Long
  Dim Message As String
  Dim pBufferLayer As ILayer
  Dim pLayer1 As ILayer
  Dim pLayer2 As ILayer
  Dim pActiveView As IActiveView
  Dim pFeatureLayer As IFeatureLayer
  Dim pFeatureSelection As IFeatureSelection
  Dim pQueryFilter As IQueryFilter
```

```
    Set pMxDoc = ThisDocument
    Set pMap = pMxDoc.FocusMap
    ' Run buffering.
    MsgBox "Select the layer to be buffered"
    Set pBufferLayer = SelectDataset
    Set pLayer1 = Buffer(pBufferLayer, pMap)
    MsgBox "Select the overlay layer"
    Set pLayer2 = SelectDataset
    Call Intersect(pLayer1, pLayer2, pMap)
    ' Query the intersect result and highlight the
    ' final selection.
    Set pActiveView = pMap
    Set pFeatureLayer = pMap.Layer(0)
    Set pFeatureSelection = pFeatureLayer
    Set pQueryFilter = New QueryFilter
    pQueryFilter.WhereClause = "Zone_ = 2"
    pActiveView.PartialRefresh esriViewGeoSelection, _
    Nothing, Nothing
    pFeatureSelection.SelectFeatures pQueryFilter, _
    esriSelectionResultNew, False
    pActiveView.PartialRefresh esriViewGeoSelection, _
    Nothing, Nothing
  End Sub
```

Start first uses a message box and *SelectDataset* to get *stream* as the input to *Buffer*. *Buffer* returns a buffer zone (*Buffer_Result*). *Start* then uses the buffer zone and *elevzone* as the inputs to *Intersect*. At the completion of the intersect operation, *Start* uses a query filter object to select areas that are in elevation zone 2. The code then highlights the selected areas by refreshing the active view.

```
    Private Function Buffer(pBufferLayer As ILayer, pMap _
    As IMap) As IFeatureLayer
      Dim pFeatureLayer As IFeatureLayer
      Dim pFeatureClass As IFeatureClass
      Dim pFCursor As IFeatureCursor
      Dim pSpatialReference As ISpatialReference
      Dim pBufWSName As IWorkspaceName
      Dim pBufDatasetName As IDatasetName
      Dim pBufFCName As IFeatureClassName
      Dim pFeatureCursorBuffer2 As IFeatureCursorBuffer2
      Dim pName As IName
      Dim pBufFC As IFeatureClass
```

```
    Dim pBufFL As IFeatureLayer
    Set pFeatureLayer = pBufferLayer
    ' Create a feature cursor.
    Set pFeatureClass = pFeatureLayer.FeatureClass
    Set pFCursor = pFeatureClass.Search(Nothing, False)
    Set pSpatialReference = pMap.SpatialReference
    ' Define the output.
    Set pBufWSName = New WorkspaceName
    pBufWSName.WorkspaceFactoryProgID = _
    "esriCore.AccessWorkspaceFactory.1"
    pBufWSName.PathName = "c:\data\chap14\Binary.mdb"
    Set pBufFCName = New FeatureClassName
    Set pBufDatasetName = pBufFCName
    Set pBufDatasetName.WorkspaceName = pBufWSName
    pBufDatasetName.Name = "Buffer_result"
    ' Perform buffering.
    ' Create a feature cursor buffer.
    Set pFeatureCursorBuffer2 = New FeatureCursorBuffer
    ' Define the feature cursor buffer.
    With pFeatureCursorBuffer2
      Set.FeatureCursor = pFCursor
      .Dissolve = True
      .ValueDistance = 200
      Set.BufferSpatialReference = pSpatialReference
      Set.DataFrameSpatialReference = pSpatialReference
      Set.SourceSpatialReference = pSpatialReference
      Set.TargetSpatialReference = pSpatialReference
    End With
    ' Run Buffer.
    pFeatureCursorBuffer2.Buffer pBufFCName
    ' Create the output layer from the output and add
    ' it to the active map.
    Set pName = pBufFCName
    Set pBufFC = pName.Open
    Set pBufFL = New FeatureLayer
    Set pBufFL.FeatureClass = pBufFC
    pBufFL.Name = "Buffer_Result"
    pMap.AddLayer pBufFL
    Set Buffer = pBufFL
End Function
```

Buffer gets the layer to be buffered and the active map as arguments from ***Start***. The code creates a feature cursor object by including all features in the input dataset and defines the workspace and name for the output. Next the code creates *pFeatureCursorBuffer2* as an instance of the *FeatureCursorBuffer* class, defines its properties, and uses the *Buffer* method to create the output referenced by *pBufFCName*. Then the code opens the name object, creates a new feature layer from the buffered feature class, and adds the layer to the active map. ***Buffer*** returns *pBufFL* as an *IFeatureLayer* to ***Start***.

```vba
Private Sub Intersect(pLayer1 As ILayer, pLayer2 As _
ILayer, pMap As IMap)
  Dim pInputLayer As IFeatureLayer
  Dim pOverlayLayer As IFeatureLayer
  Dim pInputTable As ITable
  Dim pOverlayTable As ITable
  Dim pNewWSName As IWorkspaceName
  Dim pFeatClassName As IFeatureClassName
  Dim pDatasetName As IDatasetName
  Dim pBGP As IBasicGeoprocessor
  Dim pOutputFeatClass As IFeatureClass
  Dim tol As Double
  Dim pOutputFeatLayer As IFeatureLayer
  ' Define the input and overlay tables.
  Set pInputLayer = pLayer1
  Set pOverlayLayer = pLayer2
  Set pInputTable = pInputLayer
  Set pOverlayTable = pOverlayLayer
  ' Define the output.
  Set pNewWSName = New WorkspaceName
  pNewWSName.WorkspaceFactoryProgID = _
  "esriCore.AccessWorkspaceFactory"
  pNewWSName.PathName = "c:\data\chap14\Binary.mdb"
  Set pFeatClassName = New FeatureClassName
  Set pDatasetName = pFeatClassName
  pDatasetName.Name = "Intersect_result"
  Set pDatasetName.WorkspaceName = pNewWSName
  ' Perform intersect.
  Set pBGP = New BasicGeoprocessor
  tol = 0#
  Set pOutputFeatClass = pBGP.Intersect(pInputTable, _
  False, pOverlayTable, False, tol, pFeatClassName)
```

```
     ' Create the output feature layer and add it to
     ' the active map.
     Set pOutputFeatLayer = New FeatureLayer
     Set pOutputFeatLayer.FeatureClass = pOutputFeatClass
     pOutputFeatLayer.Name = pOutputFeatClass.AliasName
     pMap.AddLayer pOutputFeatLayer
End Sub
```

Intersect gets the two layers to be intersected and the active map as arguments from *Start*. The code defines the input tables and the workspace and name of the intersect output. Next the code creates *pBGP* as an instance of the *BasicGeoprocessor* class and uses the *Intersect* method to create the output referenced by *pOutputFeat-Class*. *Intersect* then creates a feature layer from *pOutputFeatClass* and adds the layer to the active map.

```
Private Function SelectDataset() As IFeatureLayer
   ' Part 1: Prepare an Add Data dialog.
  Dim pGxDialog As IGxDialog
  Dim pGxFilter As IGxObjectFilter
  Dim pGxObjects As IEnumGxObject
  Dim pMxDoc As IMxDocument
  Dim pMap As IMap
  Dim pGxDataset As IGxDataset
  Dim pLayer As IFeatureLayer
  Set pGxDialog = New GxDialog
  Set pGxFilter = New GxFilterFeatureClasses
  ' Define the dialog's properties.
  With pGxDialog
    .AllowMultiSelect = False
    .ButtonCaption = "Add"
    Set.ObjectFilter = pGxFilter
    .StartingLocation = "c:\data\chap14\Binary.mdb"
    .Title = "Add Data"
  End With
  ' Open the dialog.
  pGxDialog.DoModalOpen 0, pGxObjects
  Set pGxDataset = pGxObjects.Next
  ' Exit sub if no dataset has been added.
  If pGxDataset Is Nothing Then
    Exit Function
```

```
   End If
   ' Add the layers to the active map.
   Set pLayer = New FeatureLayer
   Set pLayer.FeatureClass = pGxDataset.Dataset
   pLayer.Name = pLayer.FeatureClass.AliasName
   Set pMxDoc = ThisDocument
   Set pMap = pMxDoc.FocusMap
   pMap.AddLayer pLayer
   pMxDoc.ActiveView.Refresh
   pMxDoc.UpdateContents
   ' Return pLayer to the Start sub.
   Set SelectDataset = pLayer
 End Function
```

SelectDataset creates *pGxDialog* as an instance of the *GxDialog* class and defines its properties. The dialog shows only the feature classes in *Binary.mdb* and allows only one dataset to be selected. The code then creates *pLayer* as a new feature layer from the selected feature class and adds the layer to the active map. *SelectDataset* returns *pLayer* to *Start*.

14.3.2 *VectorIndexModel*

VectorIndexModel uses *soil*, *landuse*, and *depwater* as the inputs and produces a vector-based index model that shows the degree of susceptibility to ground water contamination. The three feature classes are stored in a personal geodatabase. The criterion values have been computed and are stored in the feature classes: soilrate in *soil* for the soils criterion, lurate in *landuse* for the land use criterion, and dwrate in *depwater* for the depth to water criterion. *VectorIndexModel* calculates the index value using the following expression: 3 * soilrate + lurate + dwrate. *VectorIndex-Model* then displays the index values in four classes. The index model does not apply to urban areas. Therefore, an additional task in building the model is to exclude urban areas from the analysis.

VectorIndexModel has five subs and two functions (Figure 14.2):

Start: a sub for managing the input and output layers and for calling the other subs
Intersect: a sub for performing the overlay operation
AddIndex: a sub for adding an index field to the overlay output
CalculateIndex: a sub for calculating the index values
DisplayIndexClasses: a sub for displaying the index values in four classes
FindLayer: a function for finding a specific layer
GetRGB: a function for defining a color symbol

Usage: Add *soil*, *landuse*, and *depwater* from *Index.mdb* to an active map. Import *VectorIndexModel* to Visual Basic Editor. Run the macro. The macro creates the

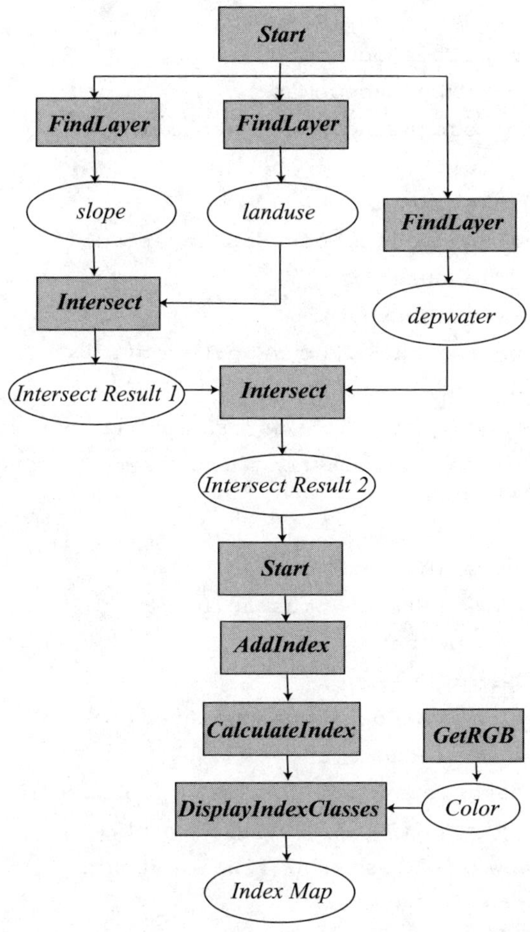

Figure 14.2   The flow chart shows the modular structure of *VectorIndexModel*.

feature classes of *Intersect_1* and *Intersect_2* and adds them to the active map. *Intersect_2* shows the index model in four classes.

```
Private Sub Start()
' Start uses the subs of FindLayer, Intersect, AddIndex,
' CalculateIndex, and DisplayIndexClasses.
  Dim pMxDoc As IMxDocument
  Dim pMap As IMap
  Dim Id As Long
  Dim Message As String
  Dim pLayer1 As ILayer
  Dim pLayer2 As ILayer
  Dim pOutName As String
```

```
    Dim pFLayer As IFeatureLayer
    Set pMxDoc = ThisDocument
    Set pMap = pMxDoc.FocusMap
    ' Run intersect using soil and landuse.
    Message = "soil"
    Id = FindLayer(Message)
    Set pLayer1 = pMxDoc.FocusMap.Layer(Id)
    Message = "landuse"
    Id = FindLayer(Message)
    Set pLayer2 = pMxDoc.FocusMap.Layer(Id)
    pOutName = "Intersect_1"
    Call Intersect(pLayer1, pLayer2, pOutName)
    ' Run intersect using the first overlay output
    ' and depwater.
    Message = "Intersect_1"
    Id = FindLayer(Message)
    Set pLayer1 = pMxDoc.FocusMap.Layer(Id)
    Message = "depwater"
    Id = FindLayer(Message)
    Set pLayer2 = pMxDoc.FocusMap.Layer(Id)
    pOutName = "Intersect_2"
    Call Intersect(pLayer1, pLayer2, pOutName)
    Set pFLayer = pMxDoc.FocusMap.Layer(0)
    ' Add a new field to the final output.
    Call AddIndex(pFLayer)
    ' Calculate the new field values.
    Call CalculateIndex(pFLayer)
    ' Display index values in classes.
    Call DisplayIndexClasses(pFLayer)
End Sub
```

Start runs the intersect operation twice to combine the three input layers of *soil*, *landuse*, and *depwater*. Names of the inputs and outputs for these operations are hard coded and are used by the ***FindLayer*** function to locate them in the active map. ***Start*** then passes the output from the second intersect operation, which is referenced by *pFLayer*, as an argument to ***AddIndex***, ***CalculateIndex***, and ***DisplayIndexClasses***. These three subs perform the sequential tasks of adding a new (index) field, calculating the field values, and displaying the field values using a class breaks renderer.

```
Private Sub Intersect(pLayer1 As ILayer, pLayer2 As _
ILayer, pOutName As String)
```

```vb
    Dim pMxDoc As IMxDocument
    Dim pInputLayer As IFeatureLayer
    Dim pInputTable As ITable
    Dim pOverlayLayer As IFeatureLayer
    Dim pOverlayTable As ITable
    Dim pNewWSName As IWorkspaceName
    Dim pFeatClassName As IFeatureClassName
    Dim pDatasetName As IDatasetName
    Dim pBGP As IBasicGeoprocessor
    Dim tol As Double
    Dim pOutputFeatClass As IFeatureClass
    Dim pOutputFeatLayer As IFeatureLayer
    Set pMxDoc = ThisDocument
    ' Define the input and overlay tables.
    Set pInputLayer = pLayer1
    Set pInputTable = pInputLayer
    Set pOverlayLayer = pLayer2
    Set pOverlayTable = pOverlayLayer
    ' Define the output.
    Set pFeatClassName = New FeatureClassName
    Set pDatasetName = pFeatClassName
    Set pNewWSName = New WorkspaceName
    pNewWSName.WorkspaceFactoryProgID = _
    "esriCore.AccessWorkspaceFactory"
    pNewWSName.PathName = "c:\data\chap14\Index.mdb"
    pDatasetName.Name = pOutName
    Set pDatasetName.WorkspaceName = pNewWSName
    ' Perform intersect.
    Set pBGP = New BasicGeoprocessor
    tol = 0#
    Set pOutputFeatClass = pBGP.Intersect(pInputTable, _
    False, pOverlayTable, False, tol, pFeatClassName)
    ' Create the output feature layer and add it to
    ' the active map.
    Set pOutputFeatLayer = New FeatureLayer
    Set pOutputFeatLayer.FeatureClass = pOutputFeatClass
    pOutputFeatLayer.Name = pOutputFeatClass.AliasName
    pMxDoc.FocusMap.AddLayer pOutputFeatLayer
End Sub
```

Intersect gets two input layers and the output name as arguments from *Start*. *Intersect* first uses the input layers to set up the input and overlay tables. Next the code defines the output's workspace and name. Then the code creates *pBGP* as an instance of the *BasicGeoprocessor* class and uses the *Intersect* method on *IBasic-Geoprocessor* to create the overlay output. Finally, the code creates a new feature layer from the overlay output and adds the layer to the active map.

```
Private Sub AddIndex(pFLayer As IFeatureLayer)
  Dim pFeatLayer As IFeatureLayer
  Dim pFClass As IFeatureClass
  Dim pField As IFieldEdit
  Set pFeatLayer = pFLayer
  Set pFClass = pFeatLayer.FeatureClass
  ' Create and define a new field.
  Set pField = New Field
  With pField
    .Name = "Total"
    .Type = esriFieldTypeDouble
    .Length = 8
  End With
  ' Add the new field.
  pFClass.AddField pField
End Sub
```

AddIndex gets the top layer in the active map (the layer from the second intersect operation) as an argument from *Start*. The code creates *pField* as an instance of the *Field* class and defines its properties including the field name of Total. Then the code uses the *AddField* method on *IFeatureClass* to add *pField* to the feature class of the layer.

```
Private Sub CalculateIndex(pFLayer As IFeatureLayer)
  Dim pFeatLayer As IFeatureLayer
  Dim pFeatClass As IFeatureClass
  Dim pFields As IFields
  Dim ii As Integer
  Dim pQueryFilter As IQueryFilter
  Dim pCursor As ICursor
  Dim pCalc As ICalculator
  Set pFeatLayer = pFLayer
  Set pFeatClass = pFeatLayer.FeatureClass
  Set pFields = pFeatClass.Fields
  ii = pFields.FindField("Total")
```

```
' Prepare a cursor for features that have lurate < 99.
Set pQueryFilter = New QueryFilter
pQueryFilter.WhereClause = "lurate < 99"
Set pCursor = pFeatClass.Update(pQueryFilter, True)
' Use the cursor to calculate the field values of Total.
Set pCalc = New Calculator
With pCalc
   Set.Cursor = pCursor
   .Expression = "([SOILRATE] * 3 + [LURATE] + [DWRATE])/250"
   .Field = "Total"
End With
pCalc.Calculate
End Sub
```

CalculateIndex gets the layer with the new field as an argument from *Start*. The code uses the *FindField* method on *IFields* to find the Total field. Then the code uses a query filter object to select those records that have lurate < 99 (i.e., non-urban land use) and save them into a feature cursor. Next the code creates *pCalc* as an instance of the *Calculator* class and defines its properties of cursor, expression, and field. Finally, *CalculateIndex* uses the *Calculate* method on *ICalculator* to populate the field values of Total.

```
Private Function FindLayer(Message As String) As Long
   Dim pMxDoc As IMxDocument
   Dim pMap As IMap
   Dim FindDoc As Variant
   Dim aLName As String
   Dim Name As String
   Dim i As Long
   Set pMxDoc = ThisDocument
   Set pMap = pMxDoc.FocusMap
   Name = Message
   For i = 0 To pMap.LayerCount - 1
     aLName = UCase(pMap.Layer(i).Name)
     If (aLName = (UCase(Name))) Then
       FindDoc = i
     End If
   Next
   FindLayer = FindDoc
End Function
```

FindLayer finds the index of the layer that matches the name from the message statement. The function makes sure that the appropriate layers are used in the intersect operations.

```vba
Private Sub DisplayIndexClasses(pFLayer As IFeatureLayer)
' DisplayIndexClasses uses the sub of GetRGB.
  Dim pMxDoc As IMxDocument
  Dim pLayer As ILayer
  Dim pGeoFeatureLayer As IGeoFeatureLayer
  Dim pClassBreaksRenderer As IClassBreaksRenderer
  Dim pFillSymbol As IFillSymbol
  Set pMxDoc = ThisDocument
  Set pLayer = pFLayer
  Set pGeoFeatureLayer = pLayer
  ' Define a class breaks renderer.
  Set pClassBreaksRenderer = New ClassBreaksRenderer
  pClassBreaksRenderer.Field = "Total"
  pClassBreaksRenderer.BreakCount = 4
  ' Hard code the symbol, break, and label for each class.
  Set pFillSymbol = New SimpleFillSymbol
  pFillSymbol.Color = GetRGBColor(245, 245, 0)
  pClassBreaksRenderer.Symbol(0) = pFillSymbol
  pClassBreaksRenderer.Break(0) = 0#
  pClassBreaksRenderer.Label(0) = "Urban Land Use"
  Set pFillSymbol = New SimpleFillSymbol
  pFillSymbol.Color = GetRGBColor(245, 175, 0)
  pClassBreaksRenderer.Symbol(1) = pFillSymbol
  pClassBreaksRenderer.Break(1) = 0.75
  pClassBreaksRenderer.Label(1) = "0.60 - 0.75"
  Set pFillSymbol = New SimpleFillSymbol
  pFillSymbol.Color = GetRGBColor(245, 125, 0)
  pClassBreaksRenderer.Symbol(2) = pFillSymbol
  pClassBreaksRenderer.Break(2) = 0.85
  pClassBreaksRenderer.Label(2) = "0.76 - 0.85"
  Set pFillSymbol = New SimpleFillSymbol
  pFillSymbol.Color = GetRGBColor(245, 0, 0)
  pClassBreaksRenderer.Symbol(3) = pFillSymbol
  pClassBreaksRenderer.Break(3) = 1#
  pClassBreaksRenderer.Label(3) = "0.86 - 1.00"
  ' Assign the renderer to the layer and refresh the map.
```

```
   Set pGeoFeatureLayer.Renderer = pClassBreaksRenderer
   pMxDoc.ActiveView.PartialRefresh esriViewGeography, _
   pLayer, Nothing
   pMxDoc.UpdateContents
End Sub
```

DisplayIndexClasses gets the layer with the populated Total field as an argument from *Start*. The code first creates *pClassBreaksRenderer* as an instance of the *ClassBreaksRenderer* class and defines its properties of field and break count. *DisplayIndexClasses* then assigns the fill symbol, break, and label for each of the four classes. The color for the simple fill symbol is obtained by entering the red (R), green (G), and blue (B) values in the *GetRGBColor* function. Finally, the code assigns *pClassBreaksRenderer* to the layer, refreshes the map, and updates the contents of the map document.

```
Private Function GetRGBColor(R As Long, G As Long, B As Long)
   Dim pColor As IRgbColor
   Set pColor = New RgbColor
   pColor.Red = R
   pColor.Green = G
   pColor.Blue = B
   GetRGBColor = pColor
End Function
```

GetRGBColor gets the input values of R, G, and B from *DisplayIndexClasses* and returns a RGB color.

14.3.3 *RasterBinaryModel*

RasterBinaryModel performs the same task as *VectorBinaryModel* but in raster format. *stream_gd* and *elevzone_gd* are the raster equivalent of *stream.shp* and *elevzone.shp*, respectively. The binary model finds areas that are in elevation zone 2 and within 200 meters of streams. In building the binary model, *RasterBinary-Model* first creates a continuous distance raster from *stream_gd*, reclassifies the distance raster, creates a descriptor of *elevzone_gd*, and then uses the reclassified distance raster and the descriptor in a raster data query. The output from the query shows selected areas as having the cell value of 1.

RasterBinaryModel has three subs and one function (Figure 14.3):

Start: a sub for managing the input and output datasets and for calling the other subs
Distance: a sub for creating a continuous distance raster from streams
QueryGrids: a sub for raster data query of two processed rasters
FindLayer: a function for finding a specific layer

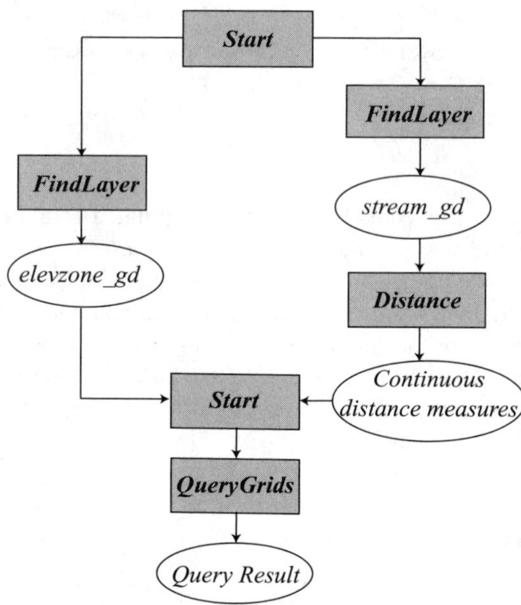

Figure 14.3   The flow chart shows the modular structure of *RasterBinaryModel*.

Usage: Add *stream_gd* and *elevzone_gd* to an active map. Import *Raster-BinaryModel* to Visual Basic Editor. Make sure that both ESRI Spatial Analyst Extension Object Library and ESRI Spatial Analyst Shared Object Library are checked in the References dialog. Run the module. The module adds two temporary rasters to the active map. One is *Distance_To_Stream*, a continuous distance raster from streams. The other is *Model*, the raster-based binary model, which should look the same as the vector-based binary model created by *VectorBinaryModel*.

```
Private Sub Start()
' Start uses the subs of FindLayer, Distance, and
' QueryGrids.
  Dim pMxDoc As IMxDocument
  Dim pMap As IMap
  Dim Id As Long
  Dim Message As String
  Dim pRasterLy As IRasterLayer
  Dim pRasterLy1 As IRasterLayer
  Dim pRasterLy2 As IRasterLayer
  Set pMxDoc = ThisDocument
  Set pMap = pMxDoc.FocusMap
  ' Run Distance.
  Message = "stream_gd"
```

```vb
    Id = FindLayer(Message)
    Set pRasterLy = pMap.Layer(Id)
    Call Distance(pRasterLy)
    ' Run QueryGrids.
    Message = "Distance_to_Stream"
    Id = FindLayer(Message)
    Set pRasterLy1 = pMap.Layer(Id)
    Message = "elevzone_gd"
    Id = FindLayer(Message)
    Set pRasterLy2 = pMap.Layer(Id)
    Call QueryGrids(pRasterLy1, pRasterLy2)
End Sub
```

Start calls the *Distance* sub to run a continuous distance measure operation from *stream_gd*. Then the code calls the **QueryGrids** sub to run a query on *elevzone_gd* and *Distance_to_Stream*, which is the output from the *Distance* sub. The result of the query shows areas that are in elevation zone 2 and within 200 meters of streams.

```vb
Private Sub Distance(pRasterLy As IRasterLayer)
    Dim pMxDoc As IMxDocument
    Dim pMap As IMap
    Dim pSourceRL As IRasterLayer
    Dim pSourceRaster As IRaster
    Dim pDistanceOp As IDistanceOp
    Dim pOutputRaster As IRaster
    Dim pOutputLayer As IRasterLayer
    Set pMxDoc = ThisDocument
    Set pMap = pMxDoc.FocusMap
    ' Define the source raster.
    Set pSourceRL = pRasterLy
    Set pSourceRaster = pSourceRL.Raster
    ' Perform the Euclidean distance operation.
    Set pDistanceOp = New RasterDistanceOp
    Set pOutputRaster = pDistanceOp.EucDistance _
    (pSourceRaster)
    ' Create the raster layer and add it to the active map.
    Set pOutputLayer = New RasterLayer
    pOutputLayer.CreateFromRaster pOutputRaster
    pOutputLayer.Name = "Distance_To_Stream"
    pMap.AddLayer pOutputLayer
End Sub
```

Distance gets the source layer for distance measures as an argument from *Start*. The code first sets *pSourceRaster* to be the raster of the stream layer. Next the code creates *pDistanceOp* as an instance of the *RasterDistanceOp* class and uses the *EucDistance* method on *IDistanceOp* to create a continuous distance raster referenced by *pOutputRaster*. Finally, the code creates a new layer from *pOutputRaster* and adds the layer to the active map.

```vba
Private Sub QueryGrids(pRasterLy1, pRasterLy2)
  Dim pMxDoc As IMxDocument
  Dim pMap As IMap
  Dim pRLayer1 As IRasterLayer
  Dim pRaster1 As IRaster
  Dim pRLayer2 As IRasterLayer
  Dim pRaster2 As IRaster
  Dim pReclassOp As IReclassOp
  Dim pRemap As IRemap
  Dim pNRemap As INumberRemap
  Dim pReclassRaster As IRaster
  Dim pFilt2 As IQueryFilter
  Dim pDesc2 As IRasterDescriptor
  Dim pLogicalOp As ILogicalOp
  Dim pOutputRaster As IRaster
  Dim pRLayer As IRasterLayer
  Set pMxDoc = ThisDocument
  Set pMap = pMxDoc.FocusMap
  ' Define the two rasters for query.
  Set pRLayer1 = pRasterLy1
  Set pRaster1 = pRLayer1.Raster
  Set pRLayer2 = pRasterLy2
  Set pRaster2 = pRLayer2.Raster
  ' Use a remap to reclassify the continuous distance
  ' raster.
  Set pReclassOp = New RasterReclassOp
  Set pNRemap = New NumberRemap
  pNRemap.MapRange 0#, 200#, 1
  pNRemap.MapRangeToNoData 200.1, 1300#
  Set pRemap = pNRemap
  Set pReclassRaster = pReclassOp.ReclassByRemap _
  (pRaster1, pRemap, False)
  ' Use the value field to create a raster descriptor
  ' from the elevzone raster.
```

```
    Set pFilt2 = New QueryFilter
    pFilt2.WhereClause = "value = 2"
    Set pDesc2 = New RasterDescriptor
    pDesc2.Create pRaster2, pFilt2, "value"
    ' Perform a logical operation.
    Set pLogicalOp = New RasterMathOps
    Set pOutputRaster = pLogicalOp.BooleanAnd _
    (pReclassRaster, pDesc2)
    ' Create the output raster layer and add it to the
    ' active map.
    Set pRLayer = New RasterLayer
    pRLayer.CreateFromRaster pOutputRaster
    pRLayer.Name = "Model"
    pMap.AddLayer pRLayer
End Sub
```

QueryGrids gets two raster layers to be queried as arguments from *Start*: one is *Distance_to_Stream* and the other is *elevzone_gd*. These two layers are processed separately prior to the query operation. *Distance_to_Stream* is a continuous raster. To isolate areas within 200 meters of streams, the code uses the *ReclassByRemap* method on *IReclassOp* to create *pReclassRaster*. The number remap has only two entries: cells within the range of 0 to 200 have the output value of one, and cells beyond 200 have the output value of no data. To isolate areas within elevation zone 2 in *elevzone_gd*, the code creates *pDesc2*, an instance of the *RasterDescriptor* class, by using a query filter object. After the two datasets are processed, the code uses the *BooleanAnd* method on *ILogicalOp* to create the logical query output referenced by *pOutputRaster*. Finally, the code creates a new layer from *pOutputRaster* and adds the layer to the active map.

```
Private Function FindLayer(Message As String) As Long
  Dim FindDoc As Variant
  Dim pMxDoc As IMxDocument
  Dim pMap As IMap
  Dim aLName As String
  Dim Name As String
  Dim i As Long
  Set pMxDoc = ThisDocument
  Set pMap = pMxDoc.FocusMap
  Name = Message
  For i = 0 To pMap.LayerCount - 1
    aLName = UCase(pMap.Layer(i).Name)
    If (aLName = (UCase(Name))) Then
      FindDoc = i
```

```
    End If
  Next
  FindLayer = FindDoc
End Function
```

FindLayer finds the index of the layer that matches the Message value. The function makes sure that the appropriate layers are entered as inputs to the subs and function.

14.3.4 *RasterIndexModel*

RasterIndexModel performs the same task as ***VectorIndexModel*** in raster format. *soil_gd, landuse_gd,* and *depwater_gd* are the raster equivalents of *soil.shp, landuse.shp,* and *depwater.shp* respectively. In building the index model, ***RasterIndexModel*** first reclassifies *landuse_gd* to exclude urban land use from the analysis. The module then performs a map algebra operation using the three rasters as the inputs. The output from the map algebra operation is the index model. The cell values of the index model range from 0.58 to 1.0. Finally, ***RasterIndexModel*** uses a color ramp renderer to display the index model in three classes.

RasterIndexModel has three subs and two functions (Figure 14.4):

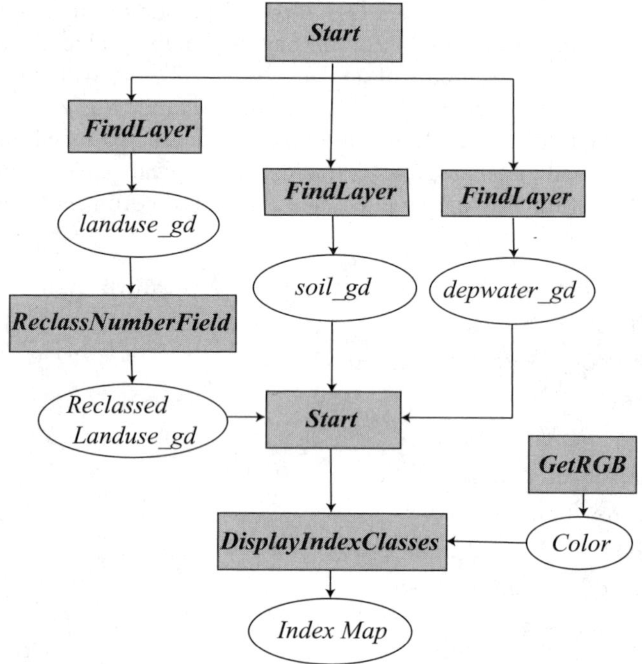

Figure 14.4 The flow chart shows the modular structure of ***RasterIndexModel***.

Start: a sub for performing the map algebra operation and for calling the other subs
ReclassNumberField: a sub for reclassifying the landuse raster
DisplayIndexClasses: a sub for displaying the index values in three classes
FindLayer: a function for finding a specific layer
GetRGB: a function for defining a color symbol

Usage: Add *soil_gd*, *landuse_gd*, and *depwater_gd* to an active map. Import ***RasterIndexModel*** to Visual Basic Editor. Make sure that both ESRI Spatial Analyst Extension Object Library and ESRI Spatial Analyst Shared Object Library are checked in the References dialog. Run the module. The module adds two temporary rasters to the active map. One is *Reclass_Landuse*, a reclassed land use raster. The other is *Model*, the raster-based index model, shown in three classes. Urban land use is treated as no data and is not included in the legend. The raster-based index model should look very similar to the vector-based index model created by ***VectorIndexModel***.

```
Private Sub Start()
' Start uses the subs of FindLayer, ReclassNumberField,
' and DisplayIndexClasses.
  Dim pMxDoc As IMxDocument
  Dim pMap As IMap
  Dim Id As Long
  Dim Message As String
  Dim pRasterLy As IRasterLayer
  Dim pMapAlgebraOp As IMapAlgebraOp
  Dim pRasterAnalysisEnv As IRasterAnalysisEnvironment
  Dim pSoilLayer As IRasterLayer
  Dim pSoilRaster As IRaster
  Dim pLanduseLayer As IRasterLayer
  Dim pLanduseRaster As IRaster
  Dim pDepwaterLayer As IRasterLayer
  Dim pDepwaterRaster As IRaster
  Dim pModel As IRaster
  Dim pModelLayer As IRasterLayer
  Set pMxDoc = ThisDocument
  Set pMap = pMxDoc.FocusMap
  ' Reclass landuse_gd to exclude urban land use from
  ' analysis.
  Message = "landuse_gd"
  Id = FindLayer(Message)
  Set pRasterLy = pMap.Layer(Id)
  Call ReclassNumberField(pRasterLy)
```

```
' Create a map algebra operation.
Set pMapAlgebraOp = New RasterMapAlgebraOp
Set pRasterAnalysisEnv = pMapAlgebraOp
' Bind the symbol R1 to soil_gd.
Message = "soil_gd"
Id = FindLayer(Message)
Set pSoilLayer = pMap.Layer(Id)
Set pSoilRaster = pSoilLayer.Raster
pMapAlgebraOp.BindRaster pSoilRaster, "R1"
' Bind the symbol R2 to reclassified landuse_gd.
Message = "Reclass_Landuse"
Id = FindLayer(Message)
Set pLanduseLayer = pMap.Layer(Id)
Set pLanduseRaster = pLanduseLayer.Raster
pMapAlgebraOp.BindRaster pLanduseRaster, "R2"
' Bind the symbol R3 to depwater_gd.
Message = "depwater_gd"
Id = FindLayer(Message)
Set pDepwaterLayer = pMap.Layer(Id)
Set pDepwaterRaster = pDepwaterLayer.Raster
pMapAlgebraOp.BindRaster pDepwaterRaster, "R3"
' Execute the map algebra operation.
Set pModel = pMapAlgebraOp.Execute _
("([R1] * 3 + [R2] + [R3])/250")
' Create the output raster layer and add it to the
' active map.
Set pModelLayer = New RasterLayer
pModelLayer.CreateFromRaster pModel
pModelLayer.Name = "Model"
pMap.AddLayer pModelLayer
Call DisplayIndexClasses(pModelLayer)
End Sub
```

Start first calls **ReclassNumberField** to reclassify urban areas in *landuse_gd* as no data, thus excluding urban areas from further analysis. Next, **Start** runs a raster map algebra operation. The operation requires that each input raster be given a "symbol." Therefore, the code calls the **FindLayer** sub to locate a layer in the active map and uses the *BindRaster* method on *IMapAlgebraOp* to bind the raster of the layer to a symbol. The R1 symbol binds *soil_gd*, the R2 symbol binds *reclass_landuse*, and the R3 symbol binds *depwater_gd*. These symbols of R1, R2, and R3 are then used in the expression that executes the map algebra operation. The result of the operation is an index model raster referenced by *pModel*. The code then creates a new layer named *Model* from *pModel* and adds the layer to the active map. **Start** concludes by calling ***DisplayIndexClasses*** to display the model layer in three classes.

```
Private Sub ReclassNumberField(pRasterLy As IRasterLayer)
  Dim pMxDoc As IMxDocument
  Dim pMap As IMap
  Dim pRaster2Ly As IRasterLayer
  Dim pGeoDs As IGeoDataset
  Dim pReclassOp As IReclassOp
  Dim pRemap As IRemap
  Dim pNRemap As INumberRemap
  Dim pOutRaster As IRaster
  Dim pReclassLy As IRasterLayer
  Set pMxDoc = ThisDocument
  Set pMap = pMxDoc.FocusMap
  ' Pass landuse_gd.
  Set pRaster2Ly = pRasterLy
  Set pGeoDs = pRaster2Ly.Raster
  ' Use a number remap to reclass landuse_gd.
  Set pReclassOp = New RasterReclassOp
  Set pNRemap = New NumberRemap
  With pNRemap
    .MapValue 20, 20
    .MapValue 40, 40
    .MapValue 45, 45
    .MapValue 50, 50
    .MapValueToNoData 99
  End With
  Set pRemap = pNRemap
  Set pOutRaster = pReclassOp.ReclassByRemap _
  (pGeoDs, pRemap, False)
  ' Create the output raster layer and add it to the
  ' active map.
  Set pReclassLy = New RasterLayer
  pReclassLy.CreateFromRaster pOutRaster
  pReclassLy.Name = "Reclass_Landuse"
  pMap.AddLayer pReclassLy
End Sub
```

ReclassNumberField gets *landuse_gd* as an argument from ***Start***. Designed to exclude urban areas from analysis, the code creates *pNRemap* as an instance of the *NumberRemap* class and uses the *MapValueToNoData* method on *INumberRemap* to assign no data to urban areas (i.e., with the map value of 99). Then the code uses the *ReclassByRemap* method on *IReclassOp* to create a reclassified raster referenced by *pOutRaster*. Finally, the code creates a new layer from *pOutRaster* and adds the layer called *Reclass_Landuse* to the active map.

```
Private Function FindLayer(Message As String) As Long
  Dim FindDoc As Variant
  Dim pMxDoc As IMxDocument
  Dim pMap As IMap
  Dim aLName As String
  Dim Name As String
  Dim i As Long
  Set pMxDoc = ThisDocument
  Set pMap = pMxDoc.FocusMap
  Name = Message
  For i = 0 To pMap.LayerCount - 1
    aLName = UCase(pMap.Layer(i).Name)
    If (aLName = (UCase(Name))) Then
      FindDoc = i
    End If
  Next
  FindLayer = FindDoc
End Function
```

FindLayer finds the index of the layer that matches the Message value. The function makes sure that the appropriate layers are used in the reclassification and map algebra operations.

```
Private Sub DisplayIndexClasses(pModelLayer As _
IRasterLayer)
' DisplayIndexClasses uses the sub of GetRGB.
  Dim pMxDoc As IMxDocument
  Dim pMap As IMap
  Dim pRLayer As IRasterLayer
  Dim pRaster As IRaster
  Dim pClassRen As IRasterClassifyColorRampRenderer
  Dim pRasRen As IRasterRenderer
  Dim pFillSymbol As IFillSymbol
  Set pMxDoc = ThisDocument
```

```
  Set pMap = pMxDoc.FocusMap
  Set pRLayer = pModelLayer
  Set pRaster = pRLayer.Raster
  ' Define a raster classify color ramp renderer.
  Set pClassRen = New RasterClassifyColorRampRenderer
  Set pRasRen = pClassRen
  Set pRasRen.Raster = pRaster
  pClassRen.ClassCount = 3
  pRasRen.Update
  ' Hard code the symbol, break, and label for each class.
  Set pFillSymbol = New SimpleFillSymbol
  pFillSymbol.Color = GetRGBColor(245, 245, 0)
  pClassRen.Symbol(0) = pFillSymbol
  pClassRen.Break(0) = 0.58
  pClassRen.Label(0) = "0.58 - 0.75"
  pFillSymbol.Color = GetRGBColor(245, 175, 0)
  pClassRen.Symbol(1) = pFillSymbol
  pClassRen.Break(1) = 0.76
  pClassRen.Label(1) = "0.76 - 0.85"
  pFillSymbol.Color = GetRGBColor(245, 125, 0)
  pClassRen.Symbol(2) = pFillSymbol
  pClassRen.Break(2) = 0.86
  pClassRen.Label(2) = "0.86 - 1.00"
  ' Assign the renderer to the layer and refresh the map.
  pRasRen.Update
  Set pRLayer.Renderer = pRasRen
  pMxDoc.ActiveView.Refresh
  pMxDoc.UpdateContents
End Sub
```

DisplayIndexClasses gets the model layer as an argument from **Start**. The code creates *pClassRen* as an instance of the *RasterClassifyColorRampRenderer* class and uses *IRasterRenderer* and *IRasterClassifyColorRampRenderer* to define the renderer's raster and class count. The code then provides the fill symbol, break, and label for each class in *pClassRen*. The **GetRGBColor** function generates the color for the fill symbol. Finally, the code assigns *pRasRen* to be the renderer of *pRLayer*, refreshes the map, and updates the contents of the map document.

```
Private Function GetRGBColor(R As Long, G As Long, _
B As Long)
  Dim pColor As IRgbColor
```

```
    Set pColor = New RgbColor
    pColor.Red = R
    pColor.Green = G
    pColor.Blue = B
    GetRGBColor = pColor
  End Function
```

GetRGBColor uses the input values of R, G, and B from ***DisplayIndexClasses*** and returns an RGB color.

B

C

M

N

S